GRAPH CLASSES

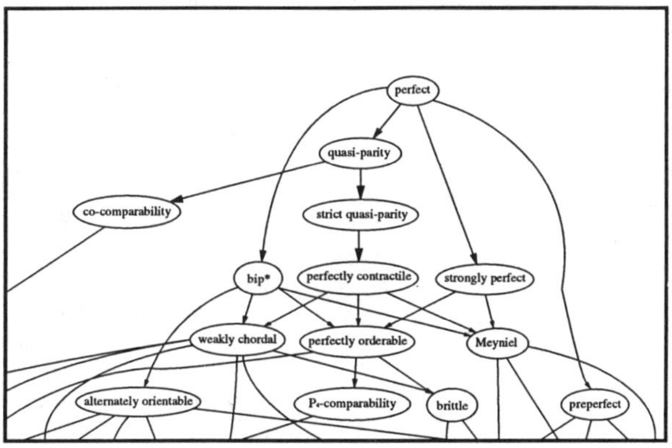

SIAM Monographs on Discrete Mathematics and Applications

The series includes advanced monographs reporting on the most recent theoretical, computational, or applied developments in the field, introductory volumes aimed at mathematicians and other mathematically motivated readers interested in understanding certain areas of pure or applied combinatorics, and graduate textbooks. The volumes are devoted to various areas of discrete mathematics and its applications.

Mathematicians, computer scientists, operations researchers, computationally oriented natural and social scientists, engineers, medical researchers and other practitioners will find the volumes of interest.

Series Volumes

Brandstädt, A., Le, V. B., and Spinrad, J. P., *Graph Classes: A Survey*
McKee, T. A. and McMorris, F. R., *Topics in Intersection Graph Theory*
Grilli di Cortona, P., Manzi, C., Pennisi, A., Ricca, F., and Simeone, B., *Evaluation and Optimization of Electoral Systems*

GRAPH CLASSES

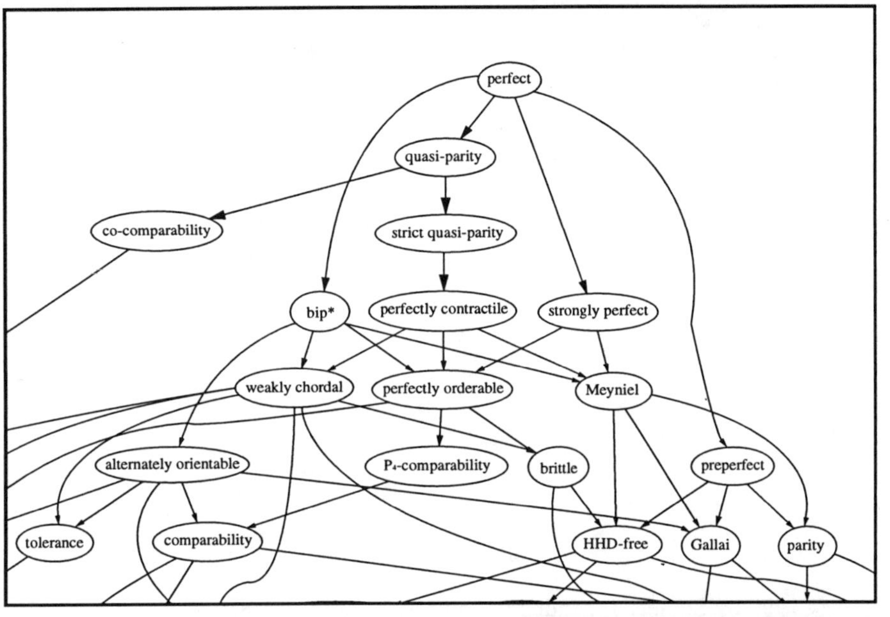

A SURVEY

Andreas Brandstädt
University of Rostock
Rostock, Germany

Van Bang Le
University of Rostock
Rostock, Germany

Jeremy P. Spinrad
Vanderbilt University
Nashville, Tennessee

siam.
Society for Industrial and Applied Mathematics Philadelphia

10 9 8 7 6 5 4 3 2 1

Library of Congress Cataloging-in-Publication Data

Brandstädt, Andreas.
Graph classes : a survey / Andreas Brandstädt, Van Bang Le, Jeremy P. Spinrad.
p. cm. -- (SIAM monographs on discrete mathematics and applications)
Includes bibliographical references and index.
ISBN 0-89871-432-X (pbk.)
1. Graph theory. I. Le, Van Bang. II. Spinrad, Jeremy P. III. Title.
IV. Series.
QA166.B73 1999
511'.5--dc21 99-11680

Contents

Preface

When dealing with special graph classes and algorithmic problems on them, a main source is the classical book of Golumbic, *Algorithmic Graph Theory and Perfect Graphs* [454]. The book, however, appeared in 1980, and since that time many interesting new classes have been introduced. Therefore, it is probably useful to have a new survey that attempts to describe the world of special graph classes with an emphasis on the algorithmic point of view.

There are many reasons for the interest in this field of research. These come from discrete mathematics as well as theoretical and practical computer science.

- Graphs are a good model for describing "real-world" and computer-science situations such as
  - interconnection and transport networks for information exchange;
  - VLSI layouts;
  - computational geometry;
  - graph drawing;
  - scheduling and partial-order problems;
  - molecular biology (DNA mappings), phylogenetic trees;
  - temporal reasoning;
  - synchronizing parallel processes;
  - sparse systems of linear equations;
  - desirable properties of relational database schemes.

The last properties are closely related to hypergraph acyclicity, Helly and tree properties of hypergraphs, and graph chordality.

- For many applications, the graphs used in the models have special properties such as
  - min-max (in)equalities of certain parameters;
  - cycles and chords;
  - separator properties;
  - distance properties;
  - elimination orderings of the vertex set;
  - composition/decomposition properties, recursive constructions;
  - intersection, containment, or overlap models or measured variants of them;
  - tree structure in many variants (reflected by related hypergraphs and matrices).

To efficiently solve such basic algorithmic problems as graph coloring, maximum independent set/maximum clique, Steiner tree, and dominating set (see [419]), it is extremely important to know the structural properties of the graphs under consideration.

A typical and classical example is the class of chordal graphs: every cycle of length at least four has at least one chord. These graphs have many different characterizations and thus appear in almost every chapter of this survey.

There are (at least) two main groups of graph properties that are helpful for designing polynomial-time algorithms:

- Graphs that fulfill certain min-max equalities such as König's theorem for bipartite graphs and Dilworth's theorem for partial orders. This approach leads to perfect graphs that have many nice algorithmic consequences.

- Graphs that generalize tree properties (trees of vertices or trees of hyperedges), since many problems are easy to solve on trees. Of course the problems again become hard on these generalizations if the graphs are "too far" from trees.

These properties are sometimes closely related — chordal graphs have both properties.

We will assume that for algorithmic problems on graphs, the input is given in some standard form such as adjacency matrices or adjacency lists, unless noted otherwise. For convenience, we will often say that an optimization problem is $\mathbb{NP}$-complete; by this, we mean that the corresponding decision problem is $\mathbb{NP}$-complete.

Algorithmic problems on graphs such as the maximum independent set problem or the minimum dominating set problem are in many cases $\mathbb{NP}$-complete even on special graph classes, but sometimes become solvable in polynomial time on certain smaller classes.

Clearly, if for graph classes $\mathcal{C}_1 \subset \mathcal{C}_2$, these problems are solvable in polynomial time on $\mathcal{C}_2$, then they are also solvable in polynomial time on $\mathcal{C}_1$, and if these problems are $\mathbb{NP}$-hard on $\mathcal{C}_1$, then they are $\mathbb{NP}$-hard on $\mathcal{C}_2$. Thus, inclusions between graph classes are interesting from the algorithmic point of view and, assuming that $\mathbb{P} \neq \mathbb{NP}$, it is an interesting problem to refine the "borderline" between $\mathbb{P}$ and $\mathbb{NP}$ for special problems and graph classes.

The field of special graph classes is developing rapidly. We describe here almost 200 classes. In the survey paper [631], Johnson remarked that "... many graph theorists have made careers out of inventing and characterizing new classes of graphs; there are by now far too many classes for a single column to survey."

Of course the survey is also restricted to the basic properties; characterizations and inclusions of a collection of graph classes that seem to be important cannot exhaustively present the current knowledge. In particular, it is not possible to discuss the complexity of selected algorithmic graph problems on as many classes as was done in [631]; for several problems there are separate monographs on the algorithmic behavior of the problem on special graph classes; see, for instance, [706] for the traveling salesman problem. It is sometimes hard to decide whether an algorithmically oriented paper using special graphs should be included in the survey. We try to include only those papers that describe interesting structural properties of the classes. We are not able to include all of these several thousand papers and apologize for the omitted ones, which could turn out to be important. The same holds true for the papers on special graph classes not included here due to space restrictions or due to our lack of knowledge. In particular, we do not describe the worlds of directed graphs, infinite graphs, and random graphs. Good sources for more information on infinite graphs are [505, 312, 1023], and a good source for random graphs is [125].

For almost 200 classes it is impossible to devote one chapter to each class, as was done in [452]. We have tried to group the material according to some basic properties, as contained in the list given above. In some cases, there are monographs dealing with much more limited graph classes; examples are the book of Fishburn [387] devoted to interval graphs and the monograph of Mahadev and Peled [762], which deals with threshold graphs.

However, these monographs do not describe the many connections between the graph classes, which are like a very complicated network and lead to many links between classes defined and characterized in completely different ways.

Since many classes have surprisingly different characterizations, they appear in different chapters. By giving a detailed index at the end we hope to give a good tool for finding the many places where some basic classes and notions occur.

We also provide a summary of the recognition complexities of various graph classes (Appendix A), and a section is devoted to containment relationships between classes (Appendix B).

As already mentioned the survey has profited very much from the book of Golumbic [452] but also from several papers containing surveys:

Johnson [628, 629, 630, 631, 632, 633, 634, 635], Möhring [786, 788], Golumbic [455], and Chvátal [209], the Ph.D. dissertations of C.T. Hoàng [553], S. Olariu [820], L.K. Stewart [1002], F.F. Dragan [321], T. Kloks [668], R. Garbe [417], C. Flotow [391], and F. Nicolai [810], the Habilitation theses of L. Babel [37], E. Dahlhaus [276], D. Kratsch [697], and V. Giakoumakis [438], and many other papers that give partial surveys.

The authors gratefully acknowledge discussions with and communications from many colleagues, among them G. Bacsó (Budapest), H.-J. Bandelt (Hamburg), H. Bodlaender (Utrecht), V. Chepoi (Marseille), A.A. Chernyak (Minsk), V. Chvátal (New Brunswick), E. Dahlhaus (Bonn), P. Damaschke (Hagen), F.F. Dragan (Kishinev), E. Eschen (Denver), F. Gavril (Haifa), V. Giakoumakis (Amiens), V. Gurvich (Moscow), K. Jansen (Trier), S. Klein (Rio de Janeiro), T. Kloks (Eindhoven), D. Kratsch (Jena), F. Maffray (Grenoble), R. McConnell (Denver), H. Müller (Jena), F. Nicolai (Duisburg), S. Olariu (Norfolk), U. Peled (Chicago), M. Penn (Haifa), P. Scheffler (Stralsund), D. Seese (Karlsruhe), R. Sritharan (Terre Haute), J.L. Szwarcfiter (Rio de Janeiro), Zs. Tuza (Budapest), W. Unger (Aachen), and V.I. Voloshin (Kishinev).

Special thanks also to W. Brauer, P.L. Hammer, E.W. Mayr, R. Shamir, G. Tinhofer, and D. de Werra for encouragement. We also would like to thank the anonymous referees for many helpful suggestions.

Jerry Spinrad acknowledges the generous support of the Vanderbilt University Research Council.

ANDREAS BRANDSTÄDT
VAN BANG LE
JEREMY P. SPINRAD

Chapter 1

Basic Concepts

1.1 Basic graph notions

This chapter introduces a few basic concepts. For others see standard textbooks or monographs such as the book of Berge [90], the book of Bondy and Murty [126], or the book of Golumbic [452].

Definition 1.1.1 *Let V be a finite set of* vertices (*or* nodes). *We usually denote the number of vertices by n, i.e., $|V| = n$. Let $\mathcal{P}(V)$ denote the power set of the set V, i.e., the set of all subsets of V, and let $\mathcal{P}_2(V)$ denote the set of all 2-element subsets of V (undirected pairs).*

> $G = (V, E)$ *is a* directed graph *if $E \subseteq V \times V$ (E is the set of* directed edges *or* arcs *of G).*
>
> $G = (V, E)$ *is an* undirected graph *if $E \subseteq \mathcal{P}_2(V)$ (E is the set of* edges *of G).*

In what follows, in most cases m denotes the number of edges of a graph, i.e., $m = |E|$. $V(G)$ is the vertex set of G and $E(G)$ is the edge set of G.

If nothing else is stated graphs are understood to be finite, undirected, and simple (i.e., without self-loops and parallel edges). We will write xy for the edge $\{x, y\}$.

Definition 1.1.2 *The* complement $\overline{G} = (V, \overline{E})$ *of a graph $G = (V, E)$ is defined by $\overline{E} = \{xy : x, y \in V$ and $x \neq y$ and $xy \notin E\}$.*

Definition 1.1.3 *Let $G = (V, E)$ be a graph.*

> *A graph $G' = (V', E')$ is a* subgraph *of G if $V' \subseteq V$ and $E' \subseteq E$.*
>
> *A subgraph $G' = (V', E')$ is an* induced subgraph *of G if $E' = \{uv : uv \in E$ and $u, v \in V'\}$. We also say that G' is induced by V' and usually write $G(V')$ for G'.*
>
> *A graph property P is* hereditary *if the property P holds for every induced subgraph of G whenever it holds for G.*

Definition 1.1.4 *Let $G = (V, E)$ be a graph.*

A sequence $(v_1, \dots, v_k)$ of pairwise distinct vertices is a path *in G if $v_1v_2, \dots, v_{k-1}v_k \in E$. These edges are called* the edges of the path. *The* length of the path *is the number $k - 1$ of its edges.*

A path $(v_1, \dots, v_k)$ is a cycle *if in addition $v_1v_k \in E$. The* edges of the cycle *are defined analogously to the edges of the path. The* length of the cycle *is the number k of its edges.*

An even (odd) cycle (path) *is a cycle (path) of even (odd) length.*

A chord *of a path (cycle) $(v_1, \dots, v_k)$ is an edge between two vertices of the path (cycle) that is not an edge of the path (cycle). A path (cycle) is* chordless *if it contains no chords.*

In what follows, we denote the chordless paths (cycles) with n vertices by P_n (C_n). Chordless cycles C_n, $n \geq 5$, are holes. *Chordless cycles C_{2n+1}, $n \geq 2$, are* odd holes. *Complements $\overline{C_n}$ of chordless cycles C_n, $n \geq 5$, are* antiholes, *complements $\overline{C_{2n+1}}$ of odd holes are* odd antiholes.

G is connected *if for all $u, v \in V$, $u \neq v$, there is a path $(v_1, \dots, v_k)$ in G connecting u and v, i.e., $\{v_1, v_k\} = \{u, v\}$.*

The distance *$d(u, v)$ between two vertices u and v is the minimum length of a path between u and v if there is one and ∞ otherwise. The* eccentricity *$ecc(v)$ of a vertex v is the maximum distance to another vertex. The* diameter *$diam(G)$ of G is the maximum eccentricity over all vertices of G, i.e., the maximum distance between two vertices (and ∞ if G is not connected).*

An induced connected subgraph $G(U)$ is isometric *(or* distance preserving*) in G if the distances in $G(U)$ are the same as in G.*

Note that for $n \geq 5$, P_{n-1} is a nonisometric subgraph in C_n.
The path P_4 plays a special role for many graph classes.

Definition 1.1.5 *Let G be a graph and P be a P_4 in G with vertices a, b, c, d and edges ab, bc, cd. Then the vertices a and d are the* endpoints *of P while b and c are the* midpoints *of P. The edges ab and cd are the* end-edges *or* wings *of P while bc is the* midedge *of P.*

Many graph algorithms are based on cutsets of graphs.

Definition 1.1.6 *Let $G = (V, E)$ be a connected graph.*

A subset $V' \subset V$ is a cutset *(or* separator*) of G if $G(V \setminus V')$ is disconnected.*

The vertex x is an articulation point *of G if $\{x\}$ is a cutset of G.*

G is 2-connected *if G has no articulation point. The maximal 2-connected subgraphs of G are the* blocks *or* 2-connected components *of G. The* k-connected components *are defined analogously.*

One of the most fundamental and algorithmically useful classes of graphs is the class of trees.

Definition 1.1.7 *Let G be a graph.*

G is a forest *if G contains no cycle.*

G is a tree *if G is a connected forest, i.e., G is connected and cycle free.*

There are many well-known properties and characterizations of trees described in standard textbooks. We mention only a few of them.

Proposition 1.1.1

(i) *Trees are maximal cycle-free graphs (i.e., adding one edge creates a cycle) and minimal connected graphs (i.e., deleting one edge destroys connectedness). Trees with n vertices have exactly $n-1$ edges. For each pair of vertices, there is exactly one path in the tree connecting them.*

(ii) *Trees and forests can be recognized in linear time using standard graph-searching methods such as depth-first search (DFS).*

(iii) *Each tree with at least two vertices has at least two leaves (i.e., vertices of degree one).*

(iv) *Trees can be generated recursively by repeatedly attaching leaves starting with the one-vertex graph (the inverse direction leads to an elimination ordering for trees by repeated deletion of leaves).*

Many algorithmic problems are efficiently solvable on trees, and this is one of the reasons for studying generalizations of trees that admit efficient algorithms. Therefore, it is not surprising that many graph classes are generalizations of trees.

Definition 1.1.8 *Let $G = (V, E)$ be a graph.*

$V' \subseteq V$ is an independent set *or* stable set *in G (or* empty subgraph *of G) if for all $u, v \in V'$, $uv \notin E$.*

$V' \subseteq V$ is a clique *in G (or* complete subgraph*) if for all $u, v \in V'$, $u \neq v$, $uv \in E$.*

A stable set (clique) S in G is maximal *if there is no stable set (clique) $S' \neq S$ in G with $S \subset S'$.*

A stable set (clique) S is maximum *if $|S|$ is the maximum possible size of a stable set (clique) in G.*

The next definition gives some standard functions that are basic for perfect graphs.

Definition 1.1.9 *Let $G = (V, E)$ be a graph.*

$$\alpha(G) = \max \{|V'| : V' \subseteq V \textit{ and } V' \textit{ is an independent set in } G\},$$

$$\omega(G) = \max \{|V'| : V' \subseteq V \textit{ and } V' \textit{ is a clique in } G\},$$

$$\chi(G) = \min \{k : \textit{there is a partition of } V \textit{ into } k \textit{ disjoint independent sets}\},$$

$$\kappa(G) = \min \{k : \textit{there is a partition of } V \textit{ into } k \textit{ disjoint cliques}\}.$$

Note that for every graph G, $\omega(G) \leq \chi(G)$ and $\alpha(G) \leq \kappa(G)$. $\chi(G)$ is often called the *chromatic number of* G, since a partition $V_1, \ldots, V_k$ of V into independent sets V_i, $i = 1, \ldots, k$, is exactly a coloring of G such that no two adjacent vertices have the same color.

Note that $\alpha(\overline{G}) = \omega(G)$ and $\chi(\overline{G}) = \kappa(G)$. It is well known that determining each of the parameters $\alpha(G)$, $\omega(G)$, $\chi(G)$, and $\kappa(G)$ is an $\mathbb{NP}$-complete problem.

One of the most important graph classes is the class of bipartite graphs.

Definition 1.1.10 *A graph G is* bipartite *if $\chi(G) \leq 2$.*

Bipartite graphs are usually written as $B = (X, Y, E)$, where X and Y denote the color classes of V in a 2-coloring of B (a *bipartition* of B). $K_{k,l}$ denotes the *complete bipartite graph* $B = (X, Y, E)$ with $|X| = k$, $|Y| = l$, and all pairs between X and Y being edges.

Theorem 1.1.1 (König [682]) *Bipartite graphs are exactly the graphs without odd cycles.*

Definition 1.1.11 *Let $G = (V, E)$ be a graph and $v \in V$.*

> $N(v) = \{u : u \in V,\ u \neq v \textit{ and } uv \in E\}$ *denotes the* (open) neighborhood *of* v,
>
> $N[v] = N(v) \cup \{v\}$ *denotes the* closed neighborhood *of* v,
>
> $N^k(v) = \{u : u \in V \textit{ and } d(u, v) = k\}$ *denotes the* kth iterated neighborhood *of* v,
>
> $D_k(v) = \{u : u \in V \textit{ and } d(u, v) \leq k\}$ *denotes the* disk *of radius* k *and center* v.
>
> *Let G^k denote the* kth power *of G, which has the same vertex set V, and $u, v \in V$ are adjacent in G^k if and only if $d(u, v) \leq k$ in G.*
>
> *For* $V' \subseteq V$ *let* $N(V') = \bigcup_{v \in V'} N(v)$ *and* $N[V'] = \bigcup_{v \in V'} N[v]$.
>
> *For* $j \geq 1$ *let* $N^{j+1}(V') = N(N^j(V'))$ *and* $N^{j+1}[V'] = N[N^j[V']]$.

In connection with vertex orderings $(v_1, \ldots, v_n)$ we frequently use the following notation.

Definition 1.1.12 *Let G be a graph. For a vertex ordering $(v_1, \ldots, v_n)$ of G let*

> $G_i = G(\{v_i, \ldots, v_n\})$ *for* $i \in \{1, \ldots, n\}$ *and let*

$N_i(v)$ ($N_i[v]$) be the open (closed) neighborhood *of v in G_i (analogously for disks).*

Definition 1.1.13 *Let $G = (V, E)$ be a graph.*

A subset $V' \subseteq V$ is a vertex cover *of G if for all edges $e \in E$, $V' \cap e \neq \emptyset$.*

Let $\tau(G)$ denote the minimum size of a vertex cover in G.

A subset $E' \subseteq E$ is a matching *in G if no two edges $e, e' \in E'$, $e \neq e'$, have a common vertex, i.e., $e \cap e' = \emptyset$.*

Let $\nu(G)$ denote the maximum size of a matching in G.

Thus, V' is a vertex cover if and only if $V \setminus V'$ is a stable set. This implies that determining the parameter $\tau(G)$ is an $\mathbb{NP}$-complete problem. Determining $\nu(G)$ can be done in polynomial time; see, e.g., the book of Lovász and Plummer [742].

One of the oldest and most well-known examples for a min-max relationship is the following.

Theorem 1.1.2 (König [682]) *For bipartite graphs B the equality $\nu(B) = \tau(B)$ holds.*

There are many generalizations of this property, which is also sometimes called the *König property.*

The following notion of join generalizes the complete bipartite subgraphs.

Definition 1.1.14 *Let $G = (V, E)$ be a graph.*

The disjoint sets $U, W \subseteq V$ form a join *denoted by $U * W$ if for all $x \in U, y \in W$, $xy \in E$.*

In bipartite graphs the complete bipartite subgraphs $K_{k,l}$ play a similar role to cliques in graphs, and are sometimes called *bicliques.*

Definition 1.1.15 *Graphs $G_1 = (V_1, E_1), G_2 = (V_2, E_2)$ are* isomorphic *($G_1 \sim G_2$) if there is a bijective function f from V_1 onto V_2 such that $uv \in E_1 \iff f(u)f(v) \in E_2$. Then f is a* graph isomorphism *between G_1 and G_2.*

Isomorphism for directed graphs can be defined analogously.

Definition 1.1.16 *Let G be a graph and let x, y be two vertices of G. Vertex x* dominates *vertex y if $N(y) \subseteq N[x]$ (or, equivalently, $N(y) \setminus \{x\} \subseteq N(x)$). If x dominates y or y dominates x then x and y are* comparable.

This binary relation was studied under the name *vicinal preorder* by Földes and Hammer.

Definition 1.1.17 (Földes, Hammer [397]) *The* Dilworth number *of a graph G is the largest number of pairwise incomparable vertices of the graph:*

$$dilw(G) = \max\{|V'| : V' \subseteq V \text{ and for all pairwise different } x, y \in V', x \text{ and } y \text{ are incomparable}\}.$$

A polynomial-time algorithm for computing the Dilworth number of a given graph is mentioned in the threshold graph chapter (chapter 13).

There is another important notion of domination in graphs.

Definition 1.1.18 *Let $G = (V, E)$ be a graph.*

> *The subset $S \subseteq V$ is a* dominating set *of G if every vertex outside S has a neighbor in S, i.e., $N[S] = V$.*
>
> *The* domination number *$\gamma(G)$ is the smallest size of a dominating set of G.*

It is well known that determining the domination number of a graph is an $\mathbb{NP}$-complete problem [419]. There are many results about domination on special graph classes. Johnson [631], Hedetniemi and Laskar [534], and Kratsch [699] survey some of them. See also the collections of survey papers on domination in [524, 525].

The domination number is bounded by the Dilworth number. In [397] it was observed that for graphs G without isolated vertices, $\gamma(G) \le \mathrm{dilw}(G)$ holds.

1.2 Chordal graphs

Now we come to a graph class of crucial importance: chordal graphs. They represent an example of a class that is a generalization of trees with many different characterizations, rich structure properties, and many important applications. These graphs appear under other names, such as *triangulated graphs, rigid-circuit graphs, monotone transitive graphs,* and *perfect elimination graphs* in the literature.

Definition 1.2.1 (Hajnal, Surányi [501]) *A graph G is* chordal *if each cycle in G of length at least 4 has at least one chord.*

Chordal graphs have nice cutset properties.

Theorem 1.2.1 (Dirac [316]) *G is chordal if and only if every minimal cutset in every induced subgraph of G is a clique.*

The following notion of simpliciality of a vertex is a straightforward generalization of leaves in a tree.

Definition 1.2.2 *Let $G = (V, E)$ be a graph.*

> *The vertex $v \in V$ is* simplicial *in G if $N(v)$ is a clique in G.*
>
> *The ordering $(v_1, \ldots, v_n)$ of the vertices of V is a* perfect elimination ordering *of G if for all $i \in \{1, \ldots, n\}$, the vertex v_i is simplicial in $G_i = G(\{v_i, \ldots, v_n\})$.*

Theorem 1.2.2 (Dirac [316], Fulkerson, Gross [409], Rose [926]) *A graph is chordal if and only if it has a perfect elimination ordering.*

A tree structure of chordal graphs is expressed in Theorem 1.2.3, for which we need the notion of an intersection graph.

Definition 1.2.3 *For a given set M of objects (for which intersection makes sense), the* intersection graph G_M *of these objects has M as vertex set, and two objects are adjacent in G_M if the intersection of the corresponding objects is nonempty.*

Theorem 1.2.3 (Walter [1077], Gavril [424], Buneman [167]) *A graph is chordal if and only if it is the intersection graph of subtrees of a tree.*

For the tree T of Theorem 1.2.3, it can be assumed that the vertices of T are the maximal cliques of G, and the subtrees T_v, $v \in V$, are defined by the occurrences of vertex v in the maximal cliques of G.

Such a tree T is called a *clique tree* of G. A natural question is how to find a clique tree for a given chordal graph G. One way to do this is a greedy strategy of finding a maximum spanning tree on the intersection graph of the maximal cliques of G, weighting the edges between maximal cliques by the size of their intersection. This approach is described by Bernstein and Goodman in [102] and was rediscovered in several papers such as [879, 972, 973]. It can be formulated in the following way (McKee [777]).

Theorem 1.2.4 *A graph G is chordal if and only if a maximum spanning-tree algorithm applied to the intersection graph of the maximal cliques of G weighted by the size of their intersections produces a clique tree for G.*

See also [414] for further research in this direction. Linear-time algorithms for constructing a clique tree are described by Blair and Peyton [109] and Hsu and Ma [597] and in a forthcoming monograph of Spinrad [994].

1.3 Basic hypergraph notions and properties

We now give some standard definitions and properties of hypergraphs.

Definition 1.3.1 *The pair $H = (V, \mathcal{E})$ is a* hypergraph *if $\mathcal{E} \subseteq \mathcal{P}(V)$. $\mathcal{E}$ is the set of* hyperedges *of the hypergraph.*

Thus, unlike edges in graphs, hyperedges are in general not of size 2. Subsequently we sometimes denote hypergraphs by their set $\mathcal{E}$ of hyperedges (in this case the vertex set is the union of the hyperedges).

Definition 1.3.2 *Let $H = (V, \mathcal{E})$ be a hypergraph.*

For a vertex $v \in V$, let $\mathcal{E}_v = \{e : e \in \mathcal{E} \text{ and } v \in e\}$ denote the set of all hyperedges containing the vertex v.

Let $\mathcal{E}^* = \{\mathcal{E}_v : v \in V\}$ and let $H^* = (\mathcal{E}, \mathcal{E}^*)$ denote the dual hypergraph *of H.*

There are several important constructions of graphs from given hypergraphs.

Definition 1.3.3 *Let $H = (V, \mathcal{E})$ be a hypergraph.*

The intersection *or* line graph *of H is defined as the graph $L(H) = (\mathcal{E}, E')$ with $ee' \in E'$ if and only if $e \neq e'$ and $e \cap e' \neq \emptyset$ (see Definition 1.2.3; here the objects are the hyperedges).*

The 2-section graph $2SEC(H)$ *of H is defined as the graph $2SEC(H) = (V, E')$ with $uv \in E'$ if and only if there is a hyperedge $e \in \mathcal{E}$ for which $u, v \in e$.*

Note that the 2-section graph is sometimes called the *representing graph.* The following properties of hypergraphs are well known.

Proposition 1.3.1 *Let H be a hypergraph.*

(i) *Taking the dual of a hypergraph twice is isomorphic to the hypergraph itself, i.e., $(H^*)^* \sim H$.*

(ii) *The line graph of the dual hypergraph is isomorphic to the 2-section graph of the hypergraph, i.e., $L(H^*) \sim 2SEC(H)$.*

Definition 1.3.4 *Let $H = (V, \mathcal{E})$ be a hypergraph.*

$\mathcal{E}$ fulfills the Helly property *if the following condition holds:*

If for any subsystem $\mathcal{E}'$ of $\mathcal{E}$, the elements of $\mathcal{E}'$ pairwise intersect, then also $\bigcap \mathcal{E}' = \bigcap_{e \in \mathcal{E}'} e \neq \emptyset$.

The hypergraph H is a Helly hypergraph *if $\mathcal{E}$ fulfills the Helly property.*

$\mathcal{E}$ is conformal *if every clique C of $2SEC(\mathcal{E})$ is contained in a hyperedge $e \in \mathcal{E}$. The hypergraph H is* conformal *if $\mathcal{E}$ is conformal.*

An important example is given again by trees: subtrees of a tree fulfill the Helly property. The following property is well known.

Proposition 1.3.2 *A hypergraph H is conformal if and only if H^* is a Helly hypergraph.*

For hypergraphs there are two important notions of smaller hypergraphs "contained" in a given hypergraph.

Definition 1.3.5 *Let $H = (V, \mathcal{E})$ be a hypergraph.*

$H' = (V', \mathcal{E}')$ is a partial hypergraph *of H if $\mathcal{E}' \subseteq \mathcal{E}$ and V' is the union of the hyperedges in $\mathcal{E}'$.*

For $V' \subseteq V$, the subhypergraph of H induced by V' *is the hypergraph $H(V') = (V', \mathcal{E}')$ with $\mathcal{E}' = \{e \cap V' : e \in \mathcal{E}\}$.*

Now we come to generalizations of trees in terms of hypergraphs.

Definition 1.3.6 *Let $H = (V, \mathcal{E})$ be a hypergraph.*

H is a hypertree *if there is a tree T with vertex set V such that all hyperedges $e \in \mathcal{E}$ induce subtrees in T.*

H is a dual hypertree *if there is a tree T with vertex set $\mathcal{E}$ such that for all vertices $v \in V$, $\mathcal{E}_v$ induces a subtree of T, i.e., the hyperedges in which v occurs form a connected set in T.*

Hypertrees were called *arboreal* hypergraphs in [90] and *subtree hypergraphs* in [744]. It is obvious that H is a hypertree if and only if H^* is a dual hypertree. Dual hypertrees play an important role in the theory of desirable properties of relational database schemes and appear there under the name *acyclic* or *α-acyclic hypergraphs*; see [370] for a survey. Note that the name *tree schemes* was proposed for these hypergraphs in [455] since acyclic hypergraphs may contain cycles. We describe this in more detail in Definitions 8.1.1 and 8.1.2 and Theorem 8.1.1.

The following characterization of hypertrees/dual hypertrees shows the close connection between chordal graphs and these hypergraphs.

Theorem 1.3.1 (Duchet [333], Flament [390], Slater [981]) *The hypergraph H is a hypertree if and only if H has the Helly property and its line graph $L(H)$ is chordal.*

Because of the dualities between hypertrees and dual hypertrees, the conformality and the Helly property, the line graph of a hypergraph and the 2-section graph of the dual hypergraph, Theorem 1.3.1 can also be expressed in many other variants by switching between a property and its dual. Thus it can be formulated for dual hypertrees in the following way.

Corollary 1.3.1 *Let $H = (V, \mathcal{E})$ be a hypergraph. Then H is a dual hypertree if and only if H is conformal and $2SEC(\mathcal{E})$ is chordal.*

We now give a construction of a hypergraph from a graph.

Definition 1.3.7 *The* clique hypergraph *of a graph G is $\mathcal{C}(G) = \{C : C$ is a maximal clique in $G\}$.*

Since for every graph G the clique hypergraph $\mathcal{C}(G)$ is conformal, we have the following equivalence.

Corollary 1.3.2 *A graph G is chordal if and only if $\mathcal{C}(G)$ is a dual hypertree.*

While induced subgraphs of chordal graphs are always chordal, i.e., chordality is a hereditary graph property, the partial hypergraphs of a dual hypertree are not necessarily dual hypertrees. The partial hypergraphs of hypertrees, however, are hypertrees, while the subhypergraphs of hypertrees are not necessarily hypertrees.

Subsequently we need further standard hypergraphs corresponding to graphs.

Definition 1.3.8 *Let $G = (V, E)$ be a graph.*

The (closed) neighborhood hypergraph *of* G *is* $\mathcal{N}(G) = \{N[v] : v \in V\}$.

The open neighborhood hypergraph *of* G *is* $\mathcal{N}_0(G) = \{N(v) : v \in V\}$.

The disk hypergraph *of* G *is* $\mathcal{D}(G) = \{D_k[v] : v \in V,\ k \text{ a positive integer}\}$.

The join-partitionable set hypergraph *of* G *is* $\mathcal{JP}(G) = \{U : U \subseteq V$ *and* U *is maximal join partitionable in* $G\}$. *Hereby a set* $U \subseteq V$ *is* join partitionable *if* U *has a partition into* U_1, U_2 *forming a join.*

For bipartite graphs there are similar constructions.

Definition 1.3.9 *Let* $B = (X, Y, E)$ *be a bipartite graph.*

Then the biclique hypergraph $\mathcal{BC}(B)$ *of* B *is the set of all maximal bicliques of* B, *i.e., the hyperedges of* $\mathcal{BC}(B)$ *are the maximal subsets of* $X \cup Y$ *inducing a complete bipartite subgraph.*

There are two one-sided neighborhood hypergraphs *of* B:

$\mathcal{N}^X(B) = \{N(y) : y \in Y\}$. *Analogously define* $\mathcal{N}^Y(B)$;

$\mathcal{N}_0(B) = \mathcal{N}^X(B) \cup \mathcal{N}^Y(B)$.

Note that $(\mathcal{N}^X(B))^*$ is isomorphic to $\mathcal{N}^Y(B)$, and the converse is also true.
The containment of vertices in hyperedges leads to the bipartite vertex-hyperedge incidence graph.

Definition 1.3.10 *Let* $H = (V, \mathcal{E})$ *be a hypergraph. Then* $\mathcal{I}(H) = (V, \mathcal{E}, I)$ *is the following bipartite vertex-hyperedge* incidence graph.

$ve \in I$ *if and only if* $v \in e$ *for* $v \in V, e \in \mathcal{E}$.

A good example is the bipartite incidence graph of the clique hypergraph.

Definition 1.3.11 *Let* G *be a graph. Then* $B_C(G) = \mathcal{I}(\mathcal{C}(G))$ *is the bipartite vertex-clique incidence graph.*

Another important example is the following construction.

Definition 1.3.12 *Let* $G = (V, E)$ *be a graph. Then* $B(G) = (V', V'', E')$ *with*

$V' = \{v' : v \in V\}$, $V'' = \{v'' : v \in V\}$,

$E' = \{v'v'' : v \in V\} \cup \{u'v'' : uv \in E\} \cup \{u''v' : uv \in E\}$,

is the bigraph *of* G *containing two copies of* G *together with edges between the two copies of each vertex and crossing edges for each edge from* E.

Note that $B(G)$ is simply the bipartite vertex-neighborhood incidence graph, i.e., $B(G) = \mathcal{I}(\mathcal{N}(G))$.

Definition 1.3.13 *Let $B = (X, Y, E)$ be a bipartite graph. Then $split_X(B) = (X, Y, E_X)$ with $E_X = E \cup \{xx' : x, x' \in X\}$, i.e., $split_X(B)$ is obtained from B by completing X to a clique.*

For algorithmic purposes the following notions for hypergraphs are of importance.

Definition 1.3.14 *Let $H = (V, \mathcal{E})$ be a hypergraph.*

A subset $V' \subseteq V$ is a transversal *of H if every hyperedge contains a vertex from V', i.e., for all $e \in \mathcal{E}$, $e \cap V' \neq \emptyset$.*

$\tau(H) = \min\ \{|V'| : V'$ is a transversal of $H\}$ is the transversal number *of H.*

A subset $\mathcal{E}' \subseteq \mathcal{E}$ is a matching *of H if for all $e, e' \in \mathcal{E}'$ with $e \neq e'$, the intersection is empty: $e \cap e' = \emptyset$.*

$\nu(H) = \max\ \{|\mathcal{E}'| : \mathcal{E}'$ is a matching *of $H\}$ is the* matching number *of H.*

H fulfills the König property *if $\tau(H) = \nu(H)$.*

Note that the notion of a transversal (matching) of a hypergraph is a generalization of the vertex cover (matching) for graphs.

Hypertrees fulfill the König property.

Proposition 1.3.3 *If H is a hypertree, then $\tau(H) = \nu(H)$.*

There are similar hypergraph generalizations of the graph parameters $\alpha(G)$ and $\kappa(G)$ ($\omega(G)$ and $\chi(G)$, respectively).

Definition 1.3.15 *Let $H = (V, \mathcal{E})$ be a hypergraph.*

A subset $V' \subseteq V$ is a packing *of H if every hyperedge contains at most one vertex from V', i.e., for all $e \in \mathcal{E}$, $|e \cap V'| \leq 1$.*

$\alpha(H) = \max\ \{|V'| : V'$ is a packing of $H\}$ is the packing number *of H.*

A subset $\mathcal{E}' \subseteq \mathcal{E}$ is a covering *of H if the union of all $e \in \mathcal{E}'$ is V.*

$\kappa(H) = \min\ \{|\mathcal{E}'| : \mathcal{E}'$ is a covering of $H\}$ is the covering number *of H.*

Note that $\alpha(H) \leq \kappa(H)$ for all hypergraphs H.

1.4 Partial orders

Binary relations on finite sets are the edge sets of corresponding directed graphs. Among all binary relations there are some that are of crucial importance. One of them is the notion of a partial order.

Definition 1.4.1 *Let V be a set.*

(i) *R is a* binary relation *on V if $R \subseteq V \times V$.*

In the following, let R be a binary relation.

(ii) *R is* reflexive *on V if for all $v \in V$, $(v, v) \in R$.*

(iii) *R is* transitive *on V if for all $u, v, w \in V$, the following holds:*

$$\text{If } (u, v) \in R \text{ and } (v, w) \in R, \text{ then } (u, w) \in R.$$

(iv) *R is* antisymmetric *on V if for all $u, v \in V$, the following holds:*

$$\text{If } (u, v) \in R \text{ and } (v, u) \in R, \text{ then } u = v.$$

(v) *R is a* partial order *on V if R has the properties* (ii), (iii), *and* (iv). *(V, R) is then a* partially ordered set (poset).

We often denote partial orders by $\leq$ instead of R : $x \leq y$ if $(x, y) \in R$, and $x < y$ if $x \leq y$ and $x \neq y$.

Definition 1.4.2 *Let $P = (V, \leq)$ be a poset.*

P is finite *if V is finite.*

P is a linear order *if for all $u, v \in V$, $u \leq v$ or $v \leq u$ holds.*

An ordering $(v_1, \ldots, v_n)$ of V is a linear extension *of P if for all $v_i \leq v_j$, $i \leq j$.*

An important undirected graph corresponding to a poset is defined as follows.

Definition 1.4.3 *Let $P = (V, \leq)$ be a finite poset. Then*

$G_P = (V, E_P)$ with $xy \in E_P$ if $x < y$ or $y < x$ is the comparability graph *of the poset P;*

$G = (V, E)$ is a comparability graph *if there is a poset P such that $G \sim G_P$.*

Undirected graphs can be oriented by assigning a direction to each edge.

Definition 1.4.4 *Let $G = (V, E)$ be an undirected graph. Then the directed graph*

$G' = (V, E')$ is an orientation *of G if for all $xy \in E$, either $(x, y) \in E'$ or $(y, x) \in E'$ and for all $(x, y) \in E'$, $xy \in E$ holds;*

G' is a transitive orientation *of G if E' is a transitive binary relation on V;*

G' is an acyclic orientation *of G if there are no directed cycles in G'.*

If $G' = (V, E')$ is a directed graph, then

$G = (V, E)$ with $xy \in E$ if and only if $(x, y) \in E'$ or $(y, x) \in E'$ is the underlying undirected graph *of G'.*

Note that by definition G is a comparability graph if and only if G has an *acyclic* transitive orientation. The following stronger fact holds.

Theorem 1.4.1 (Ghouila-Houri [435]) *A graph is a comparability graph if and only if it has a transitive orientation.*

1.5 Modular decomposition

The notion of a module in a graph is a very basic concept with many applications not only in graphs but also in posets, hypergraphs, and other structures. Möhring and Radermacher [791] give an excellent survey for the various applications of the modular decomposition, which is frequently called *substitution decomposition.* Modules in graphs are defined as follows.

Definition 1.5.1 *Let $G = (V, E)$ be a graph. The subset $M \subseteq V$ is a* module *(or* homogeneous set*) in G if for all vertices $u, v \in M$ and $w \in V \setminus M$, $uw \in E$ if and only if $vw \in E$.*

In other words, the vertices inside M are indistinguishable to every vertex outside M. The concept of a module is of such basic importance that it was rediscovered several times under several names. Thus, modules also appear in the literature under the names *autonomous sets* [791], *closed sets* [416], *stable sets, clumps, committees, externally related sets* [494], *intervals, nonsimplifiable subnetworks*, and *partitive sets.*

Definition 1.5.2 *Let $G = (V, E)$ be a graph.*

> *$M \subseteq V$ is a* trivial module *in G if $M = V$, $M = \emptyset$, or $|M| = 1$. M is a* proper module *if $M \neq V$.*
>
> *G is a* prime graph *(sometimes called* primal graph*) if G contains only trivial modules. Otherwise G is* decomposable.
>
> *Two modules M, M' are* overlapping *if the sets $M \cap M'$, $M \setminus M'$, and $M' \setminus M$ are all nonempty.*
>
> *A module M is* strong *if for every module M', the modules M and M' do not overlap, i.e., $M \cap M' = \emptyset$ or $M \subseteq M'$ or $M' \subseteq M$ holds.*
>
> *Two vertices $x, y \in V$ are* twins *if $\{x, y\}$ is a module in G. Twins x, y are* true twins *if $xy \in E$, otherwise x, y are* false twins.

There are some well-known basic properties of modules.

Proposition 1.5.1 *Let $G = (V, E)$ be a graph and let A, B be two modules of G. Then the following properties hold:*

(i) *$A \cap B$ is a module.*

(ii) *If $B \not\subseteq A$, then $A \setminus B$ is a module.*

(iii) *If $A \cap B \neq \emptyset$, then $A \cup B$ is a module.*

(iv) *For $U \subseteq V$, the set $A \cap U$ is a module of $G(U)$.*

Corollary 1.5.1 *Let U_1, U_2, and M be subsets of V. If M is a module of the two subgraphs $G(U_1 \cup M)$ and $G(U_2 \cup M)$, then M is a module of the subgraph $G(U_1 \cup U_2 \cup M)$.*

Due to condition (iii) of Proposition 1.5.1 strong modules M have a nice property: It is easy to see that a strong module M' of smallest size properly containing M is uniquely determined (if M is not maximal). This defines the parent M' of M in the *modular decomposition tree* of G.

This tree structure is of basic importance for many applications such as efficient recognition algorithms for a number of graph classes.

There are the following special cases of decomposing a graph G into modules:

> If G is not connected, then between any two connected components of G there is a join in $\overline{G}$.
>
> If $\overline{G}$ is not connected, then between any two connected components of $\overline{G}$ there is a join in G.

Thus, the *series decomposition* divides G into the connected components of $\overline{G}$, whereas the *parallel decomposition* decomposes G into its connected components.

Theorem 1.5.1 (Gallai [416], Habib [491], Habib, Maurer [494], Sumner [1003]) *Let $G = (V, E)$ be a graph with at least two vertices. Then exactly one of the following conditions holds:*

(i) *G is not connected, and it can be decomposed into its connected components* (parallel decomposition)*;*

(ii) *$\overline{G}$ is not connected, and G can be decomposed into the connected components of $\overline{G}$* (series decomposition)*;*

(iii) *G and $\overline{G}$ are connected. There is some $U \subseteq V$ and a unique partition P of V such that*

  (a) $|U| > 3$,

  (b) *$G(U)$ is a maximal prime subgraph of G,*

  (c) *for every class S of the partition P, S is a module and $|S \cap U| = 1$.*

Each vertex of G forms a leaf of the decomposition tree. Each module M of G occurring as a node in the tree contains exactly the vertices that are leaves of the subtree rooted at M.

According to Theorem 1.5.1, the decomposition tree has three kinds of nodes: parallel nodes, series nodes, and prime nodes. Prime nodes correspond to case (iii) of Theorem 1.5.1; the set of leaf descendants of a prime node corresponds to a maximal prime subgraph of G.

An important special case are the decomposition trees containing only parallel and series nodes. These are the *cotrees*, and the graphs decomposable in this way are an interesting class.

Definition 1.5.3 *A graph is a* cograph *if its modular decomposition tree contains only parallel and series nodes.*

Cographs were independently rediscovered several times and studied systematically, e.g., in [725, 249, 256]. In fact, our definition is not the original notion of a cograph, but the equivalence between different conditions mentioned in Theorem 11.3.3 are now so well known that each is used as the definition of a cograph.

Theorem 1.5.2 (Sumner [1003, 1004]) *If G is a prime graph, then G contains an induced path P_4.*

It is easy to see that the P_4 is a smallest nontrivial prime graph (graphs with 3 vertices are decomposable). From Theorem 1.5.2 it follows that the graphs not containing any induced P_4 are the cographs. We come back to this graph class in several chapters. Theorem 11.3.3, as the main theorem on cographs, collects some characterizations.

The linear-time recognition algorithm for cographs given by Corneil, Perl, and Stewart [256] has been used by McConnell and Spinrad [775] and Cournier and Habib [262] to design linear-time algorithms for constructing the modular decomposition tree. Another important linear-time algorithm is given by Dahlhaus, Gustedt, and McConnell [280].

The linear-time bound for the modular decomposition is a major breakthrough for further algorithmic applications of this tool.

The following constructions occur in many places.

Definition 1.5.4 *Let $G = (V, E)$ be a graph.*

> *For a proper homogeneous set H in G and a vertex $v \in H$, the* homogeneous reduction *$HRed(G, H, v)$ is the subgraph of G induced by $V \setminus (H \setminus \{v\})$.*
>
> *For a graph H with $V(H) \cap V(G) = \emptyset$, the* homogeneous extension *(or* substitution*) $HExt(G, H, v)$ is the graph G' with vertex set $(V \setminus \{v\}) \cup V(H)$ obtained by substituting v with H such that the vertices of H have the same neighbors in $V \setminus \{v\}$ as v in G.*

1.6 Some basic algorithms and problems

There are some basic algorithms on graphs described in several textbooks—cf., e.g., Aho, Hopcroft, and Ullman [6], Even [360], and Golumbic [452]. All these books contain classic techniques such as DFS and breadth first search (BFS). For a more recent textbook see Cormen, Leiserson, and Rivest [243]. Special variants of graph searching such as lexicographic breadth first search (LexBFS) and maximum cardinality search (MCS), which are described below, are important tools for working with a variety of graph classes.

Let $s_1 = (a_1, \ldots, a_k)$ and $s_2 = (b_1, \ldots, b_l)$ be vectors of positive integers. Then s_1 is *lexicographically smaller* than s_2 ($s_1 < s_2$) if

(i) There is an index $i \leq \min\{k, l\}$ such that $a_i < b_i$ and $a_j = b_j$ for all $j = 1, \ldots, i-1$, or

(ii) $k < l$ and $a_i = b_i$ for all $i = 1, \ldots, k$.

If $s = (a_1, \dots, a_k)$ is a vector and a is some positive integer, then $s + a$ denotes the vector $(a_1, \dots, a_k, a)$.

procedure LexBFS
Input : A graph $G = (V, E)$.
Output : A LexBFS ordering $\sigma = (v_1, \dots, v_n)$ of V.

```
begin for all v ∈ V do l(v) := ();
      for n := |V| downto 1 do
          choose a vertex v ∈ V with lexicographically maximal label l(v);
          define σ(n) := v;
          for all u ∈ V ∩ N(v) do l(u) := l(u) + n;
          V := V \ {v};
      endfor;
end.
```

If the ties in LexBFS are always resolved by choosing a vertex of highest degree then this variant of LexBFS will be called *cardinality LexBFS* (CLexBFS).

The following algorithm, called maximum cardinality search, chooses its next vertex to be adjacent to the maximum number of previously selected vertices.

procedure MCS
Input : A graph $G = (V, E)$.
Output : An MCS ordering $\sigma = (v_1, \dots, v_n)$ of V.

```
begin for n := |V| downto 1 do
          choose a vertex v ∈ V with a maximal number of numbered neighbors;
          number v by n;
          σ(n) := v;
          V := V \ {v};
      endfor;
end.
```

The following theorem, found by Rose, Tarjan, and Lueker [935] and Tarjan and Yannakakis [1025], is a basic example of the power of LexBFS and MCS.

Theorem 1.6.1 [928, 1021] *The following conditions are equivalent:*

(i) *G is chordal;*

(ii) *Every LexBFS ordering $(v_1, \dots, v_n)$ of G is a perfect elimination ordering;*

(iii) *Every MCS ordering $(v_1, \dots, v_n)$ of G is a perfect elimination ordering.*

The main algorithmic problem mentioned in this survey is to recognize a graph class, i.e., to decide whether a given graph is in the class. Appendix A at the end of the survey collects information about the time complexity of recognition problems.

An interesting generalization of the recognition problem is the following *graph sandwich problem*.

Definition 1.6.1 *For two graphs $G_1 = (V, E_1)$ and $G_2 = (V, E_2)$ such that $E_1 \subseteq E_2$, the graph $G = (V, E)$ is a* sandwich graph *for G_1, G_2 if $E_1 \subseteq E \subseteq E_2$ holds.*

The sandwich problem for a graph property Π is the following:

Input: Two graphs $G_1 = (V, E_1)$ and $G_2 = (V, E_2)$ such that $E_1 \subseteq E_2$.

Question: Is there a sandwich graph between G_1, G_2 satisfying property Π?

For some results on this problem see [459, 464, 448]. For other algorithmic graph problems see the book of Garey and Johnson [419] and the survey papers of Johnson [628, 629, 630, 631, 632, 633, 634, 635].

1.7 Some special graphs

Sometimes graph classes and their inclusions occur in this survey in bold letters—thus, for instance, **chordal** means the class of all chordal graphs. The prefix "co-" means the class of complements of those graphs: for instance **co-chordal** = $\{G : \overline{G}$ is chordal$\}$.

For vertex-disjoint graphs $G = (V, E)$ and $H = (W, F)$, the union $G \cup H$ denotes the graph $(V \cup W, E \cup F)$. For a positive integer k the disjoint union of k copies of a graph G is denoted by kG.

There are some standard graphs that appear under the following names:

P_n, $n \geq 1$, denotes the path with n vertices and $n-1$ edges.

C_n, $n \geq 3$, denotes the cycle with n vertices and n edges.

K_n, $n \geq 1$, denotes the complete graph with n vertices.

$K_n - e$ denotes the graph obtained from K_n by deleting an edge. More generally, if H is a graph such that all graphs obtained from H by deleting an edge of H are isomorphic, then we shall write $H - e$ for the graph obtained from H by deleting an edge of H. $H = K_{l,l}$ is another example for which $H - e$ is defined.

W_n, $n \geq 3$, denotes the graph $C_n * K_1$ called an *n-wheel.*

BW_n, $n \geq 3$, denotes the graph consisting of a cycle C_{2n} $(v_1, v_2, \ldots, v_{2n})$ and a central vertex v adjacent to $v_2, v_4, \ldots, v_{2n}$ called a *bipartite wheel.*

For the notion of suns see Definition 7.2.1. Other special graphs are shown in Figures 1.1 and 1.2.

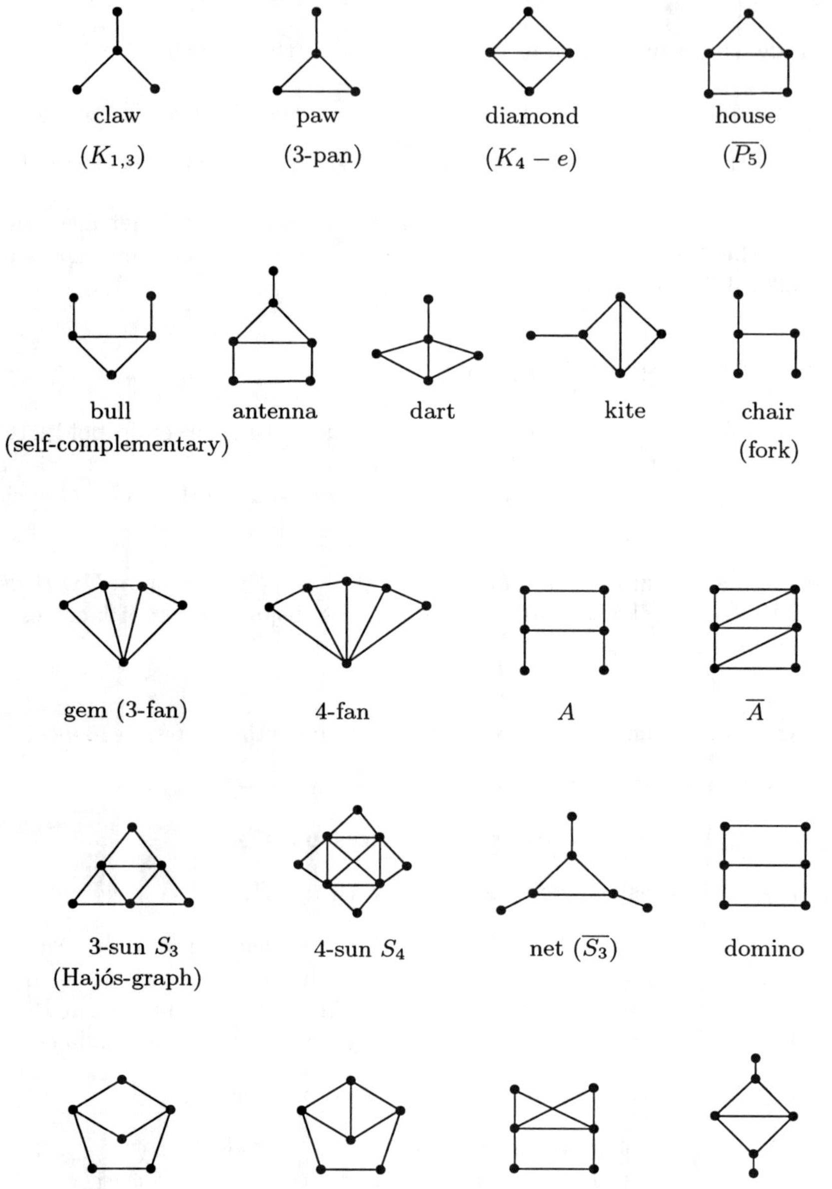

Figure 1.1: Special graphs.

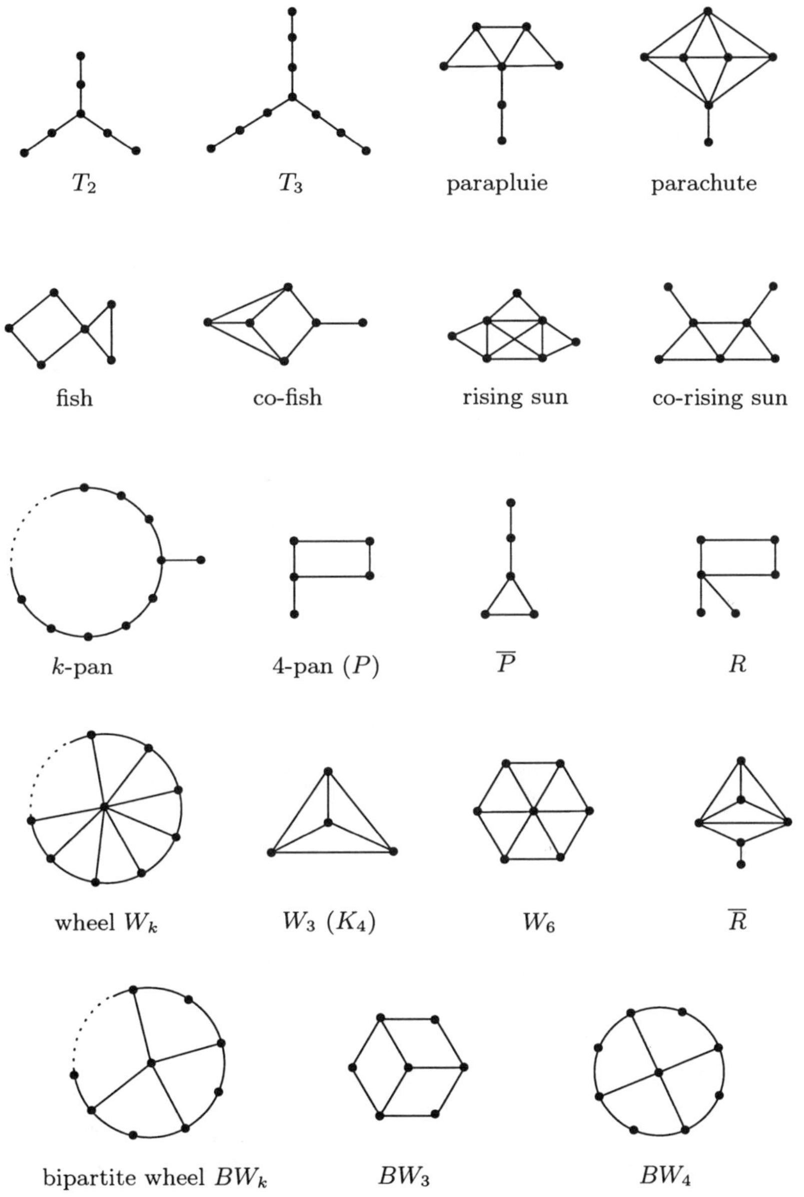

Figure 1.2: More special graphs.

Chapter 2

Perfection, Generalized Perfection, and Related Concepts

2.1 Perfect graphs and perfect graph theorems

Obviously, for every graph G, $\chi(G) \geq \omega(G)$ and $\kappa(G) \geq \alpha(G)$. This fact does not give much information about the structure of the graph G. Indeed, every graph augmented with a large clique satisfies the equality $\chi(G) = \omega(G)$. Around the early 1960s, Berge [88, 89] proposed the following notion; see also [95].

Definition 2.1.1 *A graph G is* perfect *if for all induced subgraphs H of G, $\chi(H) = \omega(H)$.*

As observed by Berge in the early 1960s [88, 89], several classical graph classes, such as bipartite graphs, chordal or triangulated graphs, comparability graphs, and complements of such graphs, are perfect. Perfect graphs are very interesting from an algorithmic point of view. While determining the clique number and the chromatic number of a graph are $\mathbb{NP}$-complete problems (even for very special graph classes [420]), they are solvable in polynomial time for perfect graphs. This fact is a deep result due to Grötschel, Lovász, and Schrijver [476] (see the matrix chapter (chapter 9)). Unfortunately, their algorithm is based on the ellipsoid method for convex programming and therefore is rather impractical.

For more background information on perfect graphs see [96, 477, 626, 677, 739, 1027]. The monograph of Golumbic [452] is an excellent reference with emphasis on the classical perfect graph classes.

We shall first give some known alternative characterizations of perfect graphs. The following one is obvious by definition.

Proposition 2.1.1 *A graph G is perfect if and only if every induced subgraph H of G has a stable set meeting all maximum cliques of H.*

The following natural weighted version of Proposition 2.1.1 is given by Boros and Gurvich [134]. Let $w(v)$ be a nonnegative weighting of the vertices v of a graph G. The weight of a set $S \subseteq V(G)$ is defined by $w(S) = \sum_{v \in S} w(v)$ and the weighted clique number $\omega(G, w)$ is the maximum of $w(C)$ for the cliques C of G.

Proposition 2.1.2 *A graph G is perfect if and only if for each nonnegative weighting $w : V(G) \to \mathbb{R}$, there exists a stable set meeting all cliques of maximum weight.*

Proposition 2.1.2 coincides with Proposition 2.1.1 when w is a 0, 1-weighting. Another generalization of Proposition 2.1.1 is based on dominating sets; see Definition 1.1.18.

Note that every maximal stable set is minimal dominating. Hammer and Maffray [512] proved the following generalization of Proposition 2.1.1.

Proposition 2.1.3 *A graph G is perfect if and only if every induced subgraph H of G has a minimal-dominating set meeting all maximum cliques of H.*

A *weighted coloring* of (G, w) consists of a family of stable sets $S_1, S_2, \ldots, S_k$ and integers $I(S_1), I(S_2), \ldots, I(S_k)$ such that for each vertex v, $w(v) \leq \sum_{v \in S_i} I(S_i)$. The weighted chromatic number $\chi(G, w)$ is the minimum of $\sum I(S_i)$ over all weighted colorings. The following characterization of perfect graphs is given in [477].

Theorem 2.1.1 *A graph G is perfect if and only if for every nonnegative weighting w, $\chi(G, w) = \omega(G, w)$.*

Definition 2.1.2 *A k-coloring with the color classes $S_1, \ldots, S_k$ of a graph G is* canonical *if for every vertex v of G the following holds: If v is in S_h, then there is a clique K containing v such that $K \cap S_i \neq \emptyset$ for $i \in \{1, \ldots, h\}$.*

As noted by Preissmann and de Werra [877], perfect graphs can be characterized in terms of canonical colorings.

Proposition 2.1.4 *A graph G is perfect if and only if every induced subgraph of G has a canonical coloring.*

Preissmann [876] pointed out that if one can optimally color every subgraph of a perfect graph G in polynomial time, then one has a polynomial algorithm to find a canonical coloring of G.

In [829], Olariu noted that the class of perfect graphs can be obtained from the class of graphs with at most two vertices by taking the "closure" of that class under certain predicates. However, since the operations are applied only if the predicates are true for all induced subgraphs, this does not imply that perfect-graph recognition is in $\mathbb{NP}$.

Among results concerning perfect graphs, perhaps the most important is the Perfect Graph Theorem (PGT) conjectured by Berge in 1961 and proved by Lovász [735] in 1972. Independently, the most difficult part of this proof was done by Fulkerson [406, 407, 408].

Theorem 2.1.2 (The Perfect Graph Theorem) *The complement of a perfect graph is perfect.*

Thus, a graph G is perfect if and only if for each induced subgraph H of G, $\alpha(H) = \kappa(H)$.

The PGT implies a number of famous min-max theorems in combinatorics. For example, it implies three well-known theorems of König on bipartite graphs, as well as Dilworth's theorem for partially ordered sets. Cameron, Edmonds, and Lovász [178] gave the following generalization of the PGT.

Theorem 2.1.3 *Let the edges of a complete graph be colored with three colors in such a way that no triangle gets three different colors. Suppose that the two monochromatic graphs formed by two of the three colors are perfect. Then so is the graph formed by the remaining color.*

In [177], Cameron and Edmonds obtained the following abstraction of Theorem 2.1.3.

Theorem 2.1.4 *Let $\mathcal{C}$ be a nonempty class of graphs closed under taking induced subgraphs, complements, and substitution. Let the edges of a complete graph be colored with three colors in such a way that no triangle gets three different colors. Suppose that the two monochromatic graphs formed by two of the three colors belong to $\mathcal{C}$. Then so does the graph formed by the remaining color.*

By the PGT and the substitution lemma (Lemma 2.4.1), the class of perfect graphs is just one such class $\mathcal{C}$ in Theorem 2.1.4.

The PGT immediately follows from the following important characterization of perfect graphs found by Lovász [736].

Theorem 2.1.5 *A graph G is perfect if and only if for each induced subgraph H of G, $\omega(H) \cdot \omega(\overline{H}) \geq |H|$.*

After Lovász [736], several proofs for Theorems 2.1.2 and 2.1.5 were given. Perz and Polewicz [868] gave an alternative proof for the PGT by considering certain norms related to graphs. Following ideas of Fulkerson [406, 407], Padberg [847] gave a proof of Theorem 2.1.5 using convex polyhedra. Zaremba [1096] showed that the inequality stated in Theorem 2.1.5 is exactly the classical Schwarz inequality for certain 0,1-vectors of the space $\mathbb{R}^n$ with a suitable norm. Recently, Gasparian [422] presented a very short and simple proof (see also Theorem 14.1.1). In contrast to all former proofs, the proof of Gasparian does not use the Lovász substitution lemma for perfect graphs (Lemma 2.4.1).

Gurvich [481] gave the following weighted version of Theorem 2.1.5.

Theorem 2.1.6 *A graph G is perfect if and only if for all nonnegative weightings x and y, $\omega(G, x) \cdot \omega(\overline{G}, y) \geq \sum_{v \in V(G)} x(v) \cdot y(v)$.*

A further possible characterization of perfect graphs could be the famous *strong perfect graph conjecture* (SPGC), posed by Berge [89] in 1961. Since the odd holes and odd antiholes are clearly imperfect, all perfect graphs are odd-hole–free and odd-antihole–free. The following notion goes back to Chvátal and Sbihi [222, 214].

Definition 2.1.3 *A graph G is* Berge *if it is odd-hole–free and odd-antihole–free, i.e., G and $\overline{G}$ contain no induced C_{2k+1} ($k \geq 2$).*

Are Berge graphs perfect? The SPGC means that the answer to this question is yes.

Conjecture 2.1.1 (The Strong Perfect Graph Conjecture) *Berge graphs are perfect.*

The validity of the SPGC would imply the PGT (and Theorem 2.1.5 as well); hence the addition "strong." The SPGC chapter (chapter 14) discusses this conjecture in detail.

There are also non-graph-theoretic characterizations of perfect graphs. As noted after Definition 2.1.1, perfect graphs have a nice feature in discrete optimization; they are exactly the graphs for which the stable polyhedron and the clique polyhedron have integral vertices [677, 477, 96]; see the matrix chapter (chapter 9).

Furthermore, perfect graphs can be characterized in terms of polynomials given by Lovász [741] as well as in terms of graph entropies by Cziszár et al. [274]. See [684] for further discussion on graph entropies and perfect graphs.

Problem 2.1.1 *Is there a polynomial-time algorithm that tests whether a graph is perfect?*

This is unknown even assuming the validity of the SPGC. Lovász [740] conjectures that a polynomial-time algorithm for recognizing perfect graphs exists; see also the paper of Knuth [677]. However, so far it is not even known whether the recognition of perfect graphs can be done by a nondeterministic polynomial-time algorithm. Cameron [175] observed that the recognition problem for perfect graphs belongs to co-$\mathbb{NP}$. Namely, if G is *not* perfect, then Lovász's characterization of perfect graphs (Theorem 2.1.5) shows that G must contain an induced partitionable subgraph; see the SPGC chapter (chapter 14). The certificate of being partitionable can be verified, by definition, in polynomial time.

We note that Wagler [1071] considered an extremal case of perfection. She calls a perfect graph G *critically perfect* if, for each edge e of G, $G-e$ is imperfect. Investigation of critically perfect graphs would certainly help us understand the structure of perfect graphs. In view of the PGT, it is interesting to characterize those graphs G such that G and $\overline{G}$ both are critically perfect.

2.2 A semistrong perfect graph theorem

In 1984, Chvátal [204] introduced the term graph P_4-isomorphism and conjectured that the perfection of a graph depends only on its P_4-structure.

Definition 2.2.1

> *Two graphs G, H on the same vertex set are* P_4-isomorphic *if for each subset S of four vertices, S induces a P_4 in G if and only if it induces a P_4 in H.*
>
> *The* P_4-structure *of a graph $G = (V, E)$ is the (4-uniform) hypergraph H having V as its vertex set and all 4-element vertex sets inducing a P_4 in G as its hyperedges.*

Chvátal conjectured that two P_4-isomorphic graphs are both perfect or both imperfect. In 1987, this conjecture was proved by Reed.

Theorem 2.2.1 **[905]** *Let G and H be two P_4-isomorphic graphs. Then G is perfect if and only if H is perfect.*

Observe that since the P_4 is self-complementary, a graph and its complement are P_4-isomorphic. Thus Theorem 2.2.1 implies the PGT. On the other hand, Chvátal [204] demonstrated that the SPGC implies Theorem 2.2.1. For this reason, Theorem 2.2.1 is often called a *semistrong perfect graph theorem.* Theorem 2.2.1 motivates the study of perfect graphs via the P_4-structure. Subsequently we will address results in this direction concerning

- decomposition schemes of perfect graphs that are derived from the P_4-structure;
- recognition of the P_4-structure of perfect graphs.

Theorem 2.2.1 also suggests the definition of classes of perfect graphs in terms of the P_4-structure only. Results from this point of view will be given in the SPGC chapter (sections 14.3 and 14.4).

Red-white decomposition theorems Let G be a graph whose vertices are colored red and white. What are the constraints on the way in which each P_4 in G may be colored, such that G is perfect if and only if the two subgraphs of G induced by all the vertices of the same color are perfect?

We classify the P_4's $abcd$ in G in the following way:

Type 1	if	a, b, c, d are red,
Type 2	if	a is white and b, c, d are red,
Type 3	if	a, c, d are red and b is white,
Type 4	if	a, b are red and c, d are white,
Type 5	if	a, c are red and b, d are white,
Type 6	if	a, d are red and b, c are white,
Type 7	if	a, d are white and b, c are red,
Type 8	if	a is red and b, c, d are white,
Type 9	if	a, c, d are white and b is red,
Type 10	if	a, b, c, d are white.

Furthermore, set

$S_1 = \{2, 3, 8, 9\}$, $\quad S_2 = \{1, 6, 7, 10\}$, $\quad S_3 = \{2, 3, 6, 7, 10\}$,
$S_4 = \{2, 5, 6, 7, 10\}$, $\quad S_5 = \{2, 5, 6, 7, 8\}$, $\quad S_6 = \{2, 7, 9\}$.

Chvátal, Lenhart, and Sbihi [222] proved that, up to "switching colors" and "complementing colors," the following theorem gives a complete answer to the question above.

Theorem 2.2.2 **[222]** *Consider a fixed i with $1 \leq i \leq 6$. If the vertices of a graph G are colored red and white in such a way that there is no P_4 of type $t \in S_i$, then G is perfect if and only if each of the two subgraphs of G induced by all the vertices of the same color is perfect.*

For the case $i = 1$, Theorem 2.2.2 was proved earlier by Chvátal and Hoàng [218]. In this case every P_4 has an even number of vertices of each color, and therefore the result is often called the *even decomposition theorem.* A special case of even decomposition is discussed by Olariu [828].

The corresponding *odd decomposition theorem* was proved by Hoàng [554]; here every P_4 has an odd number of vertices of each color. Clearly, Theorem 2.2.2 for the case $i = 2$ (independently proved by Gurvich [482]) is a generalization of the odd decomposition theorem. In [566] Hoàng and Le discussed some results related to Theorem 2.2.2.

Both even and odd decomposition theorems were generalized by Chvátal [208] as follows. Two vertices x, y of a graph are *partners* (or *siblings*) if there is a set Q such that both $Q \cup \{x\}$ and $Q \cup \{y\}$ induce a P_4 in that graph. Chvátal [210] then proved the following *partner decomposition theorem.*

Theorem 2.2.3 **[210]** *If the vertices of a graph G are colored red and white in such a way that every pair of partners has the same color, then G is perfect if and only if each of the two subgraphs of G induced by all the vertices of the same color is perfect.*

Hoàng [557] defined two graphs G and H to be *partner isomorphic* if there is a bijection $f : V(G) \to V(H)$ such that vertices x, y are partners with respect to a set Q in G if and only if $f(x)$ and $f(y)$ are partners with respect to $f(Q)$ in H. Clearly, P_4-isomorphic graphs in particular are partner isomorphic. Hence, the following theorem of Hoàng [557] generalizes Reed's semistrong perfect graph theorem.

Theorem 2.2.4 *Let G and H be two partner-isomorphic graphs. G is perfect if and only if H is perfect.*

Further discussions concerning Reed's theorem are given by Hougardy [584, 583]. Chvátal and Hoàng noted in [218] that, given a graph G, one can check in polynomial time whether or not the vertices of G can be colored red and white such that no P_4 is of type $t \in S_1$. The remaining five cases are still open.

Problem 2.2.1 *Consider a graph G and an integer $2 \leq i \leq 6$. Is there a polynomial-time algorithm that decides whether or not G has a (nontrivial) red-white coloring such that no P_4 is of type $t \in S_i$?*

See [213] for comments on this problem.

Recognizing the P_4-structure of perfect graphs Is there a polynomial-time recognition algorithm for perfect graphs? Given a graph $G = (V, E)$, Theorem 2.2.1 suggests building the P_4-structure of G. Then G is perfect if and only if there is a perfect graph $G' = (V, E')$ having the same P_4-structure as G. Therefore, to recognize perfect graphs, it is enough to recognize the P_4-structure of perfect graphs.

Problem 2.2.2 *Is there a polynomial-time algorithm for the following decision problem: Given a* 4*-uniform hypergraph* $H = (V, \mathcal{E})$*, is there a perfect graph* $G = (V, E)$ *having* H *as its* P_4*-structure?*

Problem 2.2.2 remains open. A first partial result toward this problem was made by Ding [315]. He proved that the P_4-structure of trees can be recognized in polynomial time. Based on a simpler approach, Brandstädt, Le, and Olariu [152] improved the time bound to $\mathcal{O}(|V|^2)$. Brandstädt, Le, and Olariu [150, 153] extended the approach for trees in [152] to block graphs and certain bipartite graphs. As a summary, we list currently known results in the following theorem.

Theorem 2.2.5 *There is an efficient recognition algorithm for the* P_4*-structure of*

(i) *trees* [315, 152],

(ii) *block graphs* [150],

(iii) *bipartite graphs without* 4*-pan and* C_6 [153],

(iv) *split graphs* [531],

(v) *bipartite graphs* [38],

(vi) *line graphs of bipartite graphs* [987].

Hayward, Hougardy, and Reed [531] have shown that there is a polynomial-time algorithm for deciding whether a given 4-uniform hypergraph is the P_4-structure of a graph.

Another interesting problem on the P_4-structure of perfect graphs is as follows. Given a graph G and a class $\mathcal{C}$ of perfect graphs, does G have the P_4-structure of a graph in $\mathcal{C}$? Graphs P_4-isomorphic to a member in $\mathcal{C}$ are perfect by Theorem 2.2.1 and said to be $\mathcal{C}$-*perfect*. Brandstädt and Le [151] described tree- and forest-perfect graphs; their description leads to a linear-time recognition of these graphs. Bipartite-perfect graphs are characterized in [713].

There is an interesting variant of the P_4-structure of graphs. In general, one can take the $\mathcal{F}$-*structure* for a family $\mathcal{F}$ of graphs and try to show that a graph is perfect if and only if it has the $\mathcal{F}$-structure of a perfect graph. This approach was taken by Hoàng [565] and leads to the following results.

Theorem 2.2.6 [565]

(i) *If a* C_5*-free graph* G *has the* $\{2K_2, C_4\}$*-structure of a perfect graph, then* G *is perfect.*

(ii) *If a* C_5*-free graph* G *has the* {*paw, co-paw*}*-structure of a perfect graph, then* G *is perfect.*

2.3 Sharpening perfection

The properties of perfect graphs and especially the characterizations of them given by Propositions 2.1.1 and 2.1.3 lead to some sharpenings of perfection.

Definition 2.3.1 *Let G be a graph.*

> *G is* strongly perfect *if each induced subgraph H of G has a stable set meeting all maximal cliques in H* [98].
>
> *G is* very strongly perfect *if for every induced subgraph H of G, every vertex of H belongs to a stable set of H meeting all maximal cliques in H* [555].
>
> *G is* absorbantly perfect *if every induced subgraph H of G has a minimal dominating (absorbant) set meeting all maximal cliques in H* [512].
>
> *G is* absolutely perfect *if every induced subgraph H of G has a stable set meeting all minimal vertex sets of H that are dominating the complement graph $\overline{H}$* [84].

Since these graph classes are hereditary, it is natural to look for a characterization in terms of forbidden induced subgraphs. No such characterization of these graph classes is known.

It follows immediately from the definition that all very strongly perfect graphs are strongly perfect, and since every maximal stable set is a minimal dominating set, all strongly perfect graphs are absorbantly perfect. Finally, by Proposition 2.1.3, all absorbantly perfect graphs are perfect.

The vertex sets of a graph H dominating $\overline{H}$ are called *free sets* in [84]. It is easy to see that maximal cliques have this property. Thus, absolutely perfect graphs are a special case of strongly perfect graphs. Benzaken et al. [84] show that a graph is absolutely perfect if and only if it is co-chordal.

Hoàng [555] has shown that very strongly perfect graphs are exactly those graphs in which every cycle of odd length at least five has at least two chords (known as *Meyniel graphs*; see Definition 3.5.1, Theorem 3.5.1, and related classes).

Another sharpening of perfection was proposed by Hoffman and Johnson (see [452]).

Definition 2.3.2 *Let $w(v)$ be a nonnegative weighting of the vertices of a graph G. An* interval coloring *of (G, w) maps each vertex v onto an open interval I_v of the real line of width $w(v)$ such that adjacent vertices are mapped to disjoint intervals. $\chi_{Int}(G, w) = \min\{|\cup I_v| : (I_v)_{v \in V(G)}$ is an interval coloring of $(G, w)\}$ is the* interval chromatic number *of (G, w).*

Note that for every weighting w, $\chi_{Int}(G, w) \geq \omega(G, w)$, and if $w(v) = 1$ for all vertices v, then $\chi_{Int}(G, w) = \chi(G)$, $\omega(G, w) = \omega(G)$. Hoffman and Johnson then proposed the following.

Definition 2.3.3 *A graph G is* superperfect *if for every nonnegative weighting w, $\chi_{Int}(G, w) = \omega(G, w)$.*

In [452] a characterization of superperfect graphs, due to Hoffman and Johnson, is given. By considering the 0,1-weightings of a superperfect graph, one sees easily that all superperfect graphs are perfect.

The papers [20, 669] contain some results on superperfect graphs.

2.4 Properties of perfect graphs

Nontrivial properties of perfect graphs would be helpful in finding a (good) characterization of perfect graphs. Besides the properties following trivially from the definition or Theorems 2.1.2 and 2.1.5, very few others are known.

We shall begin with a fundamental construction that was used in Lovász's proofs of Theorems 2.1.2 and 2.1.5; see [735]. For the notion of $HExt(G, H, v)$, see Definition 1.5.4.

Lemma 2.4.1 (The substitution lemma) *Let G and H be vertex disjoint perfect graphs, and let v be a vertex of G. Then the graph $HExt(G, H, v)$ resulting from substituting v with H is perfect.*

Using Lemma 2.4.1, Preissmann [876] found that perfect graphs have a coloring with a certain local property.

Proposition 2.4.1 *Let G be a perfect graph and v be a vertex of G. There exists a coloring of G with $\omega(G)$ colors such that only $\omega(N(v))$ colors appear in $N(v)$.*

This motivates the following notion.

Definition 2.4.1 *Let $G = (V, E)$ be a graph.*

> *Let C be a proper coloring of G and for a vertex $v \in V$ let $F(v, C)$ denote the number of colors of C appearing in the neighborhood $N(v)$. The coloring C is* locally perfect *if for every vertex $v \in V$, $F(v, C) = \omega(N(v))$ holds.*
>
> *G is* locally perfect *if every induced subgraph of G has a locally perfect coloring.*

Preissmann [876] proved that all K_4-free perfect graphs are locally perfect. Moreover, Meyniel graphs are locally perfect as shown by Hertz [543].

The next property of perfect graphs was found by Fournier and Las Vergnas [405]; see also [739].

Theorem 2.4.1 *Let G be a perfect graph every vertex of which has two nonadjacent neighbors. Then the maximal cliques of G can be 2-colored in such a way that every vertex is contained in cliques of both colors.*

Definition 2.4.2 [306] *A graph G is* normal *if there exist set families $\mathcal{C}$ and $\mathcal{S}$ such that*

- *the families are* cross intersecting*: for all $C \in \mathcal{C}$ and $S \in \mathcal{S}$, $C \cap S \neq \emptyset$;*

- *every member of $\mathcal{C}$ induces a clique and every member of $\mathcal{S}$ induces a stable set in G; and*
- *every vertex of G is obtainable as an intersection of members of $\mathcal{C}$ and $\mathcal{S}$.*

In [685] Körner showed the following property.

Proposition 2.4.2 *All perfect graphs are normal.*

Normal graphs are discussed in more detail by De Simone and Körner [304].

Definition 2.4.3 [98] *Let $D = (V, A)$ be a directed graph. A set $S \subseteq V$ is a* kernel *of G if it is both a* stable set, *i.e., no vertex in S has a successor in S, and an* absorbant set, *i.e., every vertex outside S has a successor in S.*

Note that a kernel in a directed clique consists simply of exactly one vertex that is a successor of all other vertices in the clique.

Definition 2.4.4 [98]

> *A* superorientation *of an undirected graph $G = (V, E)$ is a digraph $D = (V, A)$ obtained from G by orienting all its edges arbitrarily, where for some edge $uv \in E$ both arcs (u, v) and (v, u) may occur in A.*
>
> *A graph G is* kernel solvable *if for every superorientation D of G for which every clique has a kernel, D itself has a kernel.*

Berge and Duchet [99, 97] conjectured in 1983 and Boros and Gurvich [133] recently proved, using game theoretic concepts (see also [134, 135]) the following theorem; for a shorter proof, see [5].

Theorem 2.4.2 *Perfect graphs are kernel solvable.*

Earlier results had proved Theorem 2.4.2 for the following special cases: line graphs [755], Gallai graphs [754], parity graphs [336], and complements of strongly perfect graphs [112].

Berge and Duchet [99, 97] also conjectured that kernel solvability will characterize perfect graphs.

Conjecture 2.4.1 *Kernel-solvable graphs are perfect.*

Note that odd holes and odd antiholes are not kernel solvable, hence the SPGC would imply the validity of Conjecture 2.4.1. The papers [133, 755] contain partial results for this conjecture. An alternative form of the conjecture, in terms of effectivity functions, is given in [134].

The following notions are related to line graphs.

Definition 2.4.5 *Let G be a graph. Two edges $e \neq e'$ of G are*

Γ-edges *if* e *and* e' *are incident but do not form a triangle in* G*;*

Δ-edges *if they span a triangle in* G.

$\chi_\Gamma(G)$ *denotes the smallest number of colors needed to color the edges of* G *so that* Γ*-edges are distinctly colored.* $\chi_\Delta(G)$ *is defined analogously.*

Le [709, 711] proved that for all graphs G, $\chi_\Gamma(G) \leq \chi(\overline{G})$ and $\chi_\Delta(G) \leq \chi(G)$, which has the following implications for perfect graphs.

Proposition 2.4.3 *Let* G *be a perfect graph. Then* $\chi_\Gamma(G) \leq \alpha(G)$ *and* $\chi_\Delta(G) \leq \omega(G)$.

It is conjectured that the first inequality in Proposition 2.4.3 will also characterize perfect graphs; see the SPGC chapter (section 14.5).

Other interesting properties of perfect graphs are given by Cai and Corneil [173], based on Definition 2.4.6.

Definition 2.4.6 *Let* G *be a graph and* i *be a positive integer.*

A K_{i+1}*-free* k*-coloring of* G *is a partition of the vertex set of* G *into* k *subsets each of which induces a* K_{i+1}*-free subgraph of* G.

$\chi_i(G)$ *denotes the smallest number* k *for which* G *has a* K_{i+1}*-free* k*-coloring.*

$\omega_i(G)$ *denotes* $\lceil \frac{\omega(G)}{i} \rceil$.

An i-transversal *of* G *is an induced subgraph* G' *of* G *with* $\omega(G') = i$ *and* $\omega(G - G') = \omega(G) - i$.

Theorem 2.4.3 *For all positive integers* i*, all perfect graphs* G *satisfy* $\chi_i(G) = \omega_i(G)$ *and have a* $\min\{\omega(G), i\}$*-transversal.*

See also the remarks at the end of the next section.

2.5 Generalized perfection

There are many generalizations of the parameters χ, ω, α, and κ of a graph; the generalized parameters also satisfy the min-max inequalities. This leads to several concepts containing perfection as a special case.

The first such generalization is due to Hell and Roberts [538], and is based on the following notion due to Harary [519] and Sabidussi [939].

Definition 2.5.1 *Let* G *and* G' *be graphs. The* lexicographic product $G \circ G'$ *is the graph with vertex set* $V(G) \times V(G')$ *and edge set*

$$\{(g_1, g_1')(g_2, g_2') : g_1 g_2 \in E(G) \vee (g_1 = g_2 \wedge g_1' g_2' \in E(G'))\}.$$

Note that in general $G \circ G' \not\simeq G' \circ G$.

Definition 2.5.2 *For graphs G and positive integers n let*

$$\omega_n(G) = \max\{|F| : F \text{ induced subgraph of } G \circ \overline{K_n} \text{ with } \alpha(F) \leq n\} \text{ and}$$

$$\chi_n(G) = \chi(G \circ K_n).$$

Hell and Roberts [540] then proved that for a given graph G there exists n such that, for each induced subgraph H of G, $\chi_n(H) = \omega_n(H)$ (but $\chi_{n+1}(G) \neq \omega_{n+1}(G)$ is possible.) They then said that every graph is "n-perfect" for some n and therefore, "1-perfect graphs" are exactly the perfect graphs. This leads to the following definition.

Definition 2.5.3 *The* degree of perfection *of a graph G is the least positive integer n such that, for each induced subgraph H of G, $\chi_n(H) = \omega_n(H)$.*

Evidently, perfect graphs (and, by the PGT, also their complements) have degree of perfection 1. For a graph G with a higher degree of perfection, G and its complement $\overline{G}$ do not necessarily have the same degree of perfection: In [538] it is shown that the degree of perfection of C_{2k+1} is k, while that of $\overline{C_{2k+1}}$ is 2. It would be interesting to characterize those graphs G such that G and $\overline{G}$ both have the same degree of perfection n; the case $n = 2$ is of particular interest.

For the next generalization of perfect graphs see Definition 1.3.15. Let $G = (V, E)$ be a graph and let $H = (V, \mathcal{E})$ be a hypergraph on the same vertex set. Recall that a subset $\mathcal{E}' \subseteq \mathcal{E}$ is a covering of $\mathcal{E}$ if the union of all $e \in \mathcal{E}'$ is V, and the covering number $\kappa(H)$ is the smallest number of hyperedges covering H. Recall further that a subset $V' \subseteq V$ is a packing of H if every hyperedge of H contains at most one vertex from V' and the packing number $\alpha(H)$ is the largest size of a packing set of H.

Obviously, $\alpha(H) \leq \kappa(H)$ for all hypergraphs H, and for the clique hypergraph $\mathcal{C}(G)$ of a graph G, $\kappa(\mathcal{C}(G)) = \kappa(G)$ and $\alpha(\mathcal{C}(G)) = \alpha(G)$. Thus, for perfect graphs, $\alpha(\mathcal{C}(G)) = \kappa(\mathcal{C}(G))$, and the same holds for induced subgraphs $G' = G(U)$ of G and their corresponding clique subhypergraphs restricted to the vertex set U.

Chang and Nemhauser [184] considered the concept of $\mathcal{E}$-perfect graphs as described below; the notation used in [184] differs from our notation here. They assign to each induced subgraph a corresponding subhypergraph of $\mathcal{E}$, thus generalizing the special case of clique hypergraphs.

Definition 2.5.4 *Let G be a graph and let $\mathcal{E}(G)$ be a family of subsets of $V(G)$. G is* $\mathcal{E}$-perfect *if for each induced subgraph G' of G and its corresponding subhypergraph $\mathcal{E}(G')$, $\alpha(\mathcal{E}(G')) = \kappa(\mathcal{E}(G'))$.*

As already mentioned, every $\mathcal{C}(G)$-perfect graph G is perfect and vice versa. In [184], the $\mathcal{T}_k$-perfect graphs are studied; here $\mathcal{T}_k$ is the set of all trees of diameter at most k that are subgraphs of G. Among other results, graphs that are $\mathcal{T}_k$-perfect for each $k \geq 2$ and graphs that are $\mathcal{T}_k$-perfect for all even k are characterized; the latter made the assumption of the validity of the SPGC; see Theorem 7.2.3.

Note also that $\kappa(\mathcal{T}_2)$ is the *domination number* $\gamma(G)$, that is, the minimum cardinality of a dominating set of G; see Definition 1.1.18 and the subsequent remarks.

Another generalization of the parameters α and κ is due to Lovász [739].

Definition 2.5.5 *Let q be a positive integer. $\alpha_q(G)$ denotes the largest number of vertices of a graph G that induce a q-colorable subgraph. $\kappa_q(G)$ denotes the smallest number of $\sum_{j=1}^{k} \min \{|C_j|, q\}$ over all clique partitions $C_1, \ldots, C_k$ of G.*

Note that $\alpha_q(G) \leq \kappa_q(G)$ and that $\alpha_1(G) = \alpha(G)$, $\kappa_1(G) = \kappa(G)$. Lovász [739] then proposed the following.

Definition 2.5.6 *Let q be a positive integer. A graph G is q-*perfect *if for every induced subgraph H of G, $\alpha_q(H) = \kappa_q(H)$.*

Thus, 1-perfect graphs are exactly the perfect graphs. Note that a perfect graph is not necessarily q-perfect and q-perfection does not imply $(q+1)$-perfection and vice versa. However, Berge [92, 94] noted that all graphs are q-perfect for all $q \geq \chi(G)$ and proved that 2-perfect graphs are perfect.

Thus, the following classes of perfect graphs would be interesting.

Definition 2.5.7 *A graph G is* odd perfect (even perfect) *if G is q-perfect for each odd q (for each even q, respectively).*

As mentioned above, the classes of odd-perfect graphs and of even-perfect graphs are incomparable. It seems difficult to characterize these graph classes, as noted by Lovász [739] and Berge [94]. Examples of graphs that are odd *and* even perfect are the comparability graphs (by a theorem of Greene and Kleitman [471]), and their complements (by a theorem of Greene [470]). See also [93, 94, 176].

The next concept of generalized perfection is due to Scheinerman and Trenk [957]. They generalized the chromatic number and the clique number as follows.

Definition 2.5.8 *Let $\mathcal{P}$ be a nonempty, hereditary graph class. For a graph G, $\chi_{\mathcal{P}}(G)$ denotes the minimum size of a partition $V(G) = V_1 \cup \cdots \cup V_k$ such that each subgraph G_i of G induced by V_i is (isomorphic to) a member of $\mathcal{P}$. $\omega_{\mathcal{P}}(G)$ denotes the maximum of $\chi_{\mathcal{P}}(C)$ for the cliques C of G.*

Note that $\chi_{\mathcal{P}}(G) \geq \omega_{\mathcal{P}}(G)$, and if $\mathcal{P}$ is the class of edgeless graphs, then $\chi_{\mathcal{P}}(G) = \chi(G)$ and $\omega_{\mathcal{P}}(G) = \omega(G)$. Scheinerman and Trenk [964] then proposed the following.

Definition 2.5.9 *Let $\mathcal{P}$ be a nonempty, hereditary graph class. A graph G is $\mathcal{P}$-*perfect *if, for each induced subgraph H of G, $\chi_{\mathcal{P}}(H) = \omega_{\mathcal{P}}(H)$.*

For a number of specific classes $\mathcal{P}$, the $\mathcal{P}$-perfect graphs can be well characterized [957, 1028]. The classes $\mathcal{P}_n$ consisting of all K_n-free graphs (n a fixed integer) play a central role in this work. Trenk [1028] has shown that in order to characterize the $\mathcal{P}$-perfect graphs, it suffices to characterize the $\mathcal{P}_n$-perfect graphs for all $n \geq 2$. Note that the $\mathcal{P}_2$-perfect graphs are exactly the perfect graphs and the $\mathcal{P}_{i+1}$-perfect graphs are the same as the "i-perfect graphs" studied in more detail by Cai and Corneil [173]. In the same paper Cai and Corneil also considered another type of generalized perfection; the motivation is given by the characterization of perfect graphs stated in Proposition 2.1.1. Given a positive integer i, an induced subgraph G' of a graph G is an *i-transversal* if $\omega(G') = i$ and $\omega(G - G') = \omega(G) - i$ (see Definition 2.4.6).

Definition 2.5.10 *For a positive integer i, a graph G is* perfectly i-transversable *if each induced subgraph H of G admits a* $\min\{\omega(H), i\}$*-transversal of H.*

Thus, perfectly 1-transversable graphs are exactly the perfect graphs. In [173] it is shown that all perfect graphs are $\mathcal{P}_{i+1}$-perfect and perfectly i-transversable for each i; this interesting property of perfect graphs is already stated in Theorem 2.4.3.

Note that the recognition problems of $\mathcal{P}_{i+1}$-perfect graphs and perfectly i-transversable graphs ($i \geq 2$) are $\mathbb{NP}$-hard [173].

Finally, we shall note that Deuber and Zhu [307] introduced the concept of *star-perfect graphs* and *star-superperfect graphs* by generalizing the notion of interval coloring (see Definition 2.3.2) to the more general notion of *circular coloring* of weighted graphs. The class of star-perfect graphs (respectively of star-superperfect graphs) contains the perfect graphs (respectively the superperfect graphs) as a special case.

2.6 Related concepts

This section deals with some concepts related to perfect graphs. The definition of perfect graphs and the equivalent form stated after the PGT (Theorem 2.1.2) contain two conditions: First, two parameters are related by an inequality ($\chi \geq \omega$, respectively, $\kappa \geq \alpha$). The second condition is to quantify over all (induced) subgraphs. This is the idea behind all the concepts that we shall address below.

The concept of *line perfect graphs*, due to Trotter [1030], seems to be one of the first of this form.

Definition 2.6.1 *A graph G is* line perfect *if for all subgraphs H of G the maximum number of edges in a matching of H is equal to the size of a minimum cover of edges by sets of mutually adjacent edges in H.*

Notice that G is line perfect if and only if the line graph $L(G)$ is perfect in the usual sense. Line perfect graphs are perfect and characterized in [1030, 308, 755].

Let $\mathcal{C}(G)$ denote the set of all maximal cliques of a graph G. Obviously $\alpha(G) \leq |\mathcal{C}(G)|$ holds.

Definition 2.6.2 (Golumbic [451]) *A graph G is* trivially perfect *if for all induced subgraphs H of G, $\alpha(H) = |\mathcal{C}(H)|$.*

For results on this class found in [451], see Theorem 6.6.1.

Definition 2.6.3 *Let G be a graph.*

> *A* Grundy k-coloring *of G is a k-coloring such that for each color i, each vertex with color i is adjacent to at least one vertex with color j for each $j < i$.*
>
> *A k-coloring is* complete *if for each pair of different colors there are two adjacent vertices with these colors.*

The Grundy number $\gamma'(G)$ *of* G, *respectively, the* achromatic number $\psi(G)$ *of* G *is the maximum integer* k *for which* G *has a Grundy* k*-coloring, respectively, a complete* k*-coloring.*

Bodlaender [118] showed that determining the achromatic number is NP-complete for cographs and interval graphs.

Note that for any graph G, $\psi(G) \geq \gamma'(G) \geq \chi(G) \geq \omega(G)$. Christen and Selkow [197] then proposed the following.

Definition 2.6.4 *Let* G *be a graph. For distinct members* x *and* y *of* $\{\psi, \gamma', \chi, \omega\}$, G *is* xy-perfect *if for each induced subgraph* H *of* G, $x(H) = y(H)$.

The $\chi\omega$-perfect graphs are, of course, the perfect graphs. In [197], $\gamma'\omega$-perfect graphs, $\gamma'\psi$-perfect graphs, $\psi\omega$-perfect graphs, and $\psi\chi$-perfect graphs are characterized. Among other results, an interesting one stated that $\gamma'\omega$-perfect graphs and $\gamma'\chi$-perfect graphs are the same and they are exactly the P_4-free graphs, i.e., the cographs. Thus, an analogy of the PGT holds: The complement of a $\gamma'\omega$-perfect graph is $\gamma'\omega$-perfect. Let us note that no characterization of $\psi\gamma'$-perfect graphs is known.

The following notion of *neighborhood perfection* was introduced and studied by Neeralagi and Sampathkumar [809] and Lehel and Tuza [720].

Definition 2.6.5 *Let* G *be a graph.*

A neighborhood subgraph *of* G *is a subgraph induced by the* closed *neighborhood of some vertex of* G.

The neighborhood cover number $\kappa_N(G)$ *is the smallest number of vertices whose neighborhood subgraphs cover the edge set of* G.

The neighborhood independence number $\alpha_N(G)$ *is the largest number of edges of* G *such that no two of them belong to the same neighborhood subgraph.*

G *is* neighborhood perfect *if for each induced subgraph* H *of* G, $\kappa_N(H) = \alpha_N(H)$.

Determining the parameters $\kappa_N(H)$, $\alpha_N(H)$ is NP-hard [182]. Obviously, $\kappa_N(G) \geq \alpha_N(G)$. No characterization of neighborhood-perfect graphs is known. For the case of chordal neighborhood-perfect graphs see Theorem 7.2.2. The following conjecture is implicitly stated in [720, 719].

Conjecture 2.6.1 *Neighborhood-perfect graphs are perfect.*

The papers [488, 719] contain some partial results in this direction. Note that the SPGC implies Conjecture 2.6.1.

Another variant of perfection is due to Conforti, Corneil, and Mahjoub [237] studying coverings and packings of cliques of a fixed size $i \geq 2$ of a graph G. Let $\mathcal{C}_i(G)$ denote the set of all complete subgraphs K_i of G. For every $\mathcal{S} \subseteq \mathcal{C}_i(G)$, they consider parameters on $\mathcal{S}$ rather than on the graph G.

Definition 2.6.6 *Let G be a graph and $\mathcal{S} \subseteq \mathcal{C}_i(G)$.*

> *A* K_i-covering *of $\mathcal{S}$ is a subset $\mathcal{T}$ of $\mathcal{C}_{i-1}(G)$ such that each K_i in $\mathcal{S}$ contains at least one K_{i-1} in $\mathcal{T}$.*
>
> *A* K_i-packing *of $\mathcal{S}$ is a subset $\mathcal{S}'$ of $\mathcal{S}$ such that no two K_i in $\mathcal{S}'$ share a K_{i-1}.*
>
> *$c_i(\mathcal{S})$ denotes the minimum cardinality of any K_i-covering of $\mathcal{S}$.*
>
> *$p_i(\mathcal{S})$ denotes the maximum cardinality of any K_i-packing of $\mathcal{S}$.*

Note that for every graph G and set $\mathcal{S} \subseteq \mathcal{C}_i(G)$, $c_i(\mathcal{S}) \geq p_i(\mathcal{S})$. This leads to a quantification over all subsets of the set $K_i(G)$.

Definition 2.6.7 *Let $i \geq 2$ be an integer. A graph G is* K_i-perfect *if for each $\mathcal{S} \subseteq \mathcal{C}_i(G)$, $c_i(\mathcal{S}) = p_i(\mathcal{S})$.*

In [237] six different characterizations of K_i-perfect graphs are found; one of them is very similar to the SPGC. The concept of K_i-perfect graphs is flexible: In Definition 2.6.7 one could choose the quantification to be over all induced subgraphs of G or all partial subgraphs of G (instead of over all subsets of $\mathcal{C}_i(G)$.) Indeed, the *K_i-induced perfect graphs* and the *K_i-partial perfect graphs* are defined in [237]; they are not the same and each of these two new graph classes is different from the class of K_i-perfect graphs. In contrast to the K_i-perfect graphs, no characterization of K_i-induced perfect graphs or of K_i-partial perfect graphs is known.

In Definition 2.6.6, one also could choose the sets $\mathcal{H}(G)$ and $\mathcal{K}(G)$ of all subgraphs of G isomorphic to H, respectively, to K (instead of $\mathcal{C}_i(G)$ and $\mathcal{C}_{i-1}(G)$). Here H and K are two fixed graphs such that K is a subgraph of H. This leads to the concept of *K–H-perfect graphs* introduced and studied by Brown, Corneil, and Mahjoub [163]. The K_i-perfect graphs are exactly the K_{i-1}–K_i-perfect graphs. See [163] for more about K–H-perfect graphs and a possible extension of this concept.

Müller [802] considered the concept of edge perfection.

Definition 2.6.8 *Two edges are* dependent *if they have a common vertex or form a (not necessarily induced) cycle. An* edge clique *is a set of pairwise-dependent edges. An* edge-independent *set is a set of pairwise nondependent edges. The parameters α_e, ω_e, χ_e, and κ_e are defined as for α, ω, χ, and κ when "clique," respectively, "independent set," is replaced by "edge clique," respectively, "edge-independent set."*

As in the case of the parameters α, ω, χ, and κ, all graphs G satisfy $\alpha_e(G) \leq \kappa_e(G)$ and $\omega_e(G) \leq \chi_e(G)$. In general, $\chi_e(G) \neq \kappa(\overline{G})$ and $\alpha_e(G) \neq \omega_e(\overline{G})$.

Definition 2.6.9 *A graph G is*

> α_e-perfect *if for all induced subgraphs H of G, $\alpha_e(H) = \kappa_e(H)$;*
>
> ω_e-perfect *if for all induced subgraphs H of G, $\omega_e(H) = \chi_e(H)$;*
>
> e-perfect *if it is both α_e- and ω_e-perfect.*

In [802] the computations of α_e, χ_e, and κ_e are shown to be $\mathbb{NP}$-complete in general, and some classes of α_e-, ω_e-, and e-perfect graphs are given. The complexity of computing ω_e remains open.

The last concept we shall address is due to Markosjan, Gasparian, and Reed [773]. For a graph G, let $\beta(G)$ denote the maximum of the minimum degree of the induced subgraphs in G plus 1. In [774, 1015] it was shown that $\chi(G) \leq \beta(G)$.

Definition 2.6.10 *A graph G is β-*perfect *if, for all induced subgraphs H of G, $\chi(H) = \beta(H)$.*

The following interesting analogy of the SPGC is proved in [773]: G and $\overline{G}$ are β-perfect if and only if neither G nor $\overline{G}$ contains a chordless cycle of *even* length. It is pointed out that chordal graphs are exactly those graphs that are both perfect and β-perfect. The complexity of determining if a graph is β-perfect is unknown; this problem is only known to be in co-$\mathbb{NP}$.

Other interesting concepts related to perfection are *domination perfect graphs* [1006, 1097] and *irredundance perfect graphs* [542]. These concepts are discussed in the forbidden induced subgraph chapter (chapter 7). The paper [1092] contains various related concepts in terms of minimum degree or size of neighborhoods.

Chapter 3

Cycles, Chords, and Bridges

3.1 (k, l)-chordal graphs

Chordal graphs (see Definition 1.2.1) can be generalized by placing a variety of restrictions on the number and type of chords with respect to a cycle. A fairly general scheme is given in the following.

Definition 3.1.1 (Ausiello, D'Atri, Moscarini [35]) *A graph G is* (k, l)-chordal *if each cycle in G of length at least k contains at least l chords.*

Thus chordal graphs are the (4,1)-chordal graphs. Further conditions can be placed on the parity of the cycles (chords in odd cycles), the parity of the cycle distance of the end vertices of chords (odd chords), requiring crossing and/or parallel chords, requiring all these conditions for G and $\overline{G}$, and requiring these conditions in bipartite graphs (where all cycles are of even length). Thus, for example, the (5,2)-odd-crossing-chordal graphs are the graphs such that every odd cycle of length at least 5 has at least 2 crossing chords. Note that it is possible to test whether a graph contains a cycle of length $> k$ in polynomial time for any fixed k and that we can also determine whether there is any cycle of length $> k$ that has no crossing chords (parallel chords) in polynomial time [819].

We start with (5,2)-chordal graphs. These are an interesting generalization of chordal graphs. A graph G is *house-hole-domino–free* (*HHD-free*) if G contains no induced house, domino, or cycle C_k, $k \geq 5$. The following equivalence has been observed by F.F. Dragan.

Proposition 3.1.1 *A graph is* $(5, 2)$*-chordal if and only if it is HHD-free.*

The condition can be sharpened by requiring at least two crossing chords. Two chords uv, xy in a cycle C are *crossing* (*parallel, i.e., noncrossing*) if the order of these vertices in C is (u, x, v, y) ((u, x, y, v), respectively).

Definition 3.1.2 (Howorka [586]) *A graph G is* distance hereditary *if G is connected and every induced path in G is isometric.*

As already mentioned, every hole C_k, $k \geq 5$, is not distance hereditary.

Theorem 3.1.1 (Howorka [586]) *A graph has at least two crossing chords in every cycle of length at least 5 if and only if it is distance hereditary.*

These graphs have many other interesting characterizations (see, e.g., Theorems 10.1.1, 11.6.7). In particular, Theorem 10.1.1 shows that a graph G has at least two crossing chords in every cycle of length at least 5 if and only if G is house-hole-domino-gem–free (HHDG-free). It is easy to show that the graphs with at least two parallel chords in every cycle of length at least 5 can be characterized as the (house, hole, domino, $\overline{P_2 \cup P_3}, K_{3,3}, K_{3,3} - e$)-free graphs.

3.2 Chordality conditions for G and $\overline{G}$

We start with the condition that G and $\overline{G}$ are (5,1)-chordal, i.e., the graph and its complement are hole free.

Definition 3.2.1 (Hayward [526]) *A graph G is* weakly chordal (*called* weakly triangulated *in many papers) if G and $\overline{G}$ contain no induced cycle C_k, $k \geq 5$.*

As an obvious consequence, complements of weakly chordal graphs are weakly chordal. It is easy to see that chordal and co-chordal graphs are weakly chordal. Furthermore, house-hole–free graphs are weakly chordal, which refines the first inclusion and follows from the fact that every antihole $\overline{C_k}$, $k \geq 6$, contains an induced house.

Weakly chordal graphs have an interesting cutset property.

Definition 3.2.2 *In a graph, two vertices x, y are a* two-pair *if each chordless path between x and y has exactly two edges.*

Theorem 3.2.1 (Hayward, Hoàng, Maffray [530]) *A graph G is weakly chordal if and only if every induced subgraph of G either is a clique or else has a two-pair.*

Note that the question whether a pair of vertices x, y is a two-pair can be checked in polynomial time. The fastest known recognition algorithm for weakly chordal graphs has time bound $\mathcal{O}(n^4)$ and is based on Theorem 5.9.5 [997].

There is a nice characterization of hole-free graphs in terms of midedges of P_4s.

Theorem 3.2.2 (Chvátal, Rusu [221]) *A graph G contains no holes if and only if each of its induced subgraphs H with at least one edge contains an edge that is not the midedge of any P_4 in H.*

Graphs without holes of size k can be recognized in polynomial time, and the best known time bound appeared in [993].

Eschen and Sritharan [359] extend Theorem 3.2.2 to hole-free and *i-antihole-free* graphs and give some consequences for weakly chordal and chordal bipartite graphs.

The condition that G and $\overline{G}$ are chordal leads to an interesting subclass of chordal graphs.

Definition 3.2.3 (Földes, Hammer [394], Tyshkevich, Chernyak [1056]) *A graph is a* split graph *if there is a partition of its vertex set into a clique and a stable set.*

A generalization of split graphs was introduced and investigated under the name *polar graphs* in [1053, 1054, 1055] (see [762, 193]).

Theorem 3.2.3 [394] *A graph G is a split graph if and only if G and $\overline{G}$ are chordal.*

Note that split graphs have a degree-sequence characterization (see Theorem 13.3.2) that can be used to recognize them in $\mathcal{O}(n)$ time if the degree sequence is given. In Theorem 7.1.1 a forbidden induced subgraph characterization of split graphs is given, which leads to a number of interesting generalizations of split graphs.

Another example of a graph class defined in this fashion is the *Berge graphs* (Definition 2.1.3). Recall that a graph is Berge if it is odd-hole–free and odd-antihole–free. Note that each perfect graph is a Berge graph and the converse inclusion is fulfilled if and only if the SPGC holds. The connection to the SPGC gave rise to the name for this class of graphs.

3.3 Cycles and chords in bipartite graphs

It is obvious that the bipartite (4,1)-chordal graphs are exactly the cycle-free graphs, i.e., forests. Moreover, for bipartite graphs (5,1)-chordality and (6,1)-chordality are equivalent.

Definition 3.3.1 (Golumbic, Goss [456]) *A bipartite graph B is* chordal bipartite *if each cycle in B of length at least 6 has a chord (i.e., B is $(6,1)$-chordal).*

Note that chordal bipartite graphs are in general not chordal since C_4 is chordal bipartite. It is easy to see, however, that these graphs are exactly the bipartite weakly chordal graphs. In this sense, the notion "chordal bipartite" is misleading and could be replaced by "weakly chordal bipartite":

chordal bipartite = weakly chordal ∩ bipartite.

Chordal bipartite graphs have various characterizations in terms of elimination orderings, hypergraphs, and matrices. See, e.g., Theorems 5.5.5, 5.5.6, 5.9.1, Remark 5.9.1, Theorems 8.2.5, 8.2.6, Corollary 8.3.2.

A fast recognition algorithm for chordal bipartite graphs is discussed in the matrix chapter (chapter 9) based on the Γ-free (totally balanced) matrices.

In connection with relational database schemes Ausiello, D'Atri, and Moscarini [35] give an interesting generalization of chordal bipartite graphs.

Definition 3.3.2 [35] *Let $B = (X, Y, E)$ be a bipartite graph. Then*

> *B is X-*conformal *if for every set $S \subseteq Y$ with the property that all nodes of S have pairwise distance 2, there is a node $x \in X$ with $S \subseteq N(x)$. Y-*conformal *graphs are defined analogously.*

> B *is* X-chordal *if for every cycle* C *in* B *of length at least* 8, *there is a node* $x \in X$ *that is adjacent to at least two nodes in* C *whose distance in* C *is at least* 4 (*a* bridge node). Y-chordal *graphs are defined analogously.*

It is a simple observation that B is X-conformal if and only if $\mathcal{N}^X(B)$ is Helly and B is X-chordal if and only if $L(\mathcal{N}^X(B))$ is chordal. In [35] this is shown in terms of the dual notions (which justifies the terminology of Definition 3.3.2).

Theorem 3.3.1 **[35]** *Let* $B = (X, Y, E)$ *be a bipartite graph. Then*

(i) B *is* X*-conformal if and only if the hypergraph* $\mathcal{N}^Y(B)$ *is conformal;*

(ii) B *is* X*-chordal if and only if* $2SEC(\mathcal{N}^Y(B))$ *is chordal.*

Thus we have the following corollary.

Corollary 3.3.1 *The following conditions are equivalent:*

(i) B *is* X*-chordal and* X*-conformal;*

(ii) $\mathcal{N}^Y(B)$ *is a dual hypertree;*

(iii) $\mathcal{N}^X(B)$ *is a hypertree.*

An analogous proposition holds if X *and* Y *change roles.*

Dragan and Voloshin [332] give characterizations of bipartite graphs that are the incidence graphs of hypergraphs H having the property that H and its dual H^* are α-acyclic (see Theorem 8.1.1 for the equivalence of dual hypertrees with α-acyclic hypergraphs) and show a characterization of chordal bipartite graphs as the bipartite graphs for which every induced subgraph has a maximum neighborhood ordering (see Theorem 5.5.5).

In [141] the notion of star chordality (called *semichordality* there) is introduced.

Definition 3.3.3 (Brandstädt [141]) *Let* $B = (X, Y, E)$ *be a bipartite graph and* $C = (x_1, y_1, \ldots, x_k, y_k)$ *be a cycle in* B.

> C *has an* X-star (Y-star) *if there is a vertex* $x \in X$ ($y \in Y$) *such that* $x \notin \{x_1, \ldots, x_k\}$ ($y \notin \{y_1, \ldots, y_k\}$) *and there are pairwise distinct indices* $i_1, i_2, i_3 \in \{1, \ldots, k\}$ *such that* $y_{i_j}x \in E$ ($x_{i_j}y \in E$), $j \in \{1, 2, 3\}$.
>
> B *is* star chordal *if each chordless cycle in* B *of length at least* 6 *contains an* X*-star or* Y*-star.*
>
> B *is* X-star chordal *if each chordless cycle in* B *of length at least* 6 *contains an* X*-star.*

Theorem 3.3.2 **[141]** *A graph* G *is chordal if and only if* $B_C(G)$ *is* Y*-star chordal.*

Based on this characterization and LexBFS, Habib, Paul, and Viennot in [499] give a linear-time recognition of X- and Y-star–chordal graphs.

The relationship between X-chordality and X-conformality is given in the following lemma.

Lemma 3.3.1 (Behrendt, Brandstädt [75])

(i) *If B is Y-star chordal then B is Y-chordal.*

(ii) *If B is Y-chordal and Y-conformal then B is Y-star chordal.*

Another interesting special case of bipartite (k,l)-chordal graphs are the bipartite (6,2)-chordal graphs. It is easy to see that these graphs are exactly the bipartite distance-hereditary graphs.

3.4 Odd chords

The subsequently defined strongly chordal graphs are an interesting subclass of chordal graphs for many reasons. Originally they were introduced by Farber [374] as a subclass of chordal graphs for which the domination problem, which remains $\mathbb{NP}$-complete for chordal graphs and even for split graphs, can be solved efficiently.

Definition 3.4.1 **[372]** *Let G be a graph.*

A chord $x_i x_j$ in a cycle $C = (x_1, x_2, \ldots, x_{2k})$ of even length $2k$ is an odd chord *if the distance in C between x_i and x_j is odd.*

G is strongly chordal *if G is chordal and each cycle in G of even length at least 6 has an odd chord.*

Strongly chordal graphs have various characterizations in terms of elimination orderings; see Theorems 5.5.1, 5.5.2. They are also the same as the hereditary dually chordal graphs; see Theorem 5.5.4. They appear as the sun-free chordal graphs in Theorem 7.2.1. This is an important aspect, and we come back to this condition at several places in this book.

Strongly chordal graphs have characterizations in terms of β-acyclic hypergraphs and totally balanced matrices; see the hypergraph and the matrix chapters (chapters 8 and 9). The connection to Γ-free matrices is used in [850, 991] to develop fast (but not linear-time) recognition algorithms for strongly chordal graphs.

Strongly chordal graphs are closely related to chordal bipartite graphs.

Theorem 3.4.1 (Farber [372]) *A graph G is strongly chordal if and only if $B_C(G)$ is chordal bipartite.*

A similar connection is given in the following theorem.

Theorem 3.4.2 (Brandstädt [141]) *A graph G is strongly chordal if and only if $B(G)$ is chordal bipartite.*

Another connection is (see also [146]) the following theorem.

Theorem 3.4.3 (Dahlhaus [276]) *A bipartite graph $B = (X, Y, E)$ is chordal bipartite if and only if $split_X(B)$ is strongly chordal.*

All these connections have a hypergraph background; see, e.g., Corollary 8.3.2, Proposition 8.3.4.

3.5 Meyniel graphs, subclasses, and variants

If holes are avoided by requiring at least two chords in every odd cycle of length at least 5, then we again get a subclass of perfect graphs.

Definition 3.5.1 *G is a* Meyniel graph *if each cycle of odd length at least* 5 *has at least two chords.*

Meyniel [780] proved the perfection of Meyniel graphs. This has been refined later by showing inclusion in several subclasses of perfect graphs (see Appendix A). In fact, Meyniel graphs have a very special kind of perfection.

Theorem 3.5.1 (Hoàng [553, 555]) *G is a Meyniel graph if and only if G is very strongly perfect.*

This implies that Meyniel graphs are strongly perfect.

A polynomial-time recognition algorithm for Meyniel graphs is given by a so-called amalgam decomposition of Meyniel graphs into simpler ones (see Definition 11.6.2). Using this approach, Burlet and Fonlupt [169] showed that Meyniel graphs can be recognized in polynomial time. The time bound was improved to $\mathcal{O}(m^2)$ by Roussel and Rusu [932].

The observation of Preissmann that chordal graphs are locally perfect was improved by Hertz [543]; he showed that Meyniel graphs are locally perfect.

We now define two special types of Meyniel graphs. The first case is given by requiring noncrossing chords.

Definition 3.5.2 *G is a* Gallai graph *(also called* o-triangulated, i-triangulated*) if each cycle of odd length at least* 5 *has at least two noncrossing chords.*

The perfection of these graphs, which first was shown by Gallai [415] and Surányi [1009], is implied by the obvious fact that Gallai graphs are Meyniel. The recognition of these two graph classes uses similar techniques.

Theorem 3.5.2 (Burlet, Fonlupt [169]) *(see also* [1082]*) Gallai graphs can be recognized in polynomial time.*

The time bound for Gallai graph recognition has been improved by Roussel and Rusu to $\mathcal{O}(nm)$ [933]. The second case of special Meyniel graphs is given by the parity graphs.

Definition 3.5.3 *G is a* parity graph *if for any two induced paths joining the same pair of vertices the path lengths have the same parity (i.e., they are both odd or both even).*

Obviously, distance-hereditary graphs are parity graphs.

Theorem 3.5.3 (Burlet, Uhry [170]) *G is a parity graph if and only if each cycle of odd length at least* 5 *has at least two crossing chords.*

The perfection of the parity graphs, which first was shown by Olaru and Sachs [942, 840], is implied by the fact that parity graphs are Meyniel.

Burlet and Uhry [170] showed that parity graphs can be recognized in polynomial time. Recently, Cicerone and Di Stefano [224] gave a linear-time algorithm for recognizing parity graphs using the split decomposition—see section 12.3 on split decomposition—and a characterization of parity graphs given in Theorem 11.6.6.

In [745] another variant of chords in odd cycles is studied.

Definition 3.5.4 (Lubiw [745]) *A graph is* short chorded *if each odd cycle of length at least* 5 *has a* short chord, *i.e., a chord joining vertices of distance* 2 *apart in the cycle.*

Short-chorded graphs were also called *Raspail graphs*; see, e.g, Sun [1007] for the origin of this name. Obviously, short-chorded graphs are Berge graphs. Thus, if the SPGC is true, short-chorded graphs are perfect. In [1007, 745] two subclasses of short-chorded graphs are shown to be perfect.

It should be mentioned that there are other classes with chordality properties, such as comparability graphs (see Definition 1.4.3 for the definition of these graphs and Theorem 6.1.1 for such a characterization).

3.6 Bridged graphs and isometric cycles

Until this point, classes were discussed for which certain induced cycles are excluded by requiring the existence of chords. We now come to a metric version of restricting cycles. A natural generalization of chordality is to replace an edge shortcut with a path shortcut.

Definition 3.6.1 (Farber, Jamison [376], Soltan, Chepoi [984]) *Let G be a graph.*

> *A* bridge *of a cycle C is a shortest path in G joining nonconsecutive vertices of C, which is shorter than both the paths of C joining those vertices.*
>
> *G is* bridged *if every cycle C of length at least* 4 *contains two vertices connected by a bridge (i.e., the only isometric cycles in G can be of length* 3*).*

Thus a chord is a bridge of length 1. Obviously, chordal graphs are bridged. Observe that in a bridged graph every cycle of length 4 or 5 has a chord. Clearly, bridged graphs are in general not perfect, as the wheel W_7 shows. The same example shows that the property of being bridged is not hereditary.

Bridged graphs are of interest in connection with distance and convexity properties, and they can be recognized in polynomial time (see the remark after Theorem 10.5.6 and the remark after Theorem 5.8.2).

For an interesting characterization by elimination orderings see Theorem 5.8.2. For a characterization of bridged graphs in terms of convexity see Theorem 10.5.5 and of bridged and bipartite graphs (which are the hereditary modular graphs) see Theorem 10.3.8.

Chapter 4

Models and Interactions

4.1 Basic concepts

Intersection graphs (see Definition 1.2.3) are a very general notion in which objects are assigned to the nodes of a graph, and two distinct nodes are adjacent if their objects have nonempty intersection. This occurs in Definition 1.3.3 as line graphs of hypergraphs, with the important special case of line graphs of graphs (the hyperedges are now the edges of a graph); see Definition 4.2.1. Theorem 1.2.3 presents an intersection-graph characterization of chordal graphs, where the model is a set of subtrees of a tree.

In many real-world applications having a graph model this occurs in an even more general setting: The nodes of the graph are certain objects that somehow interact. This can be expressed as follows.

Definition 4.1.1 (Golumbic [454, 455]) *Let $M = \{M_1, M_2, \dots, M_n\}$ be a finite collection of nonempty sets and R be a binary and symmetric relation on M. Then this defines the following graph:*

> $G_{(M,R)} = (\{1, \dots, n\}, E_{(M,R)})$ *with* $ij \in E_{(M,R)}$ *if and only if* $(M_i, M_j) \in R$ *(since R is supposed to be symmetric, this defines an undirected graph).*
>
> *M is then called the* model *of G with respect to the* interaction *R.*

Typical examples for the relation R are

(i) *nonempty intersection*: $(M_i, M_j) \in R$ if and only if $M_i \cap M_j \neq \emptyset$;

(ii) *containment*: $(M_i, M_j) \in R$ if and only if $M_i \subseteq M_j$ or $M_j \subseteq M_i$;

(iii) *overlap*: $M_i \cap M_j \neq \emptyset$ but neither $M_i \subseteq M_j$ nor $M_j \subseteq M_i$;

(iv) *measured intersection*: the size of the intersection is not less than a given value

(see [454, 455] for a survey of this concept).

Other concepts such as visibility also give rise to graph classes with interesting models.

The close connection between containment and comparability graphs is expressed by the fact that a graph is a comparability graph if and only if it is the containment graph of some model. Therefore, special classes of containment graphs are discussed in the chapter on posets (chapter 6). The only major result on overlap is Theorem 4.7.4 on circle graphs.

It is natural to consider the question of whether every graph is the intersection graph for some model. This question is answered by the following observation of Marczewski [769]: Every graph is the intersection graph of a family of sets. Golumbic and Jamison [458] show that every graph is the edge-intersection graph of substars of a star.

Intersection classes are defined as follows.

Definition 4.1.2 (Scheinerman [953]) *A class $\mathcal{G}$ of graphs is an* intersection class *if there is a set $\mathcal{M}$ of sets such that every graph G in $\mathcal{G}$ is the intersection graph (in the sense of Definition* 1.2.3) *G_M of a family M of sets from $\mathcal{M}$ (a family allows that sets occur repeatedly).*

If $\mathcal{G}$ is an intersection class and $G \in \mathcal{G}$, then every induced subgraph of G is also in $\mathcal{G}$. Thus, not every graph class is an intersection class, e.g., the class of connected graphs is not of this type.

Theorem 4.1.1 [953] *A class $\mathcal{G}$ of graphs is an* intersection class *if and only if it satisfies the following conditions:*

(i) *If $G \in \mathcal{G}$ and G' is an induced subgraph of G, then $G' \in \mathcal{G}$;*

(ii) *If $G \in \mathcal{G}$ and G' results from G by repeatedly replacing a vertex v with an edge $v'v''$ of new vertices that have the same neighbors in the new graph as v in the original graph, then $G' \in \mathcal{G}$ (called* vertex expansion *in* [953]);

(iii) *There is a countable sequence $G_1, G_2, \ldots$ of graphs in $\mathcal{G}$ (called* composition series *in* [953]) *such that every G_i is an induced subgraph of G_{i+1} and each $G \in \mathcal{G}$ is an induced subgraph of some G_i.*

Scheinerman [954] and Quilliot [888] give further results on the topic of intersection classes.

4.2 Line graphs and generalizations

Line graphs are among the classical concepts of graph theory. They are a special case of line graphs of hypergraphs; see Definition 1.3.3.

Definition 4.2.1 *For a graph $G = (V, E)$, the graph $H = (E, E')$, with $e_1e_2 \in E'$ if and only if $e_1 \cap e_2 \neq \emptyset$ is the* line graph *(or* edge graph*) of G.*

Recall that for bipartite graphs the min-max equality from Theorem 1.1.2 for vertex cover and matching holds. This implies the following corollary.

Corollary 4.2.1 *Line graphs of bipartite graphs are perfect.*

The investigation of line graphs goes back to early papers such as Whitney [1083], where it is shown that K_3 and $K_{1,3}$ are the only connected nonisomorphic graphs with isomorphic line graphs. There is a nice characterization of line graphs in terms of forbidden subgraphs as well as other conditions—see Theorem 7.1.8—which lead to linear-time recognition.

The concept of line graphs was generalized in several directions. Hemminger and Beineke [539] and Prisner [883] give surveys on line graphs and generalizations (and on many other aspects of line graphs). Le [708] studies the following generalization of line graphs called *k-line graphs.* For a given graph G, the vertices of the k-line graph are the cliques of size k in G, and two such vertices are adjacent if they have $k-1$ common vertices in G. It is shown that odd-hole–free 3-line graphs are perfect.

Definition 4.2.2 *Let* $G = (V, E)$ *and* H *be graphs. Then the* Gallai graph $\Gamma(G)$, *the* H-line graph $L_H(G)$, *the* edge-clique graph $\mathcal{EC}(G)$, *and the* edge graph $\mathcal{E}(G)$ *of* G *have* E *as vertex set. Two edges* $e_1, e_2 \in E$ *are adjacent in*

$\Gamma(G)$ *if they share a vertex but do not lie in a common triangle of* G [707, 711, 1007];

$L_H(G)$ *if they share a vertex and lie in a common copy of* H [185];

$\mathcal{EC}(G)$ *if they lie in a common clique of* G [13];

$\mathcal{E}(G)$ *if they share a vertex or lie in a common* C_4 [53].

For the special case $H = K_3$, the K_3-line graphs are called *anti-Gallai graphs* in [709, 711] and *triangular line graphs* in [624].

There are nice applications of these notions, including the following reformulations of the four-color theorem.

Theorem 4.2.1 *The following statements are equivalent:*

(i) *For all planar graphs* G, $\chi(G) \leq 4$;

(ii) *for all planar graphs* G, $\chi(\mathcal{EC}(G)) = \omega(\mathcal{EC}(G))$ [13];

(iii) *for all planar graphs* G, $\chi(L_{K_3}(G)) = \omega(L_{K_3}(G))$ [709, 711].

Some other generalizations of line graphs are introduced in [714, 881, 882, 174].

The class of perfect graphs has an intersection model that is given by the hyperedges of normal hypergraphs—see Theorem 8.6.1—a generalization of line graphs of bipartite graphs.

4.3 Interval graphs and variants

Perhaps the best known example of intersection graphs is the interval graph. There are many applications, among them scheduling, seriation in archeology, behavioral psychology, planning, medical diagnosis and temporal reasoning in artificial intelligence, circuit design, and most recently the Human Genome Project—see [452, 459, 864, 865, 448, 646], where interval graphs (with possible side constraints) occur. A good source for the theory of interval graphs and interval orders is Fishburn's book [387].

Definition 4.3.1 (Hajós [502]) *Let $\mathcal{I} = \{I_1, I_2, \ldots, I_n\}$ be a finite collection of intervals of the real line and $G_\mathcal{I}$ its intersection graph as described in Definition* 1.2.3.

> *G is an* interval graph *if G is the intersection graph $G_\mathcal{I}$ of an interval model $\mathcal{I}$.*

The interval model defines a partial order in a natural way.

Definition 4.3.2 [502] *Let $\mathcal{I} = \{I_1, I_2, \ldots, I_n\}$ be a finite collection of intervals of the real line.*

> *The model $\mathcal{I}$ defines the partial order L as follows: $I_i \; L \; I_j$ if I_i lies entirely to the left of I_j.*
>
> *The poset $(P, <)$ is an* interval order *if there is a collection $\mathcal{I}$ of intervals such that $(P, <)$ is isomorphic to $(\mathcal{I}, L)$.*

Obviously $I_i \cap I_j = \emptyset$ if and only if $I_i \; L \; I_j$ or $I_j \; L \; I_i$. Thus, every interval graph is the complement of a comparability graph.

The partial order L can be generalized to other geometric objects between two parallel lines such as straight lines (for permutation graphs see Definition 4.7.1) or trapezoids (for trapezoid graphs see Definition 4.7.5). The poset chapter (chapter 6) describes the connections to partial orders in more detail.

Theorem 4.3.1 (Gilmore, Hoffman [444]) *Let G be a graph. The following conditions are equivalent:*

(i) *G is an interval graph;*

(ii) *G contains no induced C_4 and $\overline{G}$ is transitively orientable;*

(iii) *The maximal cliques of G can be linearly ordered such that for each vertex v, the maximal cliques containing v occur consecutively.*

Thus, every interval graph is chordal; in fact, every interval graph is a directed path graph. Booth and Lueker [127] gave a first linear-time recognition algorithm based on the linear-clique arrangement of interval graphs given in Theorem 4.3.1 and a data structure called *PQ-trees.*

Theorem 4.3.2 [127] *Interval graphs can be recognized in time $\mathcal{O}(n + m)$.*

Korte and Möhring [688] simplified the Booth–Lueker algorithm by introducing modified PQ-trees. Another simpler linear-time algorithm based on the substitution decomposition is given by Hsu and Ma [597] and by Hsu [594] that can avoid the PQ-trees. A LexBFS variant is described by Habib, Paul, and Viennot [499]. A similar approach applying a 4-sweep LexBFS algorithm is given by Corneil, Olariu, and Stewart in [254] leading to an even simpler linear-time recognition of interval graphs.

See also an efficient online recognition algorithm for interval graphs by Hsu [596]. Another characterization of interval graphs is given in Theorem 7.2.6.

The following subclasses of interval graphs are also of interest.

Definition 4.3.3 *G is a* proper interval graph *if G is an interval graph with an interval model where no two intervals $I_x, I_y \in \mathcal{I}$ properly contain each other.*

Definition 4.3.4 *G is a* unit interval graph *if G is an interval graph with an interval model $\mathcal{I}$ of unit-length intervals.*

Theorem 4.3.3 (Roberts [906]) *G is a proper interval graph if and only if G is a unit-interval graph.*

Other characterizations of proper interval graphs are given in Theorems 7.1.10 and 7.2.9.

Another subclass of interval graphs is the class of trivially perfect graphs introduced in Definition 2.6.2. Equivalently, the trivially perfect graphs are the intersection graphs of nested intervals (i.e., a set of intervals for which no pair has an overlap relationship). This follows easily from the characterization of trivially perfect graphs as the comparability graphs of rooted tree posets (see Theorem 6.6.1).

The following classes are generalizations of interval graphs.

Definition 4.3.5 (Trotter, Harary [1032], Griggs, West [472]) *A graph G is a t-*interval graph *if there is an intersection model whose objects consist of collections of t intervals, $t \geq 1$, such that G is the intersection graph of this model.*

Griggs and West [472] have shown that every graph G is a t-interval graph for some $t \leq \lceil (\Delta(G) + 1)/2 \rceil$, where $\Delta(G)$ denotes the maximum degree of G. Thus, it makes sense to define the *interval number* of a graph G to be the smallest t such that G is a t-interval graph. The interval number of graphs is unbounded, and computing the interval number is $\mathbb{NP}$-hard: West and Shmoys [1080] have shown that the recognition problem for t-interval graphs is $\mathbb{NP}$-complete for every $t \geq 2$.

The class of 2-interval graphs is of interest in the context of genetics. Trees have interval number at most 2 [1032, 472]. Planar graphs have interval number at most 3 [959].

The *unit interval number* of a graph, i.e., the case where the intervals are of unit length, is studied in [21].

Another interesting generalization of interval graphs uses k distinct parallel lines $L_1, \ldots, L_k$, and the objects are intervals $I_1, \ldots, I_k$, with interval I_j chosen from line L_j, $j \in \{1, \ldots, k\}$. This class is studied in [489].

There is an interesting bipartite variant of interval graphs called *interval bigraphs.*

Definition 4.3.6 (Harary, Kabell, McMorris [521]) *A bipartite graph $B = (X, Y, E)$ is an* interval bigraph *if there are families of intervals $\mathcal{I}_X = \{I_1, \ldots, I_r\}$, $r = |X|$, and $\mathcal{I}_Y = \{J_1, \ldots, J_s\}$, $s = |Y|$, such that $x_iy_j \in E$ if and only if $I_i \cap J_j \neq \emptyset$ for $i \in \{1, \ldots, r\}, j \in \{1, \ldots, s\}$.*

Müller [801] gives a polynomial recognition of interval bigraphs.

4.4 Tree models and variants

The model of subtrees of a tree characterizing the chordal graphs (see Theorem 1.2.3) can be varied in many ways. Farber gives a tree characterization of strongly chordal graphs in his Ph.D. thesis.

Theorem 4.4.1 (Farber [371]) *A graph G is strongly chordal if and only if G is the intersection graph of subtrees of a rooted edge-weighted tree such that for each adjacent pair of vertices x, y the following property is fulfilled: The vertices in $T_x \cap T_y$ have the smallest distance from the root in either T_x or T_y.*

Gavril studies variants of tree models in [429, 430].

As for interval graphs, one can obtain subclasses of chordal graphs by specializing both the tree and its subtrees in the intersection model. There are several interesting examples. The first are the split graphs. As the chordal graphs whose complement is chordal, the split graphs have an intersection graph representation as substars of a tree that is mentioned, e.g., in [779] and improved in [1070].

Theorem 4.4.2 *A graph G is a split graph if and only if G is the intersection graph of a set of distinct substars of a star (of the maximal cliques of G).*

The intersection model of subtrees of a tree can also be varied by taking subpaths instead of subtrees, obtaining other generalizations of interval graphs. The paths can be undirected or directed, and the intersection of two paths can be taken with respect to the vertices or edges. In this way, various cases arise that were treated in several papers in the literature. Monma and Wei [796] study these models in a systematic and unifying way. We describe here only a few cases.

Definition 4.4.1 *The graph G is an* undirected path graph (path graph, VPT-graph) *if G is the intersection graph of paths in a tree.*

Renz [904] gives a characterization of undirected path graphs, and Schäffer [947] and Gavril [428] give polynomial-time recognition algorithms for this class.

Definition 4.4.2 *G is a* directed path graph *if G is the intersection graph of a collection of directed paths in a rooted directed tree.*

From the definitions and Theorems 4.3.1 and 1.2.3 the following inclusions are obvious:

interval $\subset$ directed path $\subset$ undirected path $\subset$ chordal.

Furthermore, Farber [371] shows that

directed path $\subset$ strongly chordal.

Gavril [426] gives a polynomial-time recognition algorithm for directed path graphs. The time bound was improved by Dietz [313] to linear time.

Theorem 4.4.3 (Dietz [313]) *Directed path graphs can be recognized in linear time.*

If paths in a tree are defined to be adjacent when they share an edge instead of a vertex, one obtains the following class of graphs (Golumbic and Jamison [457, 458], Monma and Wei [796], Sysło [1013], Tarjan [1020]).

Definition 4.4.3 $G = (V, E)$ *is an* EPT graph *if* G *is the intersection graph of the following model: Let* P *be a collection of nontrivial simple paths in a tree. Then* $xy \in E$ *if and only if the corresponding paths* $P_x, P_y \in P$ *share an edge.*

Note that every cycle C_k, $k \geq 3$, is an EPT graph. Thus, in general these graphs are not perfect. The class of EPT graphs coincides with the class of *fundamental cycle graphs* of Sysło [1010, 1011].

Though the definitions are similar for undirected path and EPT graphs, these classes differ greatly; they are incomparable with respect to set inclusion [457] and furthermore their recognition complexity differs (unless $\mathbb{P} = \mathbb{NP}$).

Theorem 4.4.4 (Golumbic, Jamison [457]) *The recognition problem for EPT graphs is* $\mathbb{NP}$*-complete.*

In fact, it is even $\mathbb{NP}$-complete to decide whether a given undirected path graph is an EPT graph [457] and whether a given chordal graph is an EPT graph [1013].

It can be easily seen that G is a line graph if and only if G is the EPT graph of a star. Thus every line graph is an EPT graph. Sysło [1013] showed that every chordal EPT graph is an undirected path graph.

Theorem 4.4.5 (Golumbic, Jamison [457]) *Let* G *be a graph. The following conditions are equivalent:*

(i) *G is both an undirected path graph and an EPT graph;*

(ii) *G has an intersection representation as undirected paths on a degree-3 tree;*

(iii) *G has an EPT representation on a degree-3 tree.*

There has also been some work on intersection graphs of edge disjoint paths in a tree [852] and intersection graphs of paths in a directed graph that has a tree as its underlying undirected graph [796].

There is a generalization of the tree model where the tree is replaced by a graph having exactly one cycle [432, 429]; see also Definition 5.6.2.

4.5 Boxicity, intersection dimension, and dot product

Another way to generalize interval graphs is to make the intervals higher dimensional.

Definition 4.5.1 (Roberts [907]) *For a graph G, the* boxicity $b(G)$ *is the minimum dimension d such that G is the intersection graph of boxes in d-dimensional space.*

Every graph is representable as an intersection graph of boxes in a suitable dimension [907]. Obviously, interval graphs are the graphs of boxicity one. Quest and Wegner [885] characterize graphs of boxicity at most two. The boxicity of graphs is unbounded, and computing the boxicity of a graph is $\mathbb{NP}$-hard. The recognition problem for boxicity-k graphs is $\mathbb{NP}$-complete for every $k \geq 2$, which was shown in three steps. Cozzens [263] showed that computing the boxicity of a graph is $\mathbb{NP}$-hard. This was improved by Yannakakis [1094] to testing whether $b(G) \leq 3$ is $\mathbb{NP}$-complete and by Kratochvíl [691] to determining whether $b(G) \leq 2$ is $\mathbb{NP}$-complete.

Thomassen [1024] shows that planar graphs have boxicity at most 3, and Scheinerman [952] shows that outerplanar graphs have boxicity at most 2. For other works on boxicity of a graph see [907, 265].

An interesting generalization of boxicity called *intersection dimension* is given in [266] and studied further in [696]: The *intersection dimension of a graph $G = (V, E)$* with respect to a class $\mathcal{C}$ of graphs is the minimum k such that G is the intersection of at most k graphs on vertex set V, each of which belongs to $\mathcal{C}$.

In [696] it is shown that the intersection dimension of planar graphs with respect to the class of permutation graphs is bounded.

The intersection number is another line of generalization.

Definition 4.5.2 (Erdős, Goodman, Pósa [355]) *Let $G = (V, E)$ be an intersection graph with an* intersection representation, *i.e., a set family $\{S_v : v \in V\}$ such that $uw \in E$ if and only if $S_u \cap S_w \neq \emptyset$. The* intersection number *of G is the smallest size of the union of the sets S_v, $v \in V$, in an intersection representation.*

Note that in this definition the sets S_v, $v \in V$, can be replaced by their characteristic 0–1-vector such that $S_u \cap S_w \neq \emptyset$ if and only if the inner product fulfills $u \cdot w \geq 1$. This leads to the following notion.

Definition 4.5.3 (Fiduccia et al. [385]) *Let $G = (V, E)$ be a graph and k be a positive integer. G is a k-*dot product graph *if there is a function $f : V \to \mathbb{R}^k$ so that distinct vertices v, w are adjacent if and only if for the inner product, $f(v) \cdot f(w) \geq 1$ holds.*

The *dot product dimension* of a graph $G = (V, E)$ is the least k such that G is a k-dot product graph. It is easy to see that k can be chosen to be $k \leq |E|$ and the function f assigns to each vertex v the characteristic vector of the set of edges incident with v; see Fiduccia et al. [385]. Note that the dot product dimension of a graph is at most the intersection number of the graph.

There is also a nice connection between the concept of k-dot product graphs and what is called *local* or *implicit* representation of families of graphs in [806, 642]. In an implicit representation of a graph, a small number of bits is stored at each vertex, and the adjacency of vertices x and y can be determined using only the bits stored at x and y.

Other properties of dot product graphs can be found in [385]. If the function f is linear, the dot product graphs are the threshold graphs. Quadratic forms are more interesting.

Not surprisingly for such a general concept, many problems remain open. The recognition of k-dot product graphs is not even known to be in $\mathbb{NP}$, and the related problem of characterizing k-dot product graphs is unsolved.

4.6 Circular-arc graphs

Another generalization of interval graphs is obtained by replacing the real line by a circle and intervals by circle segments.

Definition 4.6.1 *G is a* circular-arc graph *if there is a finite collection of arcs on a circle such that G is the intersection graph of this model.*

Obviously, all circular-arc graphs are 2-interval graphs. Characterizations of circular-arc graphs were given by Tucker [1037, 1038] and Gavril [425]. The first polynomial-time recognition algorithm was given by Tucker [1044], which shows that circular-arc graphs can be recognized in $\mathcal{O}(n^3)$ steps. This time bound is improved by Hsu [595] to $\mathcal{O}(nm)$ and by Eschen and Spinrad [358] to $\mathcal{O}(n^2)$. Tucker's and Hsu's recognition algorithms are complex. Tucker divided the recognition problem into two cases, depending on whether the input graph has clique cover number two. The clique cover two case has been studied in several papers; the cleanest characterization in terms of other classes is due to Trotter and Moore [1037].

Theorem 4.6.1 [1033] *Let G be a graph with clique cover number 2. Then G is a circular-arc graph if and only if $\overline{G}$ has interval order dimension at most 2.*

Other papers relate circular-arc graphs with clique cover number 2 to two-dimensional partial orders [992] and complements of certain chordal bipartite graphs [378].

Circular-arc graph coloring has been studied extensively. Although circular-arc graphs are not necessarily perfect, Tucker [1042] stated a conjecture relating chromatic number and clique size in a circular-arc graph that was resolved by Karapetjan.

Theorem 4.6.2 (Karapetjan [644]) *The chromatic number of a circular-arc graph G is at most $3/2$ the maximum clique size of G.*

One can define proper and unit circular-arc graphs, as was seen earlier with interval graphs.

Definition 4.6.2 *Let G be a graph.*

> *G is a* proper circular-arc graph *if G is a circular-arc graph and there is a circular-arc model where no arc properly contains another.*

G is a unit circular-arc graph *if G is a circular-arc graph with an intersection model whose arcs have unit length.*

Tucker [1038, 1041], Skrien [979], and Hell, Bang-Jensen, and Huang [536] give characterizations of proper circular-arc graphs in terms of local tournaments and other conditions. A linear-time recognition algorithm based on the algorithm of [1038] for proper circular-arc graphs is given in [536, 300]. Tucker [1041] gives a characterization of which proper circular-arc graphs are unit circular-arc graphs, which implies a polynomial-time recognition algorithm for the latter class. Moreover, [1041] gives an example showing the proper inclusion of the class of unit circular-arc graphs in the class of proper circular-arc graphs.

In [67], chordal proper circular-arc graphs are characterized in terms of forbidden subgraphs (see Theorem 7.1.11). Theorem 9.1.3 gives a characterization of Helly circular-arc graphs in terms of the circular 1s property.

4.7 Permutation, circle, and trapezoid graphs and similar concepts

An intersection model $\mathcal{L}$ of straight lines between two parallel lines is described as follows.

Definition 4.7.1 (Even, Pnueli, Lempel [873, 362])

Let $\mathcal{L}_1, \mathcal{L}_2$ be two parallel lines in the plane and label n points by $1, 2, \ldots, n$ on $\mathcal{L}_1$ as well as on $\mathcal{L}_2$. The straight lines L_i connect i on $\mathcal{L}_1$ with i on $\mathcal{L}_2$.

Let $G_{\mathcal{L}} = (\{1, 2, \ldots, n\}, E_{\mathcal{L}})$ with $ij \in E_{\mathcal{L}}$ if L_i and L_j intersect (cross) each other.

A graph G is a permutation graph *if there is an intersection model $\mathcal{L}$ as described in the first condition such that $G = G_{\mathcal{L}}$.*

The name *permutation graph* comes from the fact that the points on $\mathcal{L}_1, \mathcal{L}_2$ can be seen as a permutation $\pi = \binom{1 \cdots n}{i_1 \ldots i_n}$ and $ij \in E_{\mathcal{L}}$ if and only if $(i-j)(\pi^{-1}(i) - \pi^{-1}(j)) < 0$, i.e., i and j form an inversion in π. Permutation graphs are closely related to posets—the following theorem gives such a characterization.

Theorem 4.7.1 (Dushnik, Miller [345]) *A graph G is a permutation graph if and only if G and $\overline{G}$ are comparability graphs.*

See also Theorem 6.2.2 for a similar characterization.

Theorem 4.7.2 (McConnell, Spinrad [776]) *Permutation graphs can be recognized in linear time.*

The intersection model described in the first condition of Definition 4.7.1 can be generalized as follows.

Definition 4.7.2 *G is a* circle graph *if G is the intersection graph of chords in a circle.*

Clearly every permutation graph is a circle graph, and in [452, p. 252], as an exercise, the following property is mentioned.

Proposition 4.7.1 *G is a permutation graph if and only if G is the intersection graph of chords of a circle that admits an* equator, *i.e., an additional chord that intersects every other chord.*

By considering chords in a polygon with k sides for any k instead of a circle one obtains an infinite hierarchy of graph classes between permutation and circle graphs. Elmallah and Stewart [351] studied these graphs, which are called *k-polygon graphs.*

Theorem 4.7.3 (Elmallah, Stewart [352]) *For any fixed k, k-polygon graphs can be recognized in $\mathcal{O}(4^k n^2)$ time. If k is part of the input, determining whether G is a k-polygon graph is $\mathbb{NP}$-complete.*

Theorem 4.7.4 (Gavril [423]) *G is a circle graph if and only if G is the overlap graph of intervals on the real line.*

Other characterizations of circle graphs are the following.

Theorem 4.7.5 (Fournier [404]) *G is a circle graph if and only if there exists an acyclic orientation P of G and two linear extensions L_1, L_2 of P such that the relation $F = (L_1 \cap L_2) \setminus P$ satisfies the following conditions:*

(i) *If $(x, y) \in F$ and $(y, z) \in L_1$, then $(x, z) \in F$, and*

(ii) *If $(x, y) \in L_2$ and $(y, z) \in F$, then $(x, z) \in F$.*

Definition 4.7.3 *Let $G = (V, E)$ be a graph.*

The cocycle *δ_x of a vertex $x \in V$ is the set of edges incident to x.*

A walk *P is an alternating sequence of vertices and edges $(x_1, e_1, x_2, e_2, \ldots, e_k, x_k)$ such that the edge e_i is incident to the vertices x_i and x_{i+1} for $i \in \{1, \ldots, k-1\}$.*

A w-path *is a walk such that each vertex and each edge are different from the other vertices and edges of the walk.*

A cocyclic path *is a w-path whose edges form a cocycle.*

Theorem 4.7.6 (de Fraysseix [295]) *A connected graph G is a circle graph if and only if G is the intersection graph of cocyclic paths.*

The result of de Fraysseix can also be formulated in terms of matroids. Even and Itai [361] characterize the circle graphs without isolated vertices as the *stack-sorting graphs* of a permutation whereas the permutation graphs are the *queue-sorting graphs* of a permutation.

The characterizations above did not lead to polynomial-time recognition algorithms for circle graphs. Using the split decomposition, it was shown by Bouchet [136, 137, 138], Gabor, Hsu, and Supowit [411], and Naji [808] that circle graphs can be recognized in polynomial time; see Theorem 12.3.1. This was improved to the following theorem.

Theorem 4.7.7 (Spinrad [990]) *Circle graphs can be recognized in time $\mathcal{O}(n^2)$.*

There are generalizations of circle graphs that come from intersecting geometric objects in the circle that generalize chords. In [601, 678, 679, 680] the objects are polygons, and the class is called *extended circle graphs* in [601] and *spider graphs* in [678, 679, 680]. This graph class contains not only the circle graphs but also the trapezoid, circular-arc, chordal, and series-parallel graphs. These graphs can be recognized in polynomial time [680]; if we are restricted to using polygons of size k for fixed k (e.g., intersection graphs of triangles in the circle), the recognition problem seems to be open.

Another generalization of permutation graphs are circular permutation graphs.

Definition 4.7.4 (Rotem, Urrutia [931]) *A* circular permutation diagram D_π *consists of*

two concentric circles C_1 and C_2 in the plane,

n points $1, \ldots, n$ on C_1 in the clockwise direction,

n points $c(1), \ldots, c(n)$ on C_2 in the clockwise direction,

n paths $p_1, \ldots, p_n$, respectively, from i on C_1 to $\pi(i)$ on C_2 such that two distinct paths do not intersect in more than one point.

A graph $G = (V, E)$ with $|V| = n$ is a circular permutation graph *if there is a labeling $1, \ldots, n$ of V and a permutation diagram D_π satisfying the property that $ij \in E$ if and only if p_i and p_j intersect.*

Let switch(G, v) be the graph obtained from G by interchanging neighbors and nonneighbors of v. For a vertex set X, switch(G, X) is the graph formed by performing switch(G, x) one by one for each x in X. In the following theorem, switch$(G, N(v))$ corresponds to cutting the circle along v.

Theorem 4.7.8 (Rotem, Urrutia [931]) *Let v be an arbitrary vertex in G. G is a circular permutation graph if and only if G is a comparability graph and switch$(G, N(v))$ is a permutation graph.*

Sritharan [999] gives a linear-time algorithm for the recognition of circular permutation graphs.

An intersection model that generalizes both interval graphs and permutation graphs is the following, where the objects are trapezoids between two parallel lines.

Definition 4.7.5 (Dagan, Golumbic, Pinter [275], Corneil, Kamula [246]) *G is a* trapezoid graph *if G is the intersection graph of a finite collection of trapezoids between two parallel lines.*

Clearly, interval graphs and permutation graphs are trapezoid graphs. The fastest recognition for trapezoid graphs is currently the following.

Theorem 4.7.9 (Ma, Spinrad [752]) *Trapezoid graphs can be recognized in $\mathcal{O}(n^2)$ time.*

This recognition algorithm is a refinement of an algorithm of Cogis [231] and the fact that trapezoid graphs are the complements of posets with interval order dimension two; see Lemma 6.5.1.

Cheah and Corneil [186] investigate the structure of trapezoid graphs and show that an operation called *vertex splitting* allows a trapezoid graph to be transformed into a permutation graph with special properties. This leads to an $\mathcal{O}(n^3)$ time-bounded recognition algorithm for trapezoid graphs that is easily implemented and entirely graph theoretical.

It is interesting to generalize trapezoid graphs by defining objects between more than two parallel lines.

Definition 4.7.6 (Flotow [391, 392]) *Let $L_1, \ldots, L_{d+1}$ be parallel lines in the plane.*

> *A* d-trapezoid *T is defined to be the set of all points of the plane between the borderlines defined by $d+1$ intervals (a_i, b_i) on L_i, $i \in \{1, \ldots, d+1\}$ (the case $d = 1$ is the usual case of trapezoids, whereas the case $d = 0$ is the case of intervals). Hereby the borderline is given by the polygon $a_1, a_2, \ldots, a_{d+1}, b_{d+1}, \ldots, b_1, a_1$.*
>
> *A graph is a* d-trapezoid graph *if it is the intersection graph of a family of d-trapezoids.*

It is easy to see (and mentioned in [391, 392]) that the d-trapezoid graphs are exactly the incomparability graphs of posets of interval order dimension $d+1$. This implies that their recognition is $\mathbb{NP}$-complete for $d \geq 2$. However, many algorithmic techniques known from interval and permutation graphs generalize to d-trapezoid graphs. A so-called scanline technique was developed for solving optimization problems using separator properties; see [121, 675, 676, 697, 857].

A generalization of the borderlines of d-trapezoids for arbitrarily large d are function diagrams. This notion was discovered independently by Golumbic, Rotem, and Urrutia [462] and Kratochvíl (cited in [693]).

Theorem 4.7.10 (Golumbic, Rotem, Urrutia [462]) *G is a co-comparability graph if and only if G is the intersection graph of a continuous function diagram.*

It is also interesting to restrict trapezoid graphs by intersecting triangles instead of trapezoids. This gives rise to two classes called *PI* graphs (*PI* standing for point-interval) and PI^* graphs depending on whether each triangle (a, b, c) is required to have a on line L_1 and b, c on L_2 or whether the single point can occur on either L_1 or L_2. The following proper inclusions are known [246]:

$$\textbf{permutation} \subset \textbf{PI} \subset \textbf{PI}^* \subset \textbf{trapezoid}.$$

The problems of recognizing efficiently and characterizing PI and PI^* graphs remain open.

4.8 Measured intersection

A reasonable generalization of intersection graphs is given by measuring the size of intersections.

Definition 4.8.1 (Golumbic, Monma [460]) *Let V be a finite set of vertices, $\mathcal{I} = \{I_x : x \in V\}$ a finite collection of closed intervals on a line, and $t = \{t_x : x \in V\}$ a set of positive numbers (the* tolerances*).*

$G_{\mathcal{I},t} = (V, E)$ is the graph with $xy \in E$ if and only if $|I_x \cap I_y| \geq \min\{t_x, t_y\}$ (where the length of an interval I is denoted by $|I|$).

$G = (V, E)$ is a tolerance graph *(an* interval tolerance graph*) if there exists a finite interval collection $\mathcal{I}$ and a tolerance function t on V such that $G = G_{\mathcal{I},t}$.*

The pair $(\mathcal{I}, t)$ is a tolerance representation *of G.*

A tolerance representation $(\mathcal{I}, t)$ is bounded *if $t_x \leq |I_x|$ for all $x \in V$.*

G is a bounded tolerance graph *if G is a tolerance graph that admits a bounded tolerance representation.*

The following properties are obvious for tolerance graphs:

(i) If for all $x \in V$, the tolerances have the same positive value t, i.e., $t_x = t > 0$, then G is an interval graph.

(ii) If for all $x \in V$ $t_x = |I_x|$, then G is an interval containment graph and thus a permutation graph (see Theorem 6.3.1).

Thus, all interval graphs and all permutation graphs are bounded tolerance graphs.
The fact that every bounded tolerance graph is a cocomparability graph is refined in Theorem 4.8.2. Golumbic, Monma, and Trotter [461] showed that every tolerance graph is weakly chordal.

Conjecture 4.8.1 [461] *Let G be a tolerance graph. Then G is a bounded tolerance graph if and only if G is a co-comparability graph.*

Theorem 4.8.1 (Hennig [541], Andreae, Hennig, Parra [23]) *If G is the complement of a tree T then the following conditions are equivalent:*

(i) *G is a tolerance graph;*

(ii) *G is a bounded tolerance graph;*

(iii) *T does not contain T_3 as a subtree.*

Bounded tolerance graphs have a trapezoid intersection model with trapezoids of special shape.

Theorem 4.8.2 (Bogart et al. [124], Felsner [382]) *A graph G is a bounded tolerance graph if and only if G has a trapezoid intersection model where all trapezoids are parallelograms.*

No polynomial-time recognition algorithm is known for tolerance graphs or bounded tolerance graphs.

Problem 4.8.1 *Can one recognize bounded tolerance and tolerance graphs in polynomial time?*

By analogy to interval graphs one can study unit-length interval models and interval models where no two intervals contain each other.

Definition 4.8.2 [461] *Let G be a graph.*

> *G is a* proper tolerance graph *if it admits a tolerance representation in which no interval contains another interval.*
>
> *G is a* unit tolerance graph *if it admits a tolerance representation in which every interval is of unit length.*

Golumbic, Monma, and Trotter [461] asked the question whether a graph G is a proper tolerance graph if and only if G is a unit tolerance graph. The answer is given in the following theorem.

Theorem 4.8.3 [124] *The class of unit tolerance graphs is properly contained in the class of proper tolerance graphs.*

The paper [124] contains a characterization of unit tolerance graphs as 50% tolerance graphs.

Theorem 4.8.4 [461] *The class of proper tolerance graphs is properly contained in the class of bounded tolerance graphs.*

There are interesting generalizations of tolerance graphs. One of them is the class of *multitolerance graphs* introduced by Parra [854]: It is obtained by admitting a family $\mathcal{T}$ of sets τ_v of tolerance intervals for every vertex v represented by an interval I_v. Then

$$vw \in E \iff \text{there is a } T_v \in \tau_v \text{ with } T_v \subseteq I_w \text{ or there is a } T_w \in \tau_w \text{ with } T_w \subseteq I_v.$$

By a suitable restriction bounded multitolerance graphs are defined.

Theorem 4.8.5 (Parra [854])

(i) *The trapezoid graphs are exactly the bounded multitolerance graphs.*

(ii) *The class of tolerance graphs is properly contained in the class of multitolerance graphs.*

(iii) *Multitolerance graphs are perfect.*

Tolerance graphs are a generalization of interval graphs formed by measuring the size of the intersections and by introducing vertex tolerances. Another approach based on vertex weights and tolerances leads to the following classes.

Definition 4.8.3 (Monma, Reed, Trotter [794, 795]) *A graph $G = (V, E)$ is a* threshold tolerance graph *if there is a weight function $w : V \to \mathbb{R}^+$ and a tolerance function $t : V \to \mathbb{R}^+$ such that*

$$xy \in E \text{ if and only if } w_x + w_y \geq \min(t_x, t_y).$$

Theorem 4.8.6 [794, 795] *A graph G is a threshold tolerance graph if and only if there are weight functions $a : V \to \mathbb{R}^+$, $b : V \to \mathbb{R}^+$ such that $xy \in E$ if and only if $a_x \leq b_y$ and $a_y \leq b_x$.*

Monma, Reed, and Trotter [794, 795] give several other characterizations of these graphs and show the following inclusions for the complements of them:

(i) Interval graphs are co-threshold tolerance graphs.

(ii) Co-threshold tolerance graphs are tolerance graphs.

(iii) Co-threshold tolerance graphs are strongly chordal.

A characterization in terms of vertex orderings avoiding finitely many forbidden configurations leads to a polynomial-time recognition algorithm for threshold tolerance graphs [794, 795] (and shows that the complements of these graphs are strongly chordal; see Theorem 5.5.7).

Now we come to a small but very interesting and well-characterized class of graphs.

Definition 4.8.4 (Chvátal, Hammer [216]) *A graph is a* threshold graph *if it is a threshold tolerance graph with constant tolerance function, i.e., all tolerances have the same value.*

A good reference for threshold graphs is the monograph of Mahadev and Peled [762] dealing with threshold and tolerance graphs.

Originally, threshold graphs were introduced via linear inequalities and separations of vertices in polyhedra; see Theorem 13.1.1. It turns out that threshold graphs are the P_4-free interval and co-interval graphs, or equivalently, the P_4-free split graphs (see Theorem 6.6.3). It is also known that G is a threshold graph if and only if $\text{dilw}(G) = 1$ (see Theorem 13.2.1).

Bibelnieks and Dearing [106] investigate a generalization of tolerance graphs called *neighborhood subtree tolerance graphs*, where intervals are replaced by neighborhood subtrees. These graphs are weakly chordal. The case of constant tolerance leads to intersection graphs of neighborhood subtrees on a tree that are strongly chordal.

Jacobson, McMorris, Mulder, and Scheinerman [608, 609, 610] introduce another generalization called ϕ-tolerance (S, μ)-graphs. Here uv is an edge if and only if $\mu(S_u \cap S_v) \geq \phi(t_u, t_v)$. The function ϕ can be $\min, \max, +$, etc. Thus, the tolerance graphs are the min-tolerance interval graphs (which seems to be the most interesting case). For further results on ϕ-tolerance graphs see [607].

The notion of tolerance appeared also in [302] for bipartite graphs in the following way.

Definition 4.8.5 (Derigs, Goecke, Schrader [302]) *A bipartite graph $B = (X, Y, E)$ is a* bipartite tolerance graph *if there is an $\varepsilon > 0$ and a real-valued function f on the vertex set such that $xy \in E \iff |f(x) - f(y)| \leq \varepsilon$.*

In [155] it is shown that bipartite tolerance graphs are exactly the bipartite permutation graphs.

4.9 Other geometric objects

A fairly general class of graphs is the notion of *string graphs*, i.e., the intersection graphs of curves in the plane originally introduced in [978] and studied, e.g., in [347, 689, 690, 693]. In [690] the recognition problem of these graphs is shown to be $\mathbb{NP}$-hard. As Kratochvíl [692] remarks, so far no algorithm for string-graph recognition is known (not even an inefficient one!).

Kratochvíl [691] investigates some special cases of string graphs:

> *k-DIR* is the class of intersection graphs of straight-line segments parallel with at most k directions;
>
> *PURE-k-DIR* is the class of k-DIR graphs that have a representation by straight-line segments parallel with at most k directions such that every pair of parallel segments is disjoint;
>
> *SEG* is the class of intersection graphs of straight-line segments in the plane.

It is straightforward that SEG $= \bigcup_{k=1}^{\infty}$ k-DIR. Kratochvíl and Matoušek [694] show that SEG $= \bigcup_{k=1}^{\infty}$ PURE-k-DIR.

Kratochvíl [691] shows that for $k \geq 2$, the recognition problem for k-DIR and PURE-k-DIR graphs is $\mathbb{NP}$-complete. The recognition problem for SEG graphs is only known to be in PSPACE [694].

Note that the class PURE-2-DIR is investigated in [83, 81] under the name *grid intersection graphs.*

Further papers dealing with intersection graphs of curves in the two- and three-dimensional space are [638, 639, 640, 641].

For the topic of intersection graphs of squares and rectangles in the plane there are several interesting papers; see [907, 81, 695]. There are also some papers on special algorithmic aspects; see [32, 604, 605].

In [227, 156, 766] the class of *unit disk graphs* is studied: A graph is a unit disk graph if it is the intersection graph of circles of radius one in the plane. Hereby it is assumed that tangent circles intersect. Since C_5 and K_5 are unit disk graphs this class is neither contained in the class of perfect nor in the class of planar graphs.

The recognition problem for unit disk graphs is shown to be $\mathbb{NP}$-hard by Breu and Kirkpatrick [156]. For the more general case of intersection graphs of disks with bounded diameter, Breu and Kirkpatrick [157] show that the recognition problem is $\mathbb{NP}$-hard, and for the case of disk collections of unbounded diameter the same result was given by Kratochvíl [692]. The general class has also been called circle intersection graphs; the term disk intersection graph is more precise, in that vertices are considered to be adjacent if the circle corresponding to one vertex is inside the circle corresponding to the other vertex.

The result of [156, 157] also applies to touching graphs of bounded-diameter disks while the unbounded-diameter touching graphs of disks are the planar graphs (Koebe [681]; see [943]).

4.10 Other interactions—visibility

This section deals with models based on visibility constraints. In these classes of graphs, vertices correspond to objects, and two vertices are adjacent if the straight line between the corresponding objects has certain properties. For more information on visibility representations, see O'Rourke [842, 844, 845] and Shermer [970]. The most widely studied class of graphs with this type of model is the class of visibility graphs of a polygon, also known as polygon-vertex visibility graphs.

Definition 4.10.1 *Let P be a simple polygon in standard position. The graph $G(P)$ has vertices corresponding to extreme points of P. Vertices x and y are adjacent if and only if x and y are consecutive extreme points of P, or the line between x and y is entirely inside P. G is a* visibility graph of a polygon *if $G = G(P)$ for some polygon P. In the internal visibility graph of P, vertices x and y are adjacent if and only if the line between x and y is entirely inside P.*

The problem of recognizing visibility graphs and constructing visibility polygons in polynomial time remains open; the problem is not known to be in $\mathbb{NP}$, but is known to be in PSPACE (Everett [363]). Lin and Skiena [730] give a $2^{\Omega(n)}$ lower bound on the size of a grid need to construct a polygon corresponding to a visibility graph, and a $2^{2^{O(n)}}$ upper bound is given in [465]. Resolving the question of grid size is interesting, since a singly exponential grid size would give a polynomial-space description of the polygon, implying that the recognition problem is in $\mathbb{NP}$. Even if the ordering of endpoints around the polygon is specified, no efficient algorithm is known for constructing a representation. A necessary set of conditions on visibility graphs with a specified Hamilton circuit was presented in [434]; these were shown to be insufficient in [363].

Some positive results have been achieved by placing restrictions on the polygons. In order to describe some of these constraints, we need the following definitions.

Definition 4.10.2 *A vertex p_i of a polygon P is* convex (reflex) *if the interior angle formed by p_{i-1}, p_i, p_{i+1} is less than (at least) 180 degrees. A* convex (reflex) chain of a polygon *is a consecutive subsequence of convex (reflex) vertices.*

Visibility graphs of a single convex chain correspond to complete graphs. A polygon is called a *spiral* if it consists of one reflex chain and one convex chain.

Theorem 4.10.1 (Everett, Corneil [364]) *The visibility graph of a spiral is an interval graph.*

Everett and Corneil [364] give a linear-time algorithm for recognizing visibility graphs of spiral polygons. They also show that a generalization to a class of polygons called 2-spirals, in which reflex vertices can be grouped into two chains, produces visibility graphs that are Berge.

A *tower* (also called a *funnel*) is a polygon formed by two reflex chains, plus a single edge called the *base of the tower.*

Theorem 4.10.2 (Colley, Lubiw, Spinrad [235]) *G is the internal visibility graph of a tower if and only if G is a connected bipartite permutation graph.*

The papers [235] and [196] present linear-time algorithms for recognizing visibility graphs of towers.

Definition 4.10.3 *A* convex fan *is a polygon P in which some convex vertex x is adjacent to every other vertex in the visibility graph of P. An* orthogonal convex fan (staircase polygon) *is a convex fan consisting solely of horizontal and vertical lines.*

Visibility graphs of orthogonal convex fans are studied in [2], which gives a polynomial-time recognition algorithm for the class from an arbitrary adjacency list, and a characterization in terms of forbidden configurations if the Hamilton cycle is part of the input. Colley [234, 233] showed that recognizing visibility graphs of unimonotone polygons, in which vertices are nondecreasing in the direction of some line, can be reduced to recognition of orthogonal convex fans if the Hamilton circuit is part of the input. The restricted class of staircase polygons with uniform step-length is studied in [1].

There is currently no polynomial-time recognition algorithm for visibility graphs of general convex fans. However, there have been characterizations if the order of vertices is specified; see [348, 3].

We now review some results obtained by studying restricted graph classes rather than restrictions on the polygon.

Theorem 4.10.3 (El-Gindy [348], Lin, Skiena [730]) *Every maximal outerplanar graph G is a visibility graph, and the polygon that represents G can be constructed in polynomial time* [348] *and space* [730].

There have been a number of papers dealing with which subgraphs cannot occur as induced subgraphs of a visibility graph. Everett and Corneil [365] show that there is no finite set of forbidden induced subgraphs that characterize the visibility graphs. Every tree and every convex graph (see Definition 6.2.4) is an induced subgraph of some visibility graph. There are chordal graphs [365] and bipartite graphs [970] that are not induced subgraphs of visibility graphs. A general theorem that any sufficiently dense class of graphs must contain a graph that is not an induced subgraph of a visibility graph is given in [994].

The next results deal with reconstruction of the visibility graph if one is given extra information in addition to the adjacency matrix.

Theorem 4.10.4 (Coullard, Lubiw [260]) *Suppose that we are given a weighted graph G, where $w(x, y)$ corresponds to a required Euclidean distance between a visible pair of vertices x and y in the model. A polygon meeting these requirements can be constructed in polynomial time, if such a polygon is possible.*

Theorem 4.10.4 relies on the property that every 3-connected component of a visibility graph has a 3-clique ordering, i.e., an ordering in which each vertex added is adjacent to all vertices of an existing clique of size at least 3.

If one is given the external visibility graph as well as the visibility graph of a polygon, it is possible to reconstruct the convex hull of the polygon [843, 368].

It is interesting to note that one can find a maximum clique in a visibility graph in polynomial time if the model is given as input [319]. This is one of the only natural optimization problems on a class that is known to be solvable in polynomial time if the model is given, but is open with respect to time complexity if the input is given in adjacency-list form.

Visibilities between many other types of objects have been studied; examples can be found in the references at the start of this section. One natural class, the bar visibility graphs, has a nice characterization, as given below.

Definition 4.10.4 *Consider a set of horizontal intervals in the plane, with each interval at a different height.*

> Interval x sees interval y *if there is a vertical line l of width > 0 between x and y such that l touches no intervals other than x and y.*
>
> *G is a* bar visibility graph *if vertices can be mapped to a set of intervals such that x is adjacent to y if and only if interval x sees interval y.*

Theorem 4.10.5 (Wismath [1089], Tamassia, Tollis [1019]) *A graph is a bar visibility graph if and only if it has a planar embedding with all cutpoints on the exterior face.*

The following corollary makes it obvious that bar visibility graphs can be recognized in linear time.

Corollary 4.10.1 *Let G^+ be the graph obtained from G by adding a vertex x adjacent exactly to the cutpoints of G. Then G is a bar visibility graph if and only if G^+ is planar.*

It is interesting to note that if the definition is changed to allow collinear segments, the recognition problem becomes $\mathbb{NP}$-complete [22]. Other variants studied include allowing segments in arbitrary position, two-way (horizontal and vertical) visibility, and general visibility between segments. For a summary of these results, see [844].

Chapter 5

Vertex and Edge Orderings

5.1 Perfect elimination and generalizations

In algorithmic graph theory there are many interesting examples of vertex orderings and also of edge orderings of graphs. These orderings frequently appear under the name *elimination orderings*, *elimination schemes*, or *dismantling schemes*.

Elimination orderings are an algorithmically powerful tool. Many efficient graph algorithms use such orderings, working incrementally along the given ordering from left to right or from right to left using only local properties in a "greedy" way.

Definition 5.1.1 *The ordering $(v_1, \dots, v_n)$ of the vertex set of a graph G is an* elimination ordering satisfying property P *if for all $i \in \{1, \dots, n\}$, the vertex v_i has property P in the remaining graph $G_i = G(\{v_i, \dots, v_n\})$.*

The simplest example is a vertex ordering of a tree where each vertex is a leaf in the remaining graph. Definition 1.2.2 gives a very important generalization of this concept, namely the perfect elimination orderings. Recall that Theorem 1.2.2 characterizes the graphs having a perfect elimination ordering as the chordal graphs.

Dirac [316] and Lekkerkerker and Boland [723] have shown that every incomplete chordal graph has two nonadjacent simplicial vertices. Shibata [972] generalizes this result in the following way.

Definition 5.1.2 *Let G be a graph.*

> *A maximal clique of G is a* simplicial clique *if it contains a simplicial vertex.*
>
> *Let $S(Q)$ be the set of simplicial vertices of a simplicial clique Q. A simplicial clique Q is a* boundary clique *if either $Q = S(Q)$ or there is a maximal clique different from Q including $Q \setminus S(Q)$.*

Theorem 5.1.1 (Shibata [972]) *Every incomplete chordal graph has at least two non-adjacent simplicial vertices contained in boundary cliques.*

As in trees, simplicial vertices of chordal graphs can be chosen to be of maximum distance. This fact is mentioned by Duchet [335] as a result of Laskar and Shier [704]. We refer here to two further sources; it may have been known to many researchers.

Theorem 5.1.2 (Voloshin [1068, 1069], Farber, Jamison [375])
Every chordal graph G with at least two vertices has a diametral pair of simplicial vertices: There are simplicial vertices u, v with $d(u, v) = diam(G)$.

The existence of diametral pairs of simplicial vertices suggests the following notion.

Definition 5.1.3 *The ordering $(v_1, \ldots, v_n)$ of the vertex set of a graph G is a* diametral elimination ordering satisfying property P *if for all $i \in \{1, \ldots, n\}$, the vertex v_i has property P and maximum eccentricity in the remaining graph $G_i = G(\{v_i, \ldots, v_n\})$.*

Perfect elimination orderings and the algorithms LexBFS and MCS play a crucial role in the linear-time recognition of chordal graphs; see Theorem 1.6.1. LexBFS and MCS are of interest for some other graph classes as well.

Both algorithms lead to linear-time recognition of chordal graphs; see the papers of Rose, Tarjan, and Lueker [928] and Tarjan and Yannakakis [1021]. The algorithms work in two stages—first constructing a candidate elimination ordering and then verifying that it is really a perfect elimination ordering

Note that every lexical ordering of a chordal graph is a perfect elimination ordering, as shown by Lubiw [744]. The lexical orderings, however, do not lead to linear-time recognition of chordal graphs.

In [851] other algorithms for constructing a perfect elimination ordering using DFS instead of BFS are studied. The task of generating all perfect elimination orderings of a graph is solved in [974].

Brandstädt, Chepoi, and Dragan [144] describe an algorithm called LMCS, which gives a common perfect elimination ordering of all chordal powers of a graph and is a combination of LexBFS and MCS. Unfortunately, LMCS is not linear-time bounded. Other results dealing with powers of graphs are contained in the section on powers of graphs in the distance chapter (section 10.6).

Theorem 5.1.3 [144] *Let G be a graph whose powers $G^{i_1}, \ldots, G^{i_p}$ are chordal. Then the algorithm LMCS finds a common perfect elimination ordering of the graphs $G^{i_1}, \ldots, G^{i_p}$.*

Corollary 5.1.1 [144] *If G and G^2 are chordal, then there is a common perfect elimination ordering of all powers of G.*

LexBFS itself fulfills a weaker condition.

Theorem 5.1.4 (Brandstädt, Dragan, Nicolai [149]) *Every LexBFS ordering of a chordal graph G is a common perfect elimination ordering of all odd powers of G.*

There is no analogous result for MCS. In [149] the graphs for which LexBFS also gives a common perfect elimination ordering of all chordal even powers are characterized in terms of forbidden induced subgraphs. Similar forbidden subgraph characterizations for the case of MCS are given in the same paper.

For similar results in the case of distance-hereditary graphs see Theorems 5.1.7 and 5.1.8.

In perfect elimination orderings the local property that is fulfilled by the vertex to be eliminated next is simpliciality. This property can be varied in different ways. In the following we give several examples.

Definition 5.1.4 (Nicolai [810, 811]) *Let G be a graph.*

A vertex v is 2-simplicial *if $D_2(v)$ induces a cograph in G.*

$(v_1, \ldots, v_n)$ is a 2-simplicial elimination ordering *of G if for every $i \in \{1, \ldots, n\}$, the vertex v_i is 2-simplicial in G_i. (Such orderings were called* 2-simplicial dismantling schemes *in* [810, 811].)

Theorem 5.1.5 [810, 811] *Let G be a graph. The following conditions are equivalent:*

(i) *G is distance hereditary;*

(ii) *G has a 2-simplicial elimination ordering;*

(iii) *G has a diametral 2-simplicial elimination ordering.*

The role of LexBFS for distance-hereditary graphs is similar to its use on chordal graphs.

Theorem 5.1.6 (Dragan, Nicolai [327]) *The graph G is distance hereditary if and only if every LexBFS ordering of G is a 2-simplicial elimination ordering.*

The main reason for the similarities between chordal and distance-hereditary graphs concerning elimination orderings is the following.

Theorem 5.1.7 [327] *For a distance-hereditary graph G, an ordering $(v_1, \ldots, v_n)$ is a 2-simplicial elimination ordering of G if and only if it is a perfect elimination ordering of G^2.*

Thus, every LexBFS ordering of a distance-hereditary graph is a perfect elimination ordering of G^2. In [327] even more is shown.

Theorem 5.1.8 [327] *Every LexBFS ordering of a distance-hereditary graph G is a perfect elimination ordering of every even power G^{2k}, $k \geq 1$.*

The paper [327] studies the case of all powers and characterizes those distance-hereditary graphs G by forbidden subgraphs for which LexBFS gives a common perfect elimination ordering for all powers of G, i.e., a vertex ordering $(v_1, \ldots, v_n)$ is a common perfect elimination ordering of all powers G^k of G.

The recursive generation of distance-hereditary graphs described in Theorem 11.6.7 by adding twins and pendant vertices can also be viewed as a type of elimination ordering.

5.2 Semiperfect elimination

A simplicial vertex is exactly equal to a vertex that is not the center of any P_3. By generalizing from P_3 to P_4, we can get interesting new concepts.

Definition 5.2.1 (Jamison, Olariu [616]) *Let G be a graph.*

A vertex v is semisimplicial *in G if v is midpoint of no P_4 in G.*

A vertex ordering $(v_1, \ldots, v_n)$ is a semiperfect elimination ordering *of G if for every $i \in \{1, \ldots, n\}$, the vertex v_i is semisimplicial in G_i.*

Hoàng and Reed [571] showed that G has a semiperfect elimination ordering if and only if G is P_4-simplicial, as defined in section 5.7.

Trivially, every simplicial vertex is semisimplicial and thus every perfect elimination ordering is a semiperfect elimination ordering. For chordal graphs, every LexBFS ordering gives a perfect elimination ordering. In [616] it is shown that this holds in a similar way for HHD-free graphs and semiperfect elimination orderings.

Theorem 5.2.1 **[616]** *The graph G is HHD-free if and only if every LexBFS ordering of G is an semiperfect elimination ordering.*

Note that the house, which has a semiperfect elimination ordering, has only one semisimplicial vertex, whereas every chordal graph with at least two vertices must have at least two simplicial vertices. The following theorem leads to an $\mathcal{O}(n^4)$-time recognition algorithm of HHD-free graphs; the fastest recognition algorithm is due to Hoàng and Sritharan [573] and runs in $\mathcal{O}(n^3)$ time.

Theorem 5.2.2 (Hoàng, Khouzam [565]) *If G is an HHD-free graph, then G satisfies at least one of the following conditions:*

(i) *G is a clique;*

(ii) *G contains two nonadjacent simplicial vertices;*

(iii) *G contains a homogeneous set S such that S induces a connected subgraph in $\overline{G}$.*

Condition (iii) of Theorem 5.2.2 is closely related to a property of semisimplicial vertices. The following lemma is implicitly contained in the proof of Theorem 5.2.2.

Lemma 5.2.1 (Jamison, Olariu [616], Dragan, Nicolai, Brandstädt [329]) *If a semisimplicial vertex v is not simplicial, then $N(v)$ contains a nontrivial proper module (which is given by a $\overline{G}$-connected component of $N(v)$).*

Thus, in this sense the graphs having a semiperfect elimination ordering are a kind of modular (or homogeneous) extension of chordal graphs.

There is another interesting class of graphs that can be interpreted as homogeneous extensions of chordal graphs.

Definition 5.2.2 (Olariu [826]) *A graph G is* weak bipolarizable *if G contains no induced house, hole, domino, or "A" (the HHDA-free graphs).*

Note that the domino and the "A" contain a "P" and thus the class of HHP-free graphs is properly contained in the class of HHDA-free graphs.

Theorem 5.2.3 [826] *A graph G is weak bipolarizable if and only if every induced subgraph of G is chordal or contains a proper module.*

There are other characterizations of HHD-free graphs and of HHDA-free graphs that show the deep connection to chordal graphs (see, e.g., Theorem 10.5.8 and Theorem 5.2.1).

Theorem 5.2.1 can be shown using m^3-convexity (see Definition 10.5.5) and this leads to the following corollary.

Corollary 5.2.1 [329] *In every HHD-free graph G with at least 2 vertices there is a pair of semisimplicial vertices u, v such that $d(u,v) = diam(G)$.*

For a distance-hereditary graph G, LexBFS gives a semiperfect elimination ordering not only for G but also for its powers.

Theorem 5.2.4 (Dragan, Nicolai [327]) *Every LexBFS ordering of a distance-hereditary graph G is a common semiperfect elimination ordering of all its powers G^k, $k \geq 1$.*

This implies that all powers of distance-hereditary graphs are HHD-free, which was shown already in [54]; see Theorem 10.6.14. For powers of general HHD-free graphs see Theorem 10.6.16.

Theorem 5.2.5 (Dragan, Nicolai [328]) *Every LexBFS ordering of an HHD-free graph G is a common semiperfect elimination ordering of all odd powers G^{2k+1}, $k \geq 0$.*

The case of even powers and LexBFS is described in [328] in terms of forbidden subgraphs. Replacing LexBFS by MCS leads to the following characterization.

Theorem 5.2.6 [616] *The graph G is HHP-free if and only if every MCS ordering of G is a semiperfect elimination ordering.*

The proofs of Theorems 5.2.6 and 5.2.3 are simplified in [329] by using m^3-convexity properties.

The following graph class is an interesting generalization of graphs having a semiperfect elimination ordering.

Definition 5.2.3 (Chvátal [208]) *G is a* brittle graph *if each induced subgraph H of G contains either a vertex that is not the endpoint of any P_4 of H, or a vertex that is not the midpoint of any P_4 of H.*

Since the midpoints of a P_4 in G are the endpoints of a P_4 in $\overline{G}$ we have the following. A graph G is brittle if and only if for each induced subgraph H of G there is a semisimplicial vertex in H or $\overline{H}$. Thus, Theorem 5.2.1 implies that every HHD-free graph is brittle. Brittle graphs can be recognized in $\mathcal{O}(m^2)$ [946] time or $\mathcal{O}(n^3 \log^2 n)$ [996] time.

A similar concept is interesting for chordal graphs and simplicial vertices: In an unpublished paper, Gorgos [467] characterized the class of graphs $G = (V, E)$ having an ordering $(v_1, \ldots, v_n)$ of V such that for all $i \in \{1, \ldots, n\}$, the vertex v_i is simplicial in G_i or in $\overline{G_i}$. This graph class is called *quasi triangulated* in [1098], where the following characterization by Gorgos is described. A graph G is *latticed* if each vertex of G belongs to some C_k and to some $\overline{C_l}$, $k, l \geq 4$, in G.

Theorem 5.2.7 [467] *Let G be a graph. The following conditions are equivalent:*

(i) *G is quasi triangulated;*

(ii) *no induced subgraph of G is latticed.*

Quasi-triangulated graphs coincide with the good graphs given in Definition 5.7.2.

5.3 Domination and distance-preserving elimination

In Definition 1.1.16 the notion of domination between two vertices was introduced.

Definition 5.3.1 (Dahlhaus et al. [281]) *Let G be a graph.*

A vertex ordering $(v_1, \ldots, v_n)$ *of G is a* domination elimination ordering *if for all* $i \in \{1, \ldots, n-1\}$, *there is a* $j > i$ *such that the vertex* v_i *is dominated by* v_j *in* G_i, *i.e.,* $N_i(v_i) \subseteq N_i[v_j]$.

G is a domination graph *if each induced subgraph of G has a domination elimination ordering.*

Note the close connection to cop-win orderings (dismantling schemes) (see Definition 5.8.2).

Lemma 5.3.1 [281] *If a connected graph G has a semiperfect elimination ordering, then it also has a d.e.o.*

Dahlhaus et al. [281] give a characterization of domination graphs in terms of Boolean functions. Furthermore, it is shown that brittle graphs are domination graphs and that domination graphs are weakly chordal.

MCS plays an important role in obtaining a domination elimination ordering of a given graph. Recall that Theorem 5.2.6 describes the role of MCS on HHP-free graphs. Now the class of house-hole-free (HH-free) graphs will be considered; this class is the same as house-free weakly chordal graphs.

Theorem 5.3.1 [281] *Let G be a HH-free graph. Then every ordering of the vertices of G produced by MCS is a domination elimination ordering.*

This implies that HH-free graphs are domination graphs. In [281] it is mentioned that Theorem 5.3.1 also holds for LexBFS.

Chepoi [192] introduces the notion of a distance-preserving elimination ordering.

Definition 5.3.2 *Let G be a graph. A vertex ordering $(v_1, \ldots, v_n)$ of G is a* distance-preserving elimination ordering *if for all $i \in \{1, \ldots, n\}$, G_i is an isometric subgraph of G.*

There are connections between distance-preserving elimination orderings, pseudo-peakless functions, and convexity properties studied by Chepoi [192] that are analogous to similar connections between peakless functions and perfect elimination orderings studied by Chepoi [190]. Pseudomodular (see Definition 10.3.4) and weakly modular graphs (see Definition 10.3.1) are characterized in terms of pseudopeakless functions in [192]. Furthermore Chepoi [192] shows the following result.

Theorem 5.3.2 *Let G be a pseudomodular or a house-free weakly modular graph. Then every ordering $(v_1, \ldots, v_n)$ of G produced by BFS is a distance-preserving elimination ordering.*

Corollary 5.3.1 [192] *For a graph G the following conditions are equivalent:*

(i) *For every isometric subgraph of G, the ordering of its vertices produced by BFS is distance preserving;*

(ii) *G is hereditary weakly modular.*

Further results of Chepoi [192] imply the following corollary.

Corollary 5.3.2 *Let G be a hereditary modular graph. Then every ordering of the vertices of G produced by BFS is a domination elimination ordering.*

In [192] the following modification of domination elimination orderings is defined.

Definition 5.3.3 *Let G be a graph.*

An edge xy dominates *a vertex v if $N[v] \subseteq N[x] \cup N[y]$.*

A vertex ordering $(v_1, \ldots, v_n)$ of G is an edge-domination elimination ordering *if for all $i \in \{1, \ldots, n\}$ there is an edge v_jv_k with $j > i, k > i$, which dominates the vertex v_i in G_i.*

Theorem 5.3.3 [192] *Let G be a house-free graph. Then the following conditions are equivalent:*

(i) *G and each isometric subgraph of G admit an edge-domination elimination ordering;*

(ii) *G is hereditary weakly modular;*

(iii) *For each isometric subgraph of G every ordering of G produced by BFS is an edge-domination elimination ordering.*

This leads to the following characterization for which the notion of *bipyramids* is introduced.

Definition 5.3.4 [192] *A* bipyramid *bipyr*(C_k) *is the graph consisting of a cycle* C *of length* k *and two nonadjacent vertices which are adjacent to all vertices of* C.

Theorem 5.3.4 [192] *Let* G *be a house-free graph. Then the following conditions are equivalent:*

(i) *For every isometric subgraph of* G, *every LexBFS ordering* $(v_1, \ldots, v_n)$ *is a domination elimination ordering;*

(ii) G *contains no cycle* C_k, $k \geq 5$, *and no bipyramid bipyr*(C_k), $k \geq 6$, *as an isometric subgraph;*

(iii) G *is hereditary weakly modular and contains no bipyramid bipyr*(C_k), $k \geq 6$, *as an induced subgraph.*

5.4 Maximum neighborhood orderings and generalizations

A very natural kind of elimination ordering is the following notion given by Dragan, Prisacaru, and Chepoi [331], Behrendt and Brandstädt [75], and Brandstädt et al. [146].

Definition 5.4.1 *Let* G *be a graph.*

A vertex $u \in N[v]$ *is a* maximum neighbor of v *if for all* $w \in N[v]$, $N[w] \subseteq N[u]$ *(note that* $u = v$ *is not excluded).*

A vertex ordering $(v_1, \ldots, v_n)$ *of* G *is a* maximum neighborhood ordering *of* G *if for all* $i \in \{1, \ldots, n\}$, *the vertex* v_i *has a maximum neighbor* u_i *in* G_i, *i.e., for all* $w \in N_i[v_i]$, $N_i[w] \subseteq N_i[u_i]$.

G *is* dually chordal *if* G *has a maximum neighborhood ordering.*

Note that dual chordality of graphs, unlike chordality, is not a hereditary property, i.e., induced subgraphs of dually chordal graphs are not necessarily dually chordal. For the hereditary case see Theorem 5.5.4.

The reason for the name "dually chordal graphs" is the following characterization of graphs with maximum neighborhood orderings that is part of Theorem 8.3.1:

A graph G has a maximum neighborhood ordering if and only if its clique hypergraph $\mathcal{C}(G)$ forms a hypertree.

This means duality (in the sense of hypergraphs) to chordal graphs, but unlike the case of chordal graphs, dually chordal graphs are in general not perfect, as the wheel W_5 shows. Dually chordal graphs were called *HT-graphs* by Dragan, Prisacaru, and Chepoi [331] and Dragan [321, 323].

Theorem 8.3.1 mentions the following useful characterization via neighborhood hypergraphs:

> A graph G has a maximum neighborhood ordering if and only if its neighborhood hypergraph $\mathcal{N}(G)$ forms a hypertree.

This property leads to a linear-time recognition of dually chordal graphs due to the self-duality of the neighborhood hypergraph (see Theorem 8.3.1). It follows from $L(\mathcal{N}(G)) \sim G^2$ and Theorem 1.3.1 that for dually chordal graphs G the square graph G^2 is chordal. Note that every maximum neighborhood ordering of G is a perfect elimination ordering of G^2 (see [143]). This leads to a linear-time algorithm for obtaining an maximum neighborhood ordering for a given dually chordal graph by using the *MNO-algorithm* that is described in [143]. This algorithm is based on MCS principles.

The maximum neighborhood orderings are useful for solving algorithmic problems that are related to distances such as domination-like problems, center, radius, and diameter. Their algorithmic use is studied in [143, 145, 326].

It is interesting to consider the case of chordal graphs with maximum neighborhood orderings.

Definition 5.4.2 (Dragan, Prisacaru, Chepoi [331], Moscarini [799]) *A graph G is* doubly chordal *if G is chordal and dually chordal.*

Theorem 5.4.1 [331, 799] *A graph G is doubly chordal if and only if there is a vertex elimination ordering $(v_1, \ldots, v_n)$ of G such that for all $i \in \{1, \ldots, n\}$, v_i is simplicial in G_i and has a maximum neighbor in G_i.*

An elimination ordering of a doubly chordal graph that satisfies the conditions of the previous theorem can be found in linear time [143].

As already mentioned, every maximum neighborhood ordering of G is a perfect elimination ordering of G^2. Thus Theorem 5.4.1 is a special case of Theorem 5.1.3.

There is a similar concept of maximum neighborhood ordering's for bipartite graphs discussed by Dragan, Prisacaru, and Chepoi [331], Behrendt and Brandstädt [75], and Brandstädt et al. [146].

Definition 5.4.3 *Let $B = (X, Y, E)$ be a bipartite graph.*

> *Let $B_i^Y = B(X \cup \{y_i, y_{i+1}, \ldots, y_n\})$ and $N_i(x)$ be the neighborhood of $x \in X$ in B_i^Y.*
>
> *A linear ordering $(y_1, \ldots, y_n)$ of Y is a* maximum X-neighborhood ordering *of B if for all $i \in \{1, \ldots, n\}$, there is a maximum neighbor $x_i \in N(y_i)$ of y_i, i.e., for all $x \in N(y_i)$, $N_i(x) \subseteq N_i(x_i)$. A* maximum Y-neighborhood ordering *is defined analogously.*

The next theorem shows the close connection between maximum neighborhood orderings and a chordality property of bipartite graphs.

Theorem 5.4.2 (Behrendt, Brandstädt [75]) *Let $B = (X, Y, E)$ be a bipartite graph. Then the following conditions are equivalent:*

(i) *B has a maximum X-neighborhood ordering;*

(ii) *B is X-chordal and X-conformal.*

Furthermore, $(y_1, \ldots, y_n)$ is a maximum X-neighborhood ordering of G if and only if $(y_1, \ldots, y_n)$ is a perfect elimination ordering of the 2-section graph $2SEC(\mathcal{N}^Y(G))$.

(See Theorem 3.3.1 and Corollary 3.3.1.)

A useful generalization of maximum neighborhood ordering is the following.

Definition 5.4.4 (Brandstädt, Dragan, Nicolai [148]) *Let $G = (V, E)$ be a graph with $|V| > 1$.*

> *A vertex v is h-*extremal *if (the subgraph induced by) the disk $D_2(v)$ contains a proper homogeneous dominating set, i.e., there is a proper subset $H \subset D_2(v)$ that is homogeneous in G and for which $D_2(v) \subseteq N[H]$ holds.*
>
> *A vertex ordering $(v_1, \ldots, v_n)$ is a* homogeneous elimination ordering *if for every $i = 1, \ldots, n-1$, the vertex v_i is h-extremal in $G_i = G(\{v_i, \ldots, v_n\})$.*
>
> *G is* homogeneously orderable *if G has a homogeneous elimination ordering*

This leads to a common generalization of distance-hereditary and dually chordal graphs.

Corollary 5.4.1 *Dually chordal and distance-hereditary graphs are homogeneously orderable.*

Homogeneously orderable graphs are of interest for algorithmic reasons—thus, e.g., the (cardinality) Steiner tree problem is solvable in polynomial time on homogeneously orderable graphs. This generalizes and improves the results of D'Atri, Moscarini, and Sassano [290], where a class of graphs called *homogeneous graphs* is introduced. Brandstädt, Dragan, and Nicolai [148] show that homogeneous graphs are homogeneously orderable, and the larger class of homogeneously orderable graphs seems to be the more natural one for several reasons.

The hereditarily homogeneously orderable graphs are the house-hole-domino-sun-free (HHDS-free) graphs (see Theorem 7.2.5).

5.5 Strong and simple orderings and generalizations

Strongly chordal graphs (see Definition 3.4.1) are characterized in terms of the following elimination orderings.

Definition 5.5.1 (Farber [372]) *Let G be a graph. A perfect elimination ordering of G is a* strong perfect elimination ordering *of G if the following condition is fulfilled:*

> *For each $i < j < k < l$, if v_i is adjacent to v_k and v_l, and v_j is adjacent to v_k, then v_j is adjacent to v_l.*

Farber called these orderings *strong elimination orderings* in [372]. Note that it is easy to solve the domination problem if a strong perfect elimination ordering is given as input.

Farber showed that an ordering is a strong perfect elimination ordering if and only if it fulfills a slightly different set of properties, which will be used to define new classes of graphs later in this section.

Lemma 5.5.1 *A vertex ordering $(v_1, \ldots, v_n)$ of a graph $G = (V, E)$ is a strong perfect elimination ordering if and only if it fulfills the following two conditions:*

(i) $(v_1, \ldots, v_n)$ *is a perfect elimination ordering;*

(ii) *For each $i < j$ and $k < l$ with $v_iv_l \in E$, $v_iv_k \in E$, and $v_kv_j \in E$ it follows that $v_lv_j \in E$.*

Theorem 5.5.1 [372] *A graph is strongly chordal if and only if it admits a strong perfect elimination ordering*

Anstee and Farber [24] and Hoffman, Kolen, and Sakarovich [577] give $\mathcal{O}(n^3)$-time bounded algorithms for finding an strong perfect elimination ordering of a given strongly chordal graph.

In [277] *generalized strongly chordal graphs* are introduced using only condition (ii) of Lemma 5.5.1.

Definition 5.5.2 (Dahlhaus [277]) *Let G be a graph.*

> *A* strong ordering *of G is a vertex ordering $(v_1, \ldots, v_n)$ of G fulfilling condition* (ii) *of Lemma* 5.5.1.
>
> *G is* strongly orderable *if G has a strong ordering.*

Strong orderings were called *generalized strongly perfect elimination orderings* and graphs having such orderings were called *generalized strongly chordal* in [277]. In the same paper, generalized strongly chordal graphs were shown to be weakly chordal. The greedy matching algorithm, which is known to be optimal for strongly chordal graphs, is shown to be optimal even for generalized strongly chordal graphs (see [282, 324]). Dragan [325] gives an $\mathcal{O}(nm)$-time recognition algorithm for strongly orderable graphs.

There is another characterization of strongly chordal graphs by elimination orderings of a different type.

Definition 5.5.3 *Let G be a graph.*

> *A vertex v is* simple *in G if for all $x, y \in N[v]$, $N[x] \subseteq N[y]$ or $N[y] \subseteq N[x]$, i.e., $\{N[x] : x \in N[v]\}$ is linearly ordered by inclusion.*

A simple elimination ordering *of G is a vertex ordering $(v_1, \ldots, v_n)$ such that for all $i \in \{1, \ldots, n\}$, the vertex v_i is simple in G_i.*

Theorem 5.5.2 [372] *A graph G is strongly chordal if and only if every induced subgraph of G has a simple vertex.*

Corollary 5.5.1 *A graph is strongly chordal if and only if it admits a simple elimination ordering .*

It is easy to see that a strong perfect elimination ordering is a simple elimination ordering such that $N_i[v_l] \subseteq N_i[v_k]$ whenever $i \leq l \leq k$ and $v_l, v_k \in N_i[v_i]$ (see, e.g., [181]).

For a given graph G one can find a simple elimination ordering of G in $\mathcal{O}(\min\{m \log n, n^2\})$ time (Anstee and Farber [24], Lubiw [744], Paige and Tarjan [850], Spinrad [991]) by using the connection between strongly chordal graphs and Γ-free matrices; see [744] for the notion of *doubly lexical orderings* of matrices and graphs and the following facts (see also the matrix chapter (chapter 9)).

Theorem 5.5.3 (Hoffman, Kolen, and Sakarovitch [577], Lubiw [744], Anstee, Farber [24]) *A 0-1 matrix is totally balanced if and only if every doubly lexical ordering is Γ-free.*

Obviously a simple elimination ordering is a special case of a maximum neighborhood ordering, where not only a maximum but a linear ordering of the neighborhoods of neighbors exists. Thus, strongly chordal graphs are doubly chordal.

From the characterization of strongly chordal graphs as the sun-free chordal graphs (see Theorem 7.2.1) it follows that the strongly chordal graphs are exactly the hereditarily dually chordal graphs.

Theorem 5.5.4 *A graph G is strongly chordal if and only if all induced subgraphs of G have an maximum neighborhood ordering*

In Theorem 8.2.2 an analogous property is expressed purely in terms of hypergraphs; strongly chordal graphs correspond closely to totally balanced hypergraphs and dually chordal graphs correspond closely to hypertrees.

A similar property holds for chordal bipartite graphs.

Theorem 5.5.5 (Dragan, Voloshin [332]) *A bipartite graph $B = (X, Y, E)$ is chordal bipartite if and only if every induced subgraph of B has a maximum X-neighborhood ordering (maximum Y-neighborhood ordering).*

Note that there is another characterization of chordal bipartite graphs in terms of a vertex elimination scheme that follows easily from the Γ-free matrix characterization.

Theorem 5.5.6 (Hammer, Maffray, Preissmann [513]) *A bipartite graph is chordal bipartite if and only if a vertex that is not in the center of a P_5 can be repeatedly removed until no vertex remains.*

Dragan [325] generalizes the notion of a simple elimination ordering in the following way.

Definition 5.5.4 [325] *Let G be a graph.*

> *A vertex v is* quasi simple *in G if for all $x, y \in N(v)$, $N(x) \subset N[y]$ or $N(y) \subset N[x]$.*
>
> *A* quasi-simple elimination ordering *of G is a vertex ordering $(v_1, \ldots, v_n)$ such that for all $i \in \{1, \ldots, n\}$, the vertex v_i is quasi simple in G_i.*

It is easy to see that every strong ordering of a graph G is a quasi-simple elimination ordering but not vice versa. In [325] it is shown, however, that the class of graphs having a quasi-simple elimination ordering is exactly the class of strongly orderable graphs. Dragan [325] gives further characterizations of this class in terms of elimination orderings; see Theorem 5.9.3.

Now we come to some subclasses of strongly chordal graphs. Co-threshold tolerance graphs (see Definition 4.8.3) are an interesting subclass of strongly chordal graphs characterized by the following vertex order properties.

Definition 5.5.5 (Monma, Reed, Trotter [794, 795]) *Let G be a graph.*

> *A vertex ordering $(v_1, \ldots, v_n)$ of G fulfills the P_4* rule *if for every P_4 $xywz$ in G, $x < z \iff y < w$ holds.*
>
> *A vertex ordering $(v_1, \ldots, v_n)$ of G fulfills the $2K_2$* rule *if for every $2K_2$ xy, wz in G, $x < w \iff x < z \iff y < w \iff y < z$ holds.*
>
> *A vertex ordering $(v_1, \ldots, v_n)$ of G is a* proper ordering *if for all nonadjacent vertices x, y, either x is on the left of $N(y)$ or y is on the left of $N(x)$ in the vertex ordering.*

In [795] it is observed that proper orderings are strong perfect elimination orderings as well.

Theorem 5.5.7 [794, 795] *For a graph G the following conditions are equivalent:*

(i) *G is a co-threshold tolerance graph;*

(ii) *G is strongly chordal and has a vertex ordering fulfilling the P_4 and the $2K_2$ rule;*

(iii) *G admits a proper ordering.*

The class of chordal comparability graphs, which is mentioned again in Definition 6.7.1, Theorem 6.7.1, and in the forbidden subgraph chapter (chapter 7) is a subclass of strongly chordal graphs, since no sun has a transitive orientation. It has a nice elimination ordering property.

Theorem 5.5.8 (Borie, Spinrad [130]) *Every CLexBFS ordering of a chordal comparability graph is a simple elimination ordering .*

Interval graphs and proper interval graphs can be characterized in terms of vertex orderings.

Theorem 5.5.9 (Olariu [832]) *A graph is an interval graph if and only if it has an ordering* $(v_1, \ldots, v_n)$ *such that for all* $i < j < k$*, whenever* v_iv_k *is an edge,* v_iv_j *is an edge.*

Theorem 5.5.10 (Looges, Olariu [734]) *A graph is a proper interval graph if and only if it has an ordering* $(v_1, \ldots, v_n)$ *such that for all* $i < j < k$*, whenever* v_iv_k *is an edge,* v_iv_j *and* v_jv_k *are also edges.*

5.6 Perfect orderings

Perfect elimination orderings have a nice property with respect to the following greedy coloring algorithm.

Algorithm 5.6.1 (Greedy Coloring) *Let* G *be a graph with vertex ordering* $(v_1, \ldots, v_n)$*. Color the vertices of* G *in this ordering from left to right according to the following rule:*

> *Assign to each* v_i $(1 \leq i \leq n)$ *the smallest positive integer assigned to no neighbor* v_j *of* v_i *with* $j < i$*.*

It is easy to see that for a perfect elimination ordering $(v_1, \ldots, v_n)$, Algorithm 5.6.1 yields an optimal coloring on the reverse ordering $(v_n, \ldots, v_1)$. This simple observation implies the perfection of chordal graphs.

Chvátal [205] studies the graphs that admit an ordering such that the greedy coloring algorithm (Algorithm 5.6.1) for all induced subgraphs is optimal.

Definition 5.6.1 [205] *Let* G *be a graph.*

> *A vertex ordering* $(v_1, \ldots, v_n)$ *of* G *is* perfect *if, for each induced (ordered) subgraph* F *of* G*, the number of colors used by Algorithm* 5.6.1 *on* F *coincides with the chromatic number of* F*.*
>
> G *is* perfectly orderable *if there is a perfect ordering* $(v_1, \ldots, v_n)$ *of* G*.*

For the subsequent characterization of these graphs the P_4s of the graph are important.

Theorem 5.6.1 [205] *A vertex ordering* $<$ *of a graph* G *is perfect if and only if* G *contains no* P_4 *abcd with* $a < b$ *and* $d < c$*.*

Note that this theorem implies that perfectly orderable graph recognition is in $\mathbb{NP}$. Chvátal [205, 207] proved that perfectly orderable graphs are strongly perfect and belong to **bip*** (see the definition before Conjecture 14.3.3). Moreover, they are perfectly contractile [366], hence (strict) quasi parity [551].

Theorem 5.6.2 (Middendorf, Pfeiffer [783]) *The recognition problem for the class of perfectly orderable graphs is* $\mathbb{NP}$*-complete.*

Based on the reduction used in [783], Hoàng [561] shows even more: It is $\mathbb{NP}$-complete to determine whether a given weakly chordal graph is perfectly orderable.

An ordering $<$ of the vertices of a graph G corresponds to an acyclic orientation $\vec{G}$ of $(G,<)$ such that $(x,y) \in E(\vec{G})$ if and only if $xy \in E(G)$ and $x < y$. Given an acyclic directed graph $\vec{H}$, we can construct an ordering $<$ of the underlying undirected graph G of $\vec{H}$ such that $\vec{H}$ is the corresponding orientation of $(G,<)$. Thus, Theorem 5.6.1 can be restated as follows.

Remark 5.6.1 *A graph is perfectly orderable if and only if it admits an acyclic orientation that does not contain* $\bullet\!\longrightarrow\!\bullet\!\longrightarrow\!\bullet\!\longleftarrow\!\bullet$ *(called* obstruction*) as an induced subgraph.*

There is a generalization of obstructions that leads to a concept similar to perfectly orderable graphs.

Definition 5.6.2 (Gavril, Toledano Laredo, de Werra [431]) *Let G be a directed graph.*

> *An* m-obstruction *in* G *is a chordless simple path* $x_1, x_2, \ldots, x_{m+2}$ *having arcs* $x_1 \to x_2$ *and* $x_{m+2} \to x_{m+1}$.
>
> G *is* fraternally oriented *if it has no 1-obstructions.*
>
> G *is* perfectly oriented *if it has no 2-obstructions.*

Urrutia and Gavril [1063] give an algorithm for fraternal orientation of graphs. Fraternally orientable graphs can be recognized in time $\mathcal{O}(|E||V|)$ [431]. See [429, 432] for a connection between fraternal orientations and intersection graphs of subtree models in (directed) unicyclic graphs.

More generally, graphs orientable without m-obstructions for constant m can be recognized in $\mathcal{O}(|E||V|^m)$ time [431]. This means that perfectly oriented graphs can be recognized in $\mathcal{O}(|E||V|^2)$ time in contrast to the $\mathbb{NP}$-completeness of the recognition of perfectly orderable graphs; see Theorem 5.6.2.

5.7 Special perfectly orderable graphs

The notion of perfectly orderable graphs has been a very challenging subject of research, and a natural task is to find interesting special cases of perfectly orderable graphs for which there are efficient recognition algorithms. By Remark 5.6.1, a graph is perfectly orderable if and only if it admits an acyclic orientation in which each induced P_4 is one of the following types 1, 2, 3 shown below:

`Type 1:` $\bullet\!\longleftarrow\!\bullet\!\longrightarrow\!\bullet\!\longrightarrow\!\bullet$ `Type 2:` $\bullet\!\longrightarrow\!\bullet\!\longrightarrow\!\bullet\!\longrightarrow\!\bullet$ `Type 3:` $\bullet\!\longrightarrow\!\bullet\!\longleftarrow\!\bullet\!\longrightarrow\!\bullet$

Thus, there are six very natural subclasses of perfectly orderable graphs formed by considering any nonempty proper subset of these three types of the P_4s.

Definition 5.7.1 *Let S be a nonempty, proper subset of $\{1,2,3\}$. A graph is S-*orientable *if it admits an acyclic orientation in which every P_4 is of type $t \in S$.*

Below are remarks on each of the six cases.

a. $\{1\}$-orientable graphs are also called *bipolarizable* by Hertz [548] and *Raspail* by Hoàng and Reed [571]; notice that there is a different notion of Raspail graphs in the sense of short-chorded graphs; see Definition 3.5.4. $\{1\}$-orientable graphs can be greedily recognized in time $\mathcal{O}(nMM)$ [996] and characterized by forbidden induced subgraphs, independently in [548] and [571]. This characterization of bipolarizable graphs motivated Olariu to consider the class of weak bipolarizable graphs (see Definition 5.2.2), which properly contains all bipolarizable graphs.

b. $\{2\}$-orientable graphs are also called P_4*-indifference* graphs [571]. They can be recognized in time $\mathcal{O}(n^2m)$ [894].

c. $\{3\}$-orientable graphs generalize the comparability graphs and are also called P_4*-comparability graphs* [572, 571]. They can be recognized in time $\mathcal{O}(n^4)$ [572] or $\mathcal{O}(n^2m)$ [894].

d. $\{1,2\}$-orientable graphs are also called P_4*-simplicial* [571]; they form a subclass of brittle graphs [571] (hence, all bipolarizable graphs and all P_4-indifference graphs are brittle). P_4-simplicial graphs can be recognized in $\mathcal{O}(n^3\log^2 n)$ time [996], using equivalence to semiperfectly orderable graphs (see remark after Definition 5.2.1).

e. In [561], Hoàng shows that it is $\mathbb{NP}$-complete to decide whether a weakly chordal graph is $\{1,3\}$-orientable; see the remarks after Theorem 5.6.2. Note that all P_5-free weakly chordal graphs are perfectly orderable [529].

f. The recognition complexity of $\{2,3\}$-orientable graphs is still open.

Brittle graphs (see Definition 5.2.3) form a large subclass of perfectly orderable graphs.

Definition 5.7.2 (Hoàng, Reed [571], Hoàng, Mahadev [570]) *Let G be a graph.*

An ordering $<$ on G is brittle *if for every induced subgraph H of G, either the largest vertex of $(H,<)$ is not a midpoint of any P_4 in H or the smallest vertex of $(H,<)$ is not an endpoint of any P_4 in H.*

An ordering $<$ on G is good *if for any induced subgraph H of G, either the largest vertex of $(H,<)$ is simplicial in H or the smallest vertex of $(H,<)$ is simplicial in $\overline{H}$ (cosimplicial).*

A graph G is good *if G admits a good ordering.*

Remark 5.7.1 [571, 570] *A graph is brittle if and only if it admits a brittle ordering. Every good ordering is a brittle ordering and every brittle ordering is a perfect ordering. Thus, every good graph is brittle and every brittle graph is perfectly orderable.*

It is not difficult to see that good graphs are equal to the quasi-triangulated graphs discussed in Theorem 5.2.7. This equivalence immediately gives an $\mathcal{O}(n^3)$ algorithm for recognizing good graphs.

In [570] it is shown that graphs with Dilworth number at most 3 are good and hence brittle. More subclasses of brittle graphs are given below.

Two subclasses of brittle graphs are introduced in [878], where the elimination aspect of Definition 5.2.3 is emphasized. Recall that by definition, brittle graphs admit an elimination ordering resulting from applying the following rules.

Rule 1. If there is no induced P_4 with vertex x as an endpoint, then x (and all adjacent edges) are removed.

Rule 2. If there is no induced P_4 with vertex x as a midpoint, then x (and all adjacent edges) are removed.

Definition 5.7.3 (Preissmann, de Werra, Mahadev [878]) *A graph G is* superbrittle *if at every stage of the elimination process one may eliminate every remaining vertex by Rule* 1 *or* 2 *(i.e., in every induced subgraph of G a vertex can never be a midpoint of an induced P_4 and an endpoint of an induced P_4).*

P_4 itself is a superbrittle graph since the endpoints of the P_4 are not midpoints of a P_4 and vice versa. Note that superbrittle graphs are covered in this chapter since they are a variant of brittle graphs; however, they can be defined without respect to elimination orderings as graphs in which every vertex is either not the midpoint of any P_4 or not the endpoint of any P_4.

Preissmann, de Werra, and Mahadev [878] give a characterization of superbrittle graphs in terms of forbidden induced subgraphs (see Theorem 7.1.5), which implies that superbrittle graphs are HHDA-free. Furthermore, there are more elimination properties of superbrittle graphs.

If in Rules 1 and 2, P_4 is replaced by P_3 (thus obtaining Rule $1'$ and Rule $2'$), then the class of graphs that have an elimination ordering according to Rules $1'$ and $2'$ is exactly the class of chordal graphs [878].

Definition 5.7.4 *A graph G is* superfragile *if at every stage of the elimination process one may eliminate every remaining vertex by Rule* $1'$ *or* $2'$.

Clearly, every superfragile graph is superbrittle. Preissmann, de Werra, and Mahadev [878] give a characterization of superfragile graphs by forbidden induced subgraphs that shows that superfragile graphs are trivially perfect (see Theorem 7.1.5).

Another subclass of brittle graphs is defined as follows.

Definition 5.7.5 *A graph G is* maxibrittle *if for every induced subgraph G' of G,*

there is no P_4 having a vertex of maximum degree as an endpoint and

there is no P_4 having a vertex of minimum degree as a midpoint.

Clearly, if G is maxibrittle then so is $\overline{G}$. These graphs are brittle since vertices of maximum degree may be eliminated by Rule 1, while vertices of minimum degree may be eliminated by Rule 2. In [878] a characterization of these graphs by forbidden induced subgraphs is given (see Theorem 7.1.5). In particular, maxibrittle graphs are $(C_5, P_5, \overline{P_5})$-free and thus HHD-free.

The result of replacing P_4 by P_3 in Definition 5.7.5 is described in the next remark.

Remark 5.7.2 *The graph G is trivially perfect if and only if in every induced subgraph G' of G, there is no P_3 having a vertex of maximum (respectively, minimum) degree as an endpoint (respectively, as a midpoint).*

The class of brittle graphs has been extended to a larger class, which is still a subclass of perfectly orderable graphs.

Definition 5.7.6 (Olariu [834]) *Let G be a graph.*

> *An edge ab of G is a* symmetric wing *if there exist vertices c, d, p, q such that both $abcd$ and $bapq$ are P_4s in G. (For the notion of a wing see Definition* 1.1.5.)
>
> *G is* quasi brittle *if every induced subgraph H of G contains a vertex incident with no symmetric wing in H and $\overline{H}$.*

It is easy to see that brittle graphs are quasi brittle.

Theorem 5.7.1 [834] *Quasi-brittle graphs are perfectly orderable and can be recognized in polynomial time.*

Somewhat related to brittle graphs is the following graph class.

Definition 5.7.7 *A graph is P_4-*brittle *if there is a stable set S such that either every P_4 has a midpoint in S or every P_4 has an endpoint in S.*

Clearly, the class of P_4-brittle graphs contains all bipartite graphs and all split graphs. In [566] it is shown that P_4-brittle graphs are perfectly orderable and can be recognized in polynomial time.

As already mentioned, the class of perfectly orderable graphs is characterized by the optimality of a coloring heuristic (see Theorem 5.6.1). Other coloring heuristics have been presented by Welsh and Powell [1079], Matula [774], and Cochand and de Werra [228].

Welsh and Powell choose the linear order $<$ on G as follows:

(WP) $x < y$ whenever $\deg_G(x) \geq \deg_G(y)$.

Matula defines $<$ as follows:

(M) $x < y$ whenever $\deg_H(x) \geq \deg_H(y)$ and H is the subgraph of G induced by all z with $z \leq y$.

Cochand and de Werra define $<$, using the vicinal preorder, as follows:

(V) $x < y$ whenever x dominates y and y does not dominate x.

Cochand and de Werra also choose $<$ as follows:

(N*) $x < y$ whenever $N^*(y) \subseteq N^*(x) \cup x$ and $N^*(x) \not\subseteq N^*(y) \cup y$, where $N^*(u)$ is the set of all vertices that are at distance 1 or 2 from u.

Definition 5.7.8 *Let G be a graph.*

> G *is* Welsh–Powell perfect *if every vertex ordering $<$ on G satisfying* (WP) *is perfect* [219].
>
> G *is* Matula perfect *if every vertex ordering $<$ on G satisfying* (M) *is perfect* [219].
>
> G *is* V-perfect *if every vertex ordering $<$ on G satisfying* (V) *is perfect* [228].
>
> G *is* N^*-perfect *if every vertex ordering $<$ on G satisfying* (N*) *is perfect* [228].

Chvátal et al. [219] give characterizations of graphs G whose induced subgraphs (including G itself) are Welsh–Powell perfect (Matula perfect) in terms of forbidden induced subgraphs; see the forbidden subgraph chapter (chapter 7).

From the definition it is obvious that Welsh–Powell and Matula perfect graphs are perfectly orderable. Furthermore, hereditary Matula perfect graphs are HHD-free [219].

Cochand and de Werra [228] give a forbidden induced subgraph characterization of the graphs that are hereditarily V-perfect (N^*-perfect). V-perfect graphs are Welsh–Powell perfect.

Further subclasses of perfectly orderable graphs have been discovered by Hoàng [562]; he defines a subclass of perfectly orderable graphs in the following way.

Definition 5.7.9 [562] *Let G be a graph.*

> *A vertex x is a* d-vertex *in G if for every edge yz with $y, z \notin N(x)$, y and z are comparable; i.e., $N(y) \subseteq N[z]$ or $N(z) \subseteq N[y]$.*
>
> G *is a* D-graph *if each of its induced subgraphs contains a d-vertex.*

Hoàng shows in [562] that D-graphs are perfectly orderable and can be recognized in $\mathcal{O}(nm)$ time. He further shows that the class of D-graphs contains complements of chordal bipartite graphs, complements of chordal graphs, graphs with Dilworth number at most 3, and 2-threshold graphs.

Dalang, Trotter, and de Werra introduce the class of strongly perfectly orderable graphs in [283] as follows.

Definition 5.7.10 *Let G be a graph, and let $<$ be a perfect ordering on G. Then $<$ is a* strong perfect ordering *on G if there is no induced C_4 $abcd$ with $a < b$ and $d < c$. Graphs with a strong perfect ordering are called* strongly perfectly orderable.

In [283] it is shown that all comparability graphs and all graphs with Dilworth number at most 2 are strongly perfectly orderable.

Another subclass of perfectly orderable graphs called *bipartable graphs* is given by Hertz [544] using a graph operator. Perfectly orderable line graphs and claw-free perfectly orderable graphs have been characterized by forbidden induced subgraphs by Chvátal [211, 212]. All these subclasses of perfectly orderable graphs are recognizable in polynomial time. Chvátal proved in [210] that the complement of a bipartite graph B is perfectly orderable if and only if B is chordal bipartite.

The following two classes of perfect graphs seem to be closely related to perfectly orderable graphs.

Definition 5.7.11 (Olariu [820]) *A graph G is an* opposition graph *if it satisfies one of the following two equivalent conditions.*

(i) *There is an ordering $<$ on G such that no P_4 abcd has $a < b$ and $c < d$.*

(ii) *G admits an acyclic orientation that does not contain $\bullet\longrightarrow\bullet\longrightarrow\bullet\longrightarrow\bullet$ or $\bullet\longrightarrow\bullet\longleftarrow\bullet\longrightarrow\bullet$ as an induced subgraph.*

Opposition graphs belong to **bip*** (see the definition before Conjecture 14.3.3), and hence are perfect (Olariu [820, 822]). They also belong to the class of strict quasi-parity graphs [569]. By definition, {1}-orientable graphs or bipolarizable graphs are opposition graphs. Another subclass of opposition graphs that has been considered are the Welsh–Powell opposition graphs: A graph G is a *Welsh–Powell opposition graph* if G is an opposition graph for every order satisfying (WP). Olariu and Randall [835] characterize Welsh–Powell opposition graphs and show that they and their complements are brittle.

The complexity of recognizing opposition graphs is unknown. It is also open whether there is an opposition graph that is *not* perfectly orderable!

Definition 5.7.12 (Hoàng [553]) *A graph is* alternately orientable *if it admits an orientation in which no chordless cycle of length at least* 4 *contains a directed path P_3.*

Hoàng [556] shows that alternately orientable graphs belong to **bip*** and hence are perfect. An $\mathcal{O}(nm)$ recognition algorithm for alternately orientable graphs is given in [431].

By definition, comparability graphs and chordal graphs are alternately orientable. Furthermore, the following inclusions hold:

Gallai $\subset$ **alternately orientable** [553];

2-threshold $\subset$ **alternately orientable** [553];

tolerance $\subset$ **alternately orientable** [461];

co-comparability $\cap$ **alternately orientable** $\subset$ **trapezoid** [382].

It is not known whether an alternately orientable graph always admits an *acyclic* orientation as described in Definition 5.7.12.

5.8 Cop-win orderings

Unlike chordal graphs (but similarly to dually chordal graphs), induced subgraphs of bridged graphs are not necessarily bridged. The following notion given by Jamison describes extremal vertices in bridged graphs; see [374].

Definition 5.8.1 *A vertex $v \in V$ of a bridged graph G is* suppressible *if $G(V \setminus \{v\})$ is bridged.*

Theorem 5.8.1 (Farber [374]) *Every bridged graph has a suppressible vertex. In particular, in a connected bridged graph every vertex of maximum eccentricity is suppressible.*

This is shown by using g-convexity properties of bridged graphs (see the distance property chapter (chapter 10) and Theorem 10.5.5 for g-convexity properties of bridged graphs).

Nowakowski and Winkler [818] defined a graph class by a game of a cop and a robber played on the vertices of a graph.

Definition 5.8.2 [818] *Let G be a graph.*

An ordering $(v_1, \ldots, v_n)$ is a cop-win ordering (*or* dismantling ordering) *if for each $i < n$, there is a $j > i$ such that $N_i[v_i] \subseteq N_i[v_j]$.*

A graph is cop-win *if it has a cop-win ordering.*

Note that this notion is close to the notion of a domination elimination ordering: For cop-win orderings, v_i must be a neighbor of v_j while this is not required in a domination elimination ordering. Bandelt and Prisner [65] noticed that the cop-win graphs were introduced previously by Poston [874] and Quilliot [886] under the name *dismantlable graphs* by the following recursive definition. Cop-win graphs can be recognized in $\mathcal{O}(nm)$ time as a consequence of the definition.

Definition 5.8.3

The one-vertex graph is dismantlable.

A graph G with at least two vertices is dismantlable *if there exist vertices y, z such that $N[z] \subseteq N[y]$ and $G(V \setminus \{z\})$ is dismantlable.*

Theorem 5.8.2 (Anstee, Farber [25]) *A connected graph G is bridged if and only if every isometric subgraph of G (including G itself) is dismantlable.*

Chepoi [191] showed that a dismantling scheme of a bridged graph can be computed in linear time: Every BFS ordering of the vertices of a bridged graph G is a cop-win (see also [192]). Le and Spinrad [715] use Chepoi's algorithm together with a result of Anstee and Farber [25] that G is bridged if and only if G is cop-win and contains no induced C_4 or C_5 to recognize bridged graphs in $\mathcal{O}(nMM)$ time.

5.9 Edge elimination orderings

There are several interesting examples of edge elimination orderings of graphs. The first examples here are two bipartite analogues of the chordal graphs.

Definition 5.9.1 *Let $B = (X, Y, E)$ be a bipartite graph. Then $uv \in E$ is a* bisimplicial edge *if $N(x) \cup N(y)$ induces a complete bipartite subgraph in B.*

Let $(e_1, \ldots, e_k)$ be an ordering of pairwise vertex-disjoint edges of B (not necessarily all edges of E). Let S_i be the set of endpoints of the edges $e_1, \ldots, e_i$ and let $S_0 = \emptyset$.

Definition 5.9.2 (Golumbic, Goss [456]) *Let $B = (X, Y, E)$ be a bipartite graph.*

> $(e_1, \ldots, e_k)$ *is a* perfect edge elimination ordering *for B if $B((X \cup Y) \setminus S_k)$ has no edge and each edge e_i is bisimplicial in the remaining induced subgraph $B((X \cup Y) \setminus S_{i-1})$.*
>
> *B is a* perfect elimination bipartite graph *if B admits a perfect edge elimination ordering.*

Golumbic and Goss [456] gave an $\mathcal{O}(n^5)$ recognition algorithm for perfect elimination bipartite graphs that was improved by Goh and Rotem [447] to $\mathcal{O}(n^3)$. Both these algorithms are based directly on the definition and the fact that a greedy algorithm can be used to construct the perfect edge elimination ordering.

Note that not every perfect elimination bipartite graph is chordal bipartite as the example of the 6-pan shows. However, there is a close relationship between the two conditions, as shown in the following theorem.

Theorem 5.9.1 [456] *A bipartite graph B is chordal bipartite if and only if every induced subgraph of B is perfect elimination bipartite.*

The next theorem gives a connection between chordal graphs and perfect elimination bipartite graphs. For the notation $B(G)$ see Definition 1.3.12.

Theorem 5.9.2 (Brandstädt [141]) *A graph G is chordal if and only if $B(G)$ is perfect elimination bipartite.*

There is a similar edge elimination ordering where only the edges but not their end-vertices are deleted. We call these orderings *perfect edge-without-vertex elimination orderings.*

Definition 5.9.3 *Let $B = (X, Y, E)$ be a bipartite graph and $(e_1, \ldots, e_m)$ an edge ordering. Let $B_i = (X, Y, E_i)$ with $E_0 = E$ and $E_i = E_{i-1} \setminus \{e_i\}$. The ordering $(e_1, \ldots, e_m)$ is a* perfect edge-without-vertex elimination ordering *if for all $i \in \{1, \ldots, m\}$, e_i is bisimplicial in B_{i-1}.*

The following observation was made by several researchers (among them Ma and Müller).

Remark 5.9.1 *A bipartite graph is chordal bipartite if and only if it admits a perfect edge-without-vertex elimination ordering.*

The greedy algorithm, taking any bisimplicial edge to start the ordering scheme, can be used to construct a perfect edge-without-vertex elimination ordering. Kloks and Kratsch [672] give an efficient algorithm for finding a perfect edge-without-vertex elimination ordering of a given chordal bipartite graph that uses the Γ-free ordering of the associated bipartite adjacency matrix.

Dragan [325] generalizes the notion of a bisimplicial edge in the following way.

Definition 5.9.4 *Let $G = (V, E)$ be a graph. Then $xy \in E$ is a* simplicial edge *if every two distinct vertices $u \in N(x)$, $v \in N(y)$ are adjacent.*

Obviously, simplicial edges in bipartite graphs are bisimplicial.

Theorem 5.9.3 [325] *A graph is strongly orderable if and only if it has a perfect edge-without-vertex elimination ordering.*

Benzaken et al. [84] show that the notion of a dual shelling of the edges of a graph leads to another characterization of the co-chordal graphs.

Definition 5.9.5 *Let $G = (V, E)$ be a graph. The edge ordering $(e_1, \dots, e_m)$ of G is a* dual shelling of G *if for all $k \in \{1, \dots, m\}$, the graph $(V, \{e_1, \dots, e_k\})$ has no induced subgraph isomorphic to $2K_2$.*

Theorem 5.9.4 [84] *A graph has a dual shelling if and only if it is co-chordal.*

Finally we present an edge addition scheme instead of an edge elimination scheme for dealing with weakly chordal graphs. Since the class is closed under complement this could be presented as an edge elimination scheme as well.

Theorem 5.9.5 (Spinrad, Sritharan [997]) *If the graph $G = (V, E)$ is weakly chordal and x, y is a two-pair in G, then $G' = (V, E')$ with $E' = E \cup \{xy\}$ is weakly chordal (i.e., adding edges between two-pairs maintains weak chordality).*

Thus, G is weakly chordal if and only if successively adding edges between two-pairs finally leads to a clique. Theorem 5.9.5 is used to design an $\mathcal{O}(n^4)$ recognition algorithm for weakly chordal graphs in [997].

McKee and McMorris [778] discuss such edge addition schemes in a more general setting.

Chapter 6

Posets

6.1 Partial orders and their graphs

Comparability graphs, which were introduced in Definition 1.4.3, are one of the most fundamental classes of perfect graphs. It is easy to prove that comparability graphs are perfect by introducing a height function and applying a well-known theorem due to Dilworth [314].

Theorem 6.1.1 (Ghouila-Houri [435], Gilmore, Hoffman [444]) *Let $G = (V, E)$ be a graph. Then G is a comparability graph if and only if there is no sequence $(x_1, x_2, x_3, \ldots, x_{2k+1})$ of (not necessarily distinct) vertices from V with $k \geq 2$ such that $x_i x_{i+1} \in E$ and $x_i x_{i+2} \notin E$ (cyclically).*

Gallai [416] gives a complete list of forbidden subgraphs for comparability graphs.

The books of Trotter [1031] and Golumbic [452] and the survey papers of Möhring [786, 788] contain many properties of comparability graphs and posets.

The most efficient recognition algorithms for comparability graphs work in the following manner. Given a graph G, the algorithm first solves the orientation problem, that is, a directed graph G' is formed by assigning directions to edges of G. If G is a comparability graph, then G' is transitive. However, even if G is not a comparability graph, the algorithm may produce a directed graph G'. Therefore, the recognition algorithm must perform a verification step to determine whether G' is transitive. Perhaps surprisingly the orientation phase is faster ($\mathcal{O}(n + m)$; see McConnell and Spinrad [776]) than the verification phase ($\mathcal{O}(MM)$); this distinction has caused occasional confusion in the literature. In problems where the verification problem can be done more efficiently, as arises in the recognition problems for permutation graphs, two-dimensional partial orders (see McConnell and Spinrad [776]), trapezoid graphs (see Ma and Spinrad [752]), and circular permutation graphs (see Sritharan [999]), and some optimization problems for comparability graphs (see McConnell and Spinrad [776]), comparability graph algorithms use less time than the comparability graph recognition algorithm.

Subclasses of comparability graphs can be defined by placing restrictions on the corresponding poset.

There are definitions in which posets and graphs are combined in a less direct way. For example, Barthélemy and Constantin [69] define a site as follows.

Definition 6.1.1 *A triple $T = (V, \leq, E)$ is a* site *if the following conditions are fulfilled:*

$(V, \leq)$ is a poset and (V, E) is a graph;

For all not necessarily distinct vertices $x, y, u, v \in V$, if $xy \in E$ and $x \leq u$, $y \leq v$, then $uv \in E$.

Barthélemy and Constantin [69] note that the same notion, motivated by applications in computer science, appears in [298] under the name *conflict event structure.*

6.2 Poset dimension

Definition 6.2.1 *Let $P = (V, <)$ be a poset.*

A family $P_1, \ldots, P_k$ of posets realizes *P if $a < b \Longleftrightarrow a <_i b$ for all $i \in \{1, \ldots, k\}$.*

The order dimension *$dim(P)$ of P is the smallest number k of linear extensions $L_1, \ldots, L_k$ of P that realize P.*

*A partial order P is k-*dimensional *if $dim(P) \leq k$.*

The book of Trotter [1031] is the definitive text dealing with poset dimension. The following theorem is attributed to various sources; we follow Trotter [1031], who views it as a consequence of theorems by Gallai [416].

Theorem 6.2.1 *All posets with the same comparability graph have the same dimension.*

For this reason, dimension is called an *invariant* of comparability graphs, and dimension can be viewed as a comparability graph property as well as a poset property. Of special interest are the two-dimensional posets. The following characterization is based on [345] (see Theorem 4.7.1).

Theorem 6.2.2 (Baker, Fishburn, Roberts [45]) *A graph is a permutation graph if and only if it is the comparability graph of a two-dimensional poset.*

This becomes more clear if one recalls the fact that straight lines L_i, L_j between two parallel lines do not cross each other if $L_i \; L \; L_j$ or $L_j \; L \; L_i$ in the sense of Definition 4.3.2. Thus, it is clear that permutation graphs are exactly the cocomparability graphs of the partial order L on straight lines $L_1, \ldots, L_n$ between two parallel lines, and these are just the two-dimensional posets.

There is also a characterization of two-dimensional posets in terms of a single ordering.

Theorem 6.2.3 (Dushnik, Miller [345]) *A poset has dimension at most two if and only if there is a linear extension $(v_1, \ldots, v_n)$ of the poset such that if $i < j < k$ and v_i and v_k are comparable, then v_j is comparable to at least one of v_i,v_k.*

A linear extension satisfying the condition of the previous theorem is called a *non-separating linear extension.*

It should be mentioned that the decision problem $DIM_k = \{P : \dim(P) \leq k\}$ is $\mathbb{NP}$-complete for $k = 3$ as shown by Yannakakis [1094], whereas two-dimensional posets can be recognized in $\mathcal{O}(n+m)$ time (McConnell and Spinrad [776]).

Definition 6.2.2 *A bipartite two-dimensional poset is a two-dimensional poset of height* 1.

The corresponding comparability graphs are the *bipartite permutation graphs.*

Definition 6.2.3 *Let $B = (X, Y, E)$ be a bipartite graph.*

> *An ordering $<$ of X in B has the* adjacency property *if for each vertex $y \in Y$, $N(y)$ consists of vertices that are consecutive (an interval) in the ordering $<$ of X.*
>
> *An ordering $<$ of X in B has the* enclosure property *if for each pair of $v, w \in Y$ such that $N(v) \subset N(w)$, the vertices in $N(w) \setminus N(v)$ occur consecutively in the ordering of X.*
>
> *A* strong ordering *of $X \cup Y$ consists of orderings $<_1$ of X and $<_2$ of Y such that for all $st \in E$, $s't' \in E$ with $s, s' \in X$, $t, t' \in Y$ $s <_1 s'$, and $t' <_2 t$ implies $st' \in E$ and $s't \in E$ (a crossing of edges leads to a C_4).*

Theorem 6.2.4 (Spinrad, Brandstädt, Stewart [995]) *Let B be a bipartite graph. The following conditions are equivalent:*

(i) *B is a bipartite permutation graph;*

(ii) *B admits a strong ordering;*

(iii) *B admits an ordering that has the adjacency and enclosure property.*

Spinrad, Brandstädt, and Stewart [995] give a linear-time recognition for bipartite permutation graphs that is much simpler than the general linear-time algorithm for recognizing permutation graphs.

It is obvious from Theorem 4.7.1 that bipartite permutation graphs are equal to bipartite cocomparability graphs. Just as chordal cocomparability graphs and chordal asteroidal-triple–free (AT-free) graphs are identical, since these both characterize interval graphs by Theorems 4.3.1 and 7.2.6, the following observation comes easily from Gallai's list of forbidden subgraphs of comparability graphs in [416].

Proposition 6.2.1 *G is a bipartite permutation graph if and only if G is bipartite and G contains no AT.*

In the context of bipartite permutation graphs, two other classes of bipartite graphs, namely biconvex and convex graphs, should be mentioned.

Definition 6.2.4 *A bipartite graph $B = (X, Y, E)$ is*

biconvex *if there is an ordering of X and Y that fulfills the adjacency property;*

convex *if there is an ordering of X or Y that fulfills the adjacency property.*

The following (proper) inclusions are easy to show (see [155]):

bipartite $\cap$ permutation $\subset$ biconvex $\subset$ convex $\subset$ chordal bipartite.

Both biconvex and convex graphs can be recognized in linear time by using PQ-trees.

The rightmost inclusion is refined by Müller [801], showing that

convex $\subset$ interval bigraph $\subset$ chordal bipartite.

There is an intriguing connection between planar graphs and dimension. For any graph G, we can create a height-one poset P with vertices at one level and edges at the other, such that $v \leq e$ in P if and only if v is an endpoint of e. This poset is called the *vertex/edge inclusion poset* of G.

Theorem 6.2.5 (Schnyder [960]) *A graph is planar if and only if its vertex/edge inclusion poset has dimension at most* 3.

For more results in this direction see, e.g., de Fraysseix and Ossona de Mendez [296], where another characterization of planar graphs in terms of 2-colorability of a graph is given. The chordal comparability graphs are another example of a graph class having bounded dimension; see Theorem 6.7.1.

6.3 Containment graphs

The models chapter (chapter 4) dealt with classes of intersection graphs, in which vertices are mapped to objects, with the vertices defined to be adjacent if and only if the corresponding objects have a nonempty intersection. In a containment-graph representation, vertices map to objects, and vertices are adjacent if and only if one object contains the other. This may be either an undirected graph, or a directed graph with an edge from x to y if the object corresponding to x contains the object that corresponds to y. Although containment graphs have not been studied as extensively as intersection graphs, a number of classes of containment graphs have interesting characterizations. Urrutia [1062] gives a survey of classes of containment graphs; the topic is also covered by Trotter [1031].

It is not hard to see that the containment graphs are exactly equal to the comparability graphs; in fact, every comparability graph is the containment graph of substars of a star.

A characterization of containment classes similar to the characterization of intersection classes (see Definition 4.1.1) is given in [463].

The containment orders are defined analogously, i.e., given a family $\mathcal{F}$ of objects, a poset $P = (V, <)$ is an $\mathcal{F}$-order if there is a function assigning an object $F(v)$ to each vertex from V such that $x \leq y$ if and only if $F(x) \subseteq F(y)$. In the next example the objects are intervals on the real line.

Definition 6.3.1 *G is an* interval containment graph *if its vertices can be mapped to intervals on the real line such that vertices x and y are adjacent if and only if one of the corresponding intervals contains the other.*

Theorem 6.3.1 (Dushnik, Miller [345]) *G is an interval containment graph if and only if G is a permutation graph.*

Just as interval graphs have been generalized in a number of ways, we can get new classes by generalizing the interval containment model. Thus, circular-arc containment graphs are the containment graphs of arcs on the circle.

Theorem 6.3.2 (Nirkhe [812]) *G is a circular-arc containment graph if and only if G is a circular permutation graph.*

Generalizing intervals to multidimensional boxes preserves the relationship to dimension.

Theorem 6.3.3 (Golumbic [453]) *G is a containment graph of k-dimensional isooriented boxes if and only if G is the comparability graph of a $2k$-dimensional poset.*

Containment graphs of circles (or disks) have received quite a bit of attention. A well-known conjecture that every finite three-dimensional poset is a circle containment order was strongly refuted by the following theorem.

Theorem 6.3.4 (Felsner, Fishburn, Trotter [383]) *There are three-dimensional posets that are not the containment orders of spheres in any dimension.*

Theorem 6.3.4 implies the earlier result that there are infinite three-dimensional partial orders that are not circle orders [958]. The two-dimensional posets, however, are circle orders. The relationship between planar graphs and three-dimensional posets stated in Theorem 6.2.5 also applies to planar graphs and circle orders.

Theorem 6.3.5 (Scheinerman [955]) *G is planar if and only if the vertex/edge inclusion poset of G is a circle order.*

The generalization to containment graphs of spheres in higher dimensions has an interesting interpretation. Any event has a location and a time at which the event occurred. For two given events A and B, we can determine whether A could have known about B, in the sense that the time between A and B was large enough so that a message traveling at the speed of light could travel the distance between A and B. Orders that come from this relationship are called *causality orders.*

Theorem 6.3.6 (Scheinerman [956]) *If events take place in k-dimensional space, causality orders are exactly equal to containment orders of k-dimensional spheres.*

Regular n-gon orders and containment graphs of convex n-gons for all values of n are studied in [1062].

Another class of containment graphs (or containment orders) that has received attention is defined by angles in the plane. Each vertex is viewed as a light source sent out at a particular angle from an origin, with an edge between vertices if the area illuminated by one vertex contains the area illuminated by the other.

Theorem 6.3.7 (Fishburn, Trotter [389]) *Every four-dimensional poset is an angle order.*

6.4 Series-parallel posets

Series-parallel posets are defined by the recursive application of the following operations (see Theorem 1.5.1).

Definition 6.4.1 *Let $P_1 = (V_1, <_1)$, $P_2 = (V_2, <_2)$ be finite posets with $V_1 \cap V_2 = \emptyset$.*

The parallel composition *$P_1 + P_2 = (V, <_+)$ of P_1, P_2 is defined by $V = V_1 \cup V_2$, and $u < v$ if and only if ($u, v \in V_1$ and $u <_1 v$) or ($u, v \in V_2$ and $u <_2 v$).*

The series composition *$P_1 * P_2 = (V, <_*)$ of P_1, P_2 is defined by $V = V_1 \cup V_2$, and $u < v$ if and only if ($u, v \in V_1$ and $u <_1 v$) or ($u, v \in V_2$ and $u <_2 v$) or ($u \in V_1$ and $v \in V_2$).*

A poset $P = (V, <)$ is series parallel *if $|V| = 1$ or P is obtained by $P_1 + P_2$ or $P_1 * P_2$ of smaller series-parallel posets P_1, P_2, i.e., the class of series-parallel posets is the smallest class which contains the one-element poset and is closed under the operations $+$ and $*$.*

Corollary 6.4.1 *A graph is a cograph if and only if it is the comparability graph of a series-parallel poset.*

Jung [635] characterized cographs as the comparability graphs of multitrees.

Note that a series-parallel poset defines a decomposition tree that has nice properties and is algorithmically very helpful. The class of cographs should not be confused with the class of series-parallel graphs given in Definition 11.2.3.

Since series-parallel orders are two-dimensional (see [788]), cographs are special permutation graphs.

Valdes, Tarjan, and Lawler [1064] give an $\mathcal{O}(n+m)$-time recognition algorithm to determine whether an input poset is a series-parallel poset, and also test in the same amount of time whether the transitive closure of the poset is a series-parallel poset. This paper also explains the relationship between series-parallel posets, which they call *vertex series-parallel digraphs*, and two-terminal series-parallel digraphs.

Theorem 6.4.1 [1064] *An acyclic multidigraph with a single source and a single terminal is two-terminal series parallel if and only if its line digraph is the transitive reduction of a series-parallel poset.*

6.5 Interval orders and semiorders

Let $\mathcal{I} = (I_j)_{j=1}^n$ be a finite collection of intervals of the real line and let $(\mathcal{I}, L)$ be the poset given in Definition 4.3.2, i.e., $I_i \; L \; I_j$ if and only if I_i is completely on the left of I_j.

It is easy to see that the comparability graphs of interval orders are exactly the cointerval graphs; see Theorem 4.3.1.

Theorem 6.5.1 (Papadimitriou, Yannakakis [853]) *P is an interval order if and only if the downsets of the vertices are totally ordered by set containment.*

Theorem 6.5.1 implies a linear-time algorithm for recognizing interval orders. This is strengthened in [417], where a linear-time algorithm to test whether the transitive closure of a digraph is an interval order is presented.

The following containment relationships are not obvious.

Theorem 6.5.2 (Fishburn [388], Fishburn, Trotter [389])

(i) *Every interval order is a circle order* [388].

(ii) *Every interval order is an angle order* [389].

One can define interval order dimension analogously to order dimension.

Definition 6.5.1 *Let $P = (V, <)$ be a poset.*

The interval order dimension *$idim(P)$ of P is the smallest number k of interval orders that realize P.*

Thus interval orders have interval-order dimension 1.

Theorem 6.5.3 (Habib, Kelly, Möhring [493]) *Interval order dimension is a comparability graph invariant.*

Definition 6.5.2 *A poset $P = (V, <)$ is a* trapezoid order *if there is a collection of trapezoids $\mathcal{T} = (T_j)_{j=1}^{n}$ between two parallel lines such that $P \sim (\mathcal{T}, L)$ for the relation L given in Definition* 4.3.2.

It is easy to see that trapezoid orders are exactly the posets of interval order 2 and the trapezoid graphs are co-comparability graphs of such posets.

Proposition 6.5.1 (Dagan, Golumbic, Pinter [275]) *G is a trapezoid graph if and only if G is the incomparability graph of a partial order having interval order dimension at most* 2.

Let $P = (V, <)$ be a partial order and $x < y$ and $z < t$ be a $2 + 2$ in P. The pairs (x, t) and (z, y) are then the *diagonals* of the $2 + 2$. Now the following *incomparability graph F_P* for P is defined:

The vertices of F_P are the pairs (x, y) and (y, x) with incomparable x and y;

Two vertices of F_P are connected by an edge if and only if they are the diagonals of a common $2 + 2$ in P.

Lemma 6.5.1 (Cogis [231]) *The interval order dimension of P is at most 2 if and only if F_P is bipartite.*

The most efficient algorithm for recognizing posets with interval dimension 2, given by Ma and Spinrad [752], takes $\mathcal{O}(n^2)$ time by reducing the interval order dimension problem to the poset dimension problem; see Theorem 4.7.9.

An interesting subclass of interval orders are the semiorders described in [452].

Definition 6.5.3 (Luce [748]) *The binary relation P is a* semiorder *if the following conditions hold:*

P is irreflexive;

If $(x, y) \in P$ and $(z, w) \in P$, then $(x, w) \in P$ or $(z, y) \in P$;

If $(x, y) \in P$ and $(y, z) \in P$, then $(x, w) \in P$ or $(w, z) \in P$.

Note that there is a different notion of semiorders given in [435] in connection with comparability graphs.

Theorem 6.5.4 (Roberts [906]) *Let $G = (V, E)$ be a graph. The following conditions are equivalent:*

(i) *There exists a real-valued function $f : V \to \mathbb{R}$ satisfying for all distinct vertices $x, y \in V$: $xy \in E \Leftrightarrow |f(x) - f(y)| < 1$;*

(ii) *There exists a semiorder (V, P) such that $\overline{E} = P + P^{-1}$;*

(iii) *$\overline{G}$ is a comparability graph and every transitive orientation of $\overline{G} = (V, \overline{E})$ is a semiorder;*

(iv) *G is a proper interval graph;*

(v) *G is a unit interval graph.*

The first condition of Theorem 6.5.4 has been used as a model of preference relations by Luce [748].

Rabinovitch has shown that the maximum dimension of a semiorder is three [889], while interval orders can have arbitrarily high dimension [890].

Another class of posets that has been mentioned in the context of preference is the class of weak orders.

Definition 6.5.4 *A poset P is a* weak order *if every member x of the poset can be assigned a number $w(x)$ such that (x, y) is in P if and only if $w(x) < w(y)$.*

Trenk [1029] has generalized this notion by viewing a weak order as a structure in which incomparable vertices must occur at a single level of the poset and introducing a notion of incomparable vertices occurring at levels that differ by at most k.

Definition 6.5.5 *A poset P is* k-weak *if one can assign a level $l(x)$ to every member x of P such that the following conditions are satisfied:*

> *If (x, y) is in P, then $l(x) < l(y)$;*
>
> *If x and y are incomparable in P, then $|l(x) - l(y)| \le k$.*

Trenk [1029] gives a polynomial-time algorithm for recognizing k-weak orders for all fixed k.

6.6 Arborescence orders and threshold orders

Definition 6.6.1 *A poset $(V, <)$ is an* arborescence order *if for all $x \in V$, $\{y : y < x\}$ is a linearly ordered subset.*

Theorem 6.6.1 (Wolk [1090], Golumbic [451]) *Let G be a graph. The following conditions are equivalent:*

(i) *G is the comparability graph of an arborescence order;*

(ii) *G contains no induced P_4 and C_4;*

(iii) *G is trivially perfect.*

An even more special kind of poset is the following, which is closely related to threshold graphs; see Definition 4.8.4.

Definition 6.6.2 *A* threshold order *is a partial order $P = (V, <)$ with the property*

> *$V = \{v_1, \ldots, v_n\}$ and there are weights $w_1, \ldots, w_n$ and a threshold t with the property $C \subseteq V$ is a maximal chain of P if and only if $\sum_{v_i \in C} w_i \le t$.*

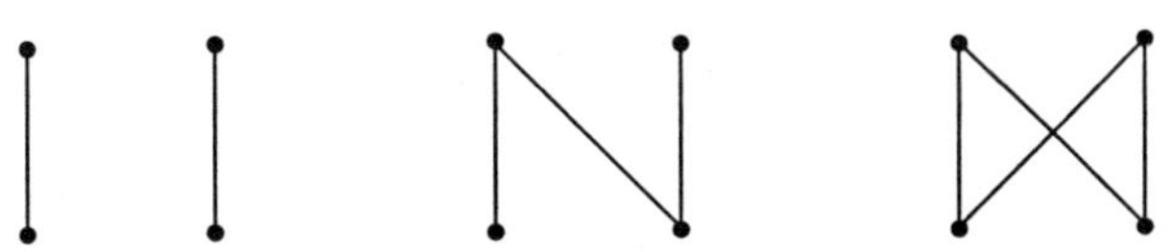

Figure 6.1: Forbidden suborders in threshold orders.

Theorem 6.6.2 (Möhring [788]) *Let P be a poset. The following conditions are equivalent:*

(i) *P is a threshold order;*

(ii) *P does not contain any of the partial orders given in Figure 6.1;*

(iii) *P is series parallel and each node of its canonical decomposition tree has at most one son that is not a leaf.*

Theorem 6.6.3 (Chvátal, Hammer [216]) *Let G be a graph. The following conditions are equivalent:*

(i) *G is a threshold graph;*

(ii) *G is the comparability graph of a threshold order;*

(iii) *G is $(2K_2, C_4, P_4)$-free.*

Thus, the class of threshold graphs is closed under complement and a graph G is threshold if and only if G and its complement are trivially perfect.

6.7 Comparability graphs with other restrictions

In this section, we discuss subclasses of comparability graphs we get if we require the comparability graphs to have other graph theoretic properties. The starting point for this research is the equivalence between permutation graphs and graphs that are both comparability graphs and co-comparability graphs, from Theorem 4.7.1.

Definition 6.7.1 *G is a* chordal comparability graph *if G is chordal and a comparability graph.*

Theorem 6.7.1 (Ma, Spinrad [751], Kierstead, Qin, Trotter [654]) *The maximum dimension of a transitive orientation of a chordal comparability graph is 4.*

The result of Theorem 6.7.1 consists of two parts: the upper bound 4 is given in [751], whereas the lower bound is given in [654].

Ma and Spinrad [751] study a number of problems on chordal comparability graphs. Chordal comparability graphs can be recognized in $\mathcal{O}(n+m)$ time as shown by Hsu and Ma [598].

Definition 6.7.2 *G is a* weakly chordal comparability graph *if G is both weakly chordal and a comparability graph.*

Weakly chordal comparability graphs have a good characterization in terms of the following bipartite transformation. Let $\vec{G}$ be a directed graph. $B(\vec{G})$ is the undirected bipartite graph defined by creating two vertices, x_1 and x_2, for each vertex x of $\vec{G}$, and adding an edge from u_1 to v_2 if and only if there is an edge from u to v in $\vec{G}$.

Theorem 6.7.2 (Eschen et al. [357]) *Let G be a comparability graph and $\vec{G}$ be any transitive orientation of G. Then G is a weakly chordal comparability graph if and only if $B(\vec{G})$ is chordal bipartite.*

The characterization above leads to an $\mathcal{O}(n^2)$ algorithm for recognizing weakly chordal comparability graphs.

Definition 6.7.3 *G is a* circular-arc comparability graph *if and only if G is both a circular-arc graph and a comparability graph.*

The dimension of every transitive orientation of a circular-arc comparability graph is at most 4, as mentioned by Trotter [1031]; it is not known whether this bound is tight.

6.8 Posets and diagrams

Definition 6.8.1 *Let $P = (V, <)$ be a poset.*

> *Element $x \in V$ is* covered *by $y \in V$ if $x < y$ and there is no $z \in V$ such that $x < z < y$ in P.*
>
> *The* covering graph *of a poset $(P, <)$ is the undirected graph $G = (V, E)$ such that xy is in E if x covers y or y covers x.*

Theorem 6.8.1 (Brightwell [158]) *It is $\mathbb{NP}$-complete to determine whether a graph G is a covering graph for some poset.*

Definition 6.8.2 *A* diagram (Hasse diagram) *D of a poset $P = (V, <)$ is a representation of the covering graph of P in the plane such that if $x < y$, then x is below y on the plane.*

The diagram is normally viewed as a directed graph. A number of types of posets have been defined by placing restrictions on the diagram. One type of restriction involves forbidden subgraphs in the diagram. The most important such class is the class of N-free posets, defined as follows.

Definition 6.8.3 *A poset P is* N-free *if the diagram of P does not contain four vertices a, b, c, d such that the set of edges in the diagram between a, b, c, d is exactly equal to $\{(a, b), (c, b), (c, d)\}$.*

Note that being N-free is not a hereditary property of posets. If P is an arbitrary poset, a poset P_s can be created by taking each comparability $x < y$ and creating a new vertex z such that $x < z < y$. This process is called *subdividing all edges* and will always create an N-free poset.

The following characterizations of N-free posets are due to Grillet [475], Leclerc and Monjardet [717], and Habib and Jegou [494]; cf. Möhring [788].

Theorem 6.8.2 [473, 716, 492] *Let $P = (V, <)$ be a poset. The following conditions are equivalent:*

(i) *P is an N-free poset;*

(ii) *Each maximal chain of P meets each maximal antichain of P;*

(iii) *The diagram of P is a line digraph;*

(iv) *There is an arc diagram for P that contains no dummy edges;*

(v) *If the immediate predecessors of a pair of vertices have a nonempty intersection, the immediate predecessors are identical.*

Habib and Jegou [492] also characterize N-free posets by an operation that generalizes the series and parallel operations of Definition 6.4.1. Hereby a *source* (*sink*) of a poset is a minimal (maximal) element.

Definition 6.8.4 *Let $P_1 = (V_1, <_1)$, $P_2 = (V_2, <_2)$ be posets. Let X_1 be a subset of the sources of P_1, and let Y_2 be a subset of the sinks of P_2.*

The quasi–series-parallel composition *is defined by $V = V_1 \cup V_2$, and $u < v$ if and only if one of the following conditions holds:*

$(u, v \in V_1$ and $u <_1 v)$ or $(u, v \in V_2$ and $u <_2 v)$;

$u \in V_1$, $v \in V_2$, and there exist some $x_1 \in X_1$ and $y_1 \in Y_2$ such that $u \leq_1 x_1$ and $y_2 \leq_2 v$.

Theorem 6.8.3 (Habib, Jegou [492]) *A poset $P = (V, <)$ is N-free if and only if P can be built from single-element posets using quasi–series-parallel operations.*

The jump-number problem, that is, the problem of finding a linear extension that minimizes the number of consecutive unrelated elements, was solved in polynomial time for N-free posets by Sysło [1014]; cf. [788]. Sysło [1014] has generalized this result by solving the jump-number problem in polynomial time on a poset if the poset admits an arc diagram with a constant number of dummy edges.

Definition 6.8.5 *A poset is* treelike *if its covering graph is a tree.*

The number of linear extensions can be counted efficiently for treelike posets [33].

A planar drawing of a poset P is a representation of the diagram of P such that no edges of the diagram cross. Note that this is not equivalent to saying that the covering graph is planar, since edges of the diagram must go upward in the drawing, i.e., if $x < y$, then x must be placed below y in the plane and the line from x to y must be drawn upward in the plane.

Definition 6.8.6 *A* planar poset *is a poset that has a planar drawing.*

In general, planar posets may have arbitrarily large dimension [649], but this becomes false if certain restrictions are added.

Theorem 6.8.4 *A lattice is planar if and only if it has dimension at most* 2.

Trotter credits Theorem 6.8.4 to K. Baker; see [1031].

Theorem 6.8.5 (Trotter, Moore [1034]) *If a planar poset has a minimum, it has dimension at most* 3.

A number of characterizations of planar posets are presented in [421]; the paper also resolved an open question regarding the complexity of the recognition problem for the class.

Theorem 6.8.6 (Garg, Tamassia [421]) *Planar-poset recognition is* $\mathbb{NP}$*-complete.*

However, single-source planar posets can be recognized in linear time [103].

Chapter 7

Forbidden Subgraphs

Many graph classes have definitions or nice characterizations in terms of a finite or infinite collection of forbidden subgraphs.

Definition 7.0.7 *Let $\mathcal{F}$ be a family of graphs. A graph G is $\mathcal{F}$-*free *if G contains no induced subgraph isomorphic to a member of $\mathcal{F}$.*

Many inclusions between graph classes follow immediately from forbidden subgraph definitions or characterizations.

Examples of the case of finitely many forbidden subgraphs are the trivially perfect graphs, which are the graphs containing no induced P_4 and no induced C_4 (see Theorem 6.6.1) and the cographs—Theorem 11.3.3 shows that G is a cograph if and only if G contains no induced P_4. Characterizations of this type will be handled in the first section.

There are some other interesting graph classes that are defined by forbidding an infinite collection of induced subgraphs, such as the induced cycles $\{C_k : k \geq 4\}$, which define the chordal graphs. The graphs characterized in this fashion will be discussed in the second section.

The famous concept of minors is yet another example of forbidden subgraph characterization. One important example is Kuratowski's theorem on planar graphs. Forbidden minor characterizations are treated in the third section. The last section describes some forbidden ordered induced subgraphs.

7.1 Finitely many forbidden induced subgraphs

7.1.1 One forbidden induced subgraph

There are several important graph classes characterized by a single forbidden induced subgraph. Definition 1.5.3 introduces cographs, which have many properties, discussed in more detail in Theorem 11.3.3. A graph G is a cograph if and only if G contains no induced P_4.

Claw-free (i.e., $K_{1,3}$-free) graphs and triangle-free (i.e., K_3-free) graphs are other examples of graph classes defined in terms of a single forbidden subgraph.

Despite considerable work dealing with claw-free and triangle-free graphs no nontrivial recognition algorithm has been developed for dense graphs. The fastest recognition algorithm uses n matrix multiplications for claw-free graphs and is thus of $\mathcal{O}(nMM)$-time complexity, while triangle-free graphs can be recognized in $\mathcal{O}(MM)$ time.

Other interesting special cases of graphs not containing a certain forbidden subgraph are the *bull-free*, *tetrahedron-free* (or *K_4-free*), *paw-free* (or 3-pan–free), *dart-free*, and *diamond-free* graphs.

For all of these graph classes the SPGC is known to be true; see the SPGC chapter (chapter 14).

7.1.2 $P_4, C_4, 2K_2$, and other subgraphs

An example of the case of finitely many forbidden subgraphs is the class of trivially perfect graphs, which are the graphs containing no induced P_4 and no induced C_4 (see Theorem 6.6.1). There are several other combinations of small forbidden subgraphs that characterize graph classes of interest.

Theorem 7.1.1 (Földes, Hammer [394]) *G is a split graph if and only if G contains no induced $2K_2$, C_4 and no C_5.*

An interesting generalization of split graphs is the class of $(C_4, 2K_2)$-free graphs, which is obviously closed under complement and is studied by Blázsik et al. [111]. Maffray and Preissmann [759] give a linear-time recognition of $(C_4, 2K_2)$-free graphs, which they call *pseudosplit graphs*, using a degree-sequence characterization that extends that of split graphs.

There are domination results for graphs that are (P_4, C_4)-free or, more generally, (P_5, C_5)-free (see Theorems 11.3.4 and 11.3.5 and the subsequent remarks).

In [111] it is shown that the $(C_4, 2K_2)$-free graphs contain at most n maximal cliques. Fouquet et al. [402] show this property to be valid even for the larger class of (C_4, P_5)-free graphs. Note that there are other classes defined by forbidden induced subgraphs having this property: Kloks [667] shows this for $(K_{1,3}, W_4)$-free graphs.

Recall that due to Theorem 6.6.3, a graph G is a threshold graph if and only if G and its complement are trivially perfect. In terms of forbidden subgraphs, this means that G is threshold if and only if G contains no induced $2K_2$, C_4, and no P_4. We summarize some of the relationships between these graph classes below.

Corollary 7.1.1

threshold = split ∩ cograph;

threshold = cograph ∩ interval ∩ co-interval;

threshold = trivially perfect ∩ co-trivially perfect.

The class **interval ∩ co-interval** has a characterization in terms of finitely many forbidden induced subgraphs.

Theorem 7.1.2 (Benzaken, Hammer, de Werra [87]) *A graph G and its complement $\overline{G}$ is an interval graph if and only if G contains no induced $2K_2, C_4, C_5, S_3, \overline{S_3}$, and no rising sun and its complement.*

Note that

interval $\cap$ co-interval = split $\cap$ permutation.

For a connection between this class and the Dilworth number see Theorem 13.2.4.

Theorem 7.1.3 (Földes, Hammer [395]) *If G is a split graph, then G is a comparability graph if and only if G contains no induced S_3, $\overline{S_3}$, and no co-rising sun.*

Corollary 7.1.2 (Golumbic [452]) *A split graph G is a comparability graph if and only if G is superperfect.*

Hoffman has shown that

comparability $\subset$ superperfect ([575], see [452]).

For some graph classes it is convenient to define forbidden *configurations* instead of forbidden subgraphs.

Definition 7.1.1 *Let $G = (V, E)$ be a graph.* G contains the configuration $H = (E_1, E_2)$ *if $E_1 \subseteq E$ and $E_2 \subseteq \overline{E}$.*

A good example is the class of threshold graphs: A graph $G = (V, E)$ is a threshold graph if and only if G does not contain the configuration $H = (\{ab, cd\}, \{ac, bd\})$ for vertices $a, b, c, d \in V$.

Matroidal graphs are characterized in Theorem 13.4.1 in terms of a forbidden 5-vertex configuration and the C_5. Matrogenic graphs are characterized in Theorem 13.4.2 in terms of the same configuration.

Hammer, Mahadev, and Peled [516] defined the class of strict 2-threshold graphs; see Definition 13.1.4. Mahadev and Peled [761] characterize this class in terms of eight forbidden configurations.

7.1.3 $C_5, P_5, \overline{P_5}, C_6, \overline{C_6}, P_6, \overline{P_6}$, and other subgraphs

The graphs C_5, P_5, and $\overline{P_5}$ play a fundamental role in several forbidden subgraph characterizations.

Theorem 7.1.4 (Hoàng [553]) *G and $\overline{G}$ are Meyniel graphs if and only if G contains no induced C_5, P_5 and no $\overline{P_5}$ (the house).*

As a consequence, these graphs are HHD-free, which implies that they are weakly chordal.

Hereditary Welsh–Powell opposition graphs are the P-free Meyniel and co-Meyniel graphs; see Olariu and Randall [835]. Chvátal et al. [219] characterize hereditary Welsh–Powell perfect graphs (see Definition 5.7.8) in terms of 17 forbidden induced subgraphs, including $C_5, C_6, \overline{C_6}, P_6, \overline{P_6}$.

Definition 7.1.2 (Hayward [527]) *A graph is a* murky graph *if it contains no induced* C_5, P_6, *and* $\overline{P_6}$.

Obviously, these graphs generalize $(C_5, P_5, \overline{P_5})$-free graphs. Hayward shows in [527] that murky graphs are perfect. Hertz simplifies the proof of this inclusion in [549].

There are some classes defined by elimination properties (see Definitions 5.7.3–5.7.5) that are characterized by finitely many forbidden induced subgraphs including C_5, P_5, and $\overline{P_5}$.

Theorem 7.1.5 (Preissmann, de Werra, Mahadev [878]) *Let G be a graph.*

(i) *G is superfragile if and only if it contains no induced* C_4, P_4 *and no induced dart;*

(ii) *G is maxibrittle if and only if it contains no induced* $C_5, P_5, \overline{P_5}$, *fish, co-fish;*

(iii) *G is superbrittle if and only if it contains no induced* $C_5, P_5, \overline{P_5}, A, \overline{A}$, *parapluie, parachute.*

Corollary 7.1.3

superfragile ⊂ trivially perfect;

maxibrittle ⊂ Meyniel ∩ co-Meyniel;

superbrittle ⊂ Meyniel ∩ co-Meyniel.

The threshold-signed graphs described in Definition 13.2.1 have a forbidden induced subgraph characterization: Benzaken, Hammer, and de Werra [86] describe such a characterization in terms of 6 forbidden configurations or, equivalently, 21 forbidden induced subgraphs including the C_5, P_5, $\overline{P_5}$, $S_3, \overline{S_3}$, fish, and co-fish. This implies that threshold-signed graphs are maxibrittle.

Definition 5.7.8 describes two other classes introduced by elimination properties: the V-perfect and the N^*-perfect graphs. Cochand and de Werra [228] give a forbidden induced subgraph characterization of the graphs that are hereditary V-perfect in terms of 11 subgraphs, including the $C_5, C_6, \overline{C_6}, P_6, \overline{P_6}$. This implies that hereditary V-perfect graphs are murky graphs.

Furthermore, [228] contains a characterization of the hereditary N^*-perfect graphs in terms of 13 forbidden induced subgraphs, including the $C_5, C_6, C_7, C_8, P_8, \overline{P_5}$, and gem. Thus, hereditary N^*-perfect graphs are house-hole-gem–free (HHG-free).

Another group of graph classes characterized by finitely many forbidden induced subgraphs including the C_5, P_5, and $\overline{P_5}$ is defined by having few P_4s: For the notion of P_4-reducible graphs, see Definition 11.4.1, and for the notion of P_4-sparse graphs, see Definition 11.4.2.

Theorem 7.1.6 (Jamison, Olariu [620], Giakoumakis, Vanherpe [440]) *Let G be a graph.*

(i) *G is* P_4*-sparse if and only if G contains no induced* $C_5, P_5, \overline{P_5}, P, \overline{P}$, *fork, and cofork.*

(ii) *G is P_4-reducible if and only if G is P_4-sparse and $(S_3, \overline{S_3})$-free.*

A slight generalization of these notions is given by admitting the existence of a C_5.

Definition 7.1.3 (Giakoumakis [436], Giakoumakis, Vanherpe [440]) *Let G be a graph.*

G is extended P_4-sparse if G contains no induced $P_5, \overline{P_5}, P, \overline{P}$, fork, and cofork.

G is extended P_4-reducible if G is extended P_4-sparse and $(S_3, \overline{S_3})$-free.

In [401] the following class is introduced.

Definition 7.1.4 (Fouquet, Giakoumakis [401]) *A graph is* semi–P_4-sparse *if it contains no induced $P_5, \overline{P_5}$ and no kite.*

Note that the complements of these graphs are exactly the $(P_5, \overline{P_5})$-free fork-free graphs.

Olariu [831] gives an interesting characterization of house- and bull-free graphs, which implies a result on *homogeneously representable interval graphs,* the subclass of P_5-free bull-free interval graphs characterized in [980, 445] (see [833], where the same result is obtained by studying sources in comparability graphs).

7.1.4 $K_{1,3}$ and other subgraphs

Interest in claw-free graphs came particularly from the fact that they generalize line graphs, as can be seen in Theorem 7.1.9, and the existence of efficient algorithms for the maximum independent-set problem on line graphs given by Minty [784] (weighted case) and Sbihi [945] (unweighted case; see a simplified version in the book of Lovász and Plummer [742]).

De Simone and Sassano [306] give a polynomial algorithm for the maximum independent-set problem on bull-free chair-free graphs. Alexejew [14] claims to have a polynomial algorithm for the same problem on general chair-free graphs.

Ebenegger, Hammer, and de Werra [346] introduced a general method called *struction,* which is used to solve the maximum independent-set problem for some classes such as claw-free net-free graphs in [515]. Struction is a method that transforms a given graph G into a graph G' such that $\alpha(G') = \alpha(G) - 1$. In general, the size of G' is greater than the size of G, but if we forbid some subgraphs, repeating the construction leads to an efficient algorithm. Hertz [550] generalizes the results of [515].

Claw-free net-free graphs have a remarkable property.

Theorem 7.1.7 (Duffus, Gould, Jacobson [343]) *If G is claw-free and net-free, then*

(i) *if G is connected, then G has a Hamiltonian path;*

(ii) *if G is 2-connected, then G has a Hamiltonian circuit.*

See [968] for more of such properties. Property (i) was discovered independently in [285].

The line graphs (see Definition 4.2.1) are characterized by a larger number of forbidden induced subgraphs (including the claw) as well as by other conditions described, e.g., by Harary [520]. For this characterization we need the following notion: A triangle T of a graph G is *odd* if there is a vertex of G adjacent to an odd number of vertices of T.

Theorem 7.1.8 (Krausz [701], Van Rooij, Wilf [1066], Beineke [76, 77]) *Let G be a graph. The following conditions are equivalent:*

(i) *G is a line graph;*

(ii) *The edges of G can be partitioned into complete subgraphs such that no node lies in more than two of the subgraphs;*

(iii) *G is claw-free and if two odd triangles have a common edge, then the subgraph induced by their vertices is a K_4;*

(iv) *G contains none of the nine forbidden graphs of Figure 7.1 as an induced subgraph.*

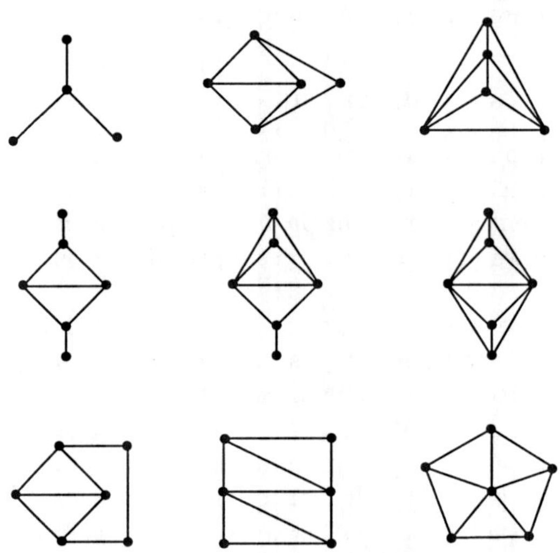

Figure 7.1: Forbidden induced subgraphs of line graphs.

Šoltés [986] gives a forbidden induced subgraph characterization of line graphs that uses a smaller number of subgraphs.

Corollary 7.1.4 *G is a line graph of a bipartite graph if and only if G contains no claw, diamond, or odd hole as an induced subgraph.*

The linear-time recognition algorithm of Roussopoulos [934] for line graphs is based on condition (ii) and the linear-time algorithm of Lehot [721] is based on conditions (iii) and (iv) of Theorem 7.1.9. See [299] for further recognition algorithms.

The domino graphs characterized in Theorem 11.1.16 are an interesting subclass of claw-free graphs.

We now turn our attention to classes characterized by chordality plus a finite number of forbidden induced subgraphs including the claw. For proper interval graphs there is such a characterization.

Theorem 7.1.9 (Wegner [1078]) *A chordal graph is a proper interval graph if and only if it contains no induced $K_{1,3}$, S_3 and $\overline{S_3}$.*

This implies that an interval graph is a proper interval graph if and only if it contains no induced $K_{1,3}$. Wegner [1078] called these graphs *indifference graphs.* Roberts [907] showed that the indifference graphs are exactly the proper interval graphs. See Theorem 7.2.9 for a characterization of proper interval graphs in terms of astral triples.

Hell, Bang-Jensen, and Huang [538] applied the result of Wegner (Theorem 7.1.9) to chordal proper circular-arc graphs.

Theorem 7.1.10 [536, 67] *A chordal graph is a proper circular-arc graph if and only if it contains no induced $K_{1,3}$ and no $\overline{S_3}$.*

7.1.5 Other examples

Definition 7.1.5 *Let G be a graph.*

> *A* local complementation *of G at vertex v consists in replacing the induced subgraph $G(N(v))$ by its complement graph.*
>
> *The graphs G_1 and G_2 are* locally equivalent *if there is a finite sequence of local complementations transforming G_1 into G_2.*

Theorem 7.1.11 (Bouchet [139]) *G is a circle graph if and only if no graph locally equivalent to G contains an induced subgraph isomorphic to the W_5 or W_7 or BW_3.*

Definition 7.1.6 *A graph G is* domination perfect *if the domination number $\gamma(H)$ and the independent domination number $\iota(H)$ (i.e., the minimum size of an independent dominating set) coincide for all induced subgraphs H of G (including G itself).*

Zverovich and Zverovich [1097] give a characterization of domination perfect graphs in terms of 17 forbidden induced subgraphs.

Henning [542] defines irredundance perfect graphs in a similar way by equality between the domination number and the irredundance number. For irredundance number 2, Henning characterizes the irredundance perfect graphs in terms of 22 forbidden induced subgraphs.

7.2 Infinitely many forbidden induced subgraphs

7.2.1 Comparability graphs and variants

In his seminal paper on comparability graphs [416], Gallai gave a forbidden induced subgraph characterization of the class, which has been used in many other papers on comparability graphs. A widely available source for the list of forbidden subgraphs is the book [1031] of Trotter.

The forbidden subgraph characterization of comparability graphs has many consequences for particular classes. We give some examples below.

Borie and Spinrad [130] mention a forbidden induced subgraph characterization of chordal comparability graphs. Theorem 7.1.3 characterizes the split comparability graphs. The great influence of Gallai's result on modular decomposition is discussed in the decomposition chapter (chapter 12).

7.2.2 Chordality and suns

Strongly chordal graphs have a characterization in terms of the following forbidden induced subgraphs.

Definition 7.2.1 (Farber [372], Chang [180])

> *A* sun (*or* trampoline) *is a chordal graph G on $2n$ vertices for some $n \geq 3$ whose vertex set can be partitioned into two sets, $W = \{w_1, \ldots, w_n\}$, $U = \{u_1, \ldots, u_n\}$, such that W is independent and for each i and j, w_j is adjacent to u_i if and only if $i = j$ or $i \equiv j + 1 \pmod{n}$.*
>
> *A* complete sun *is a sun G in which $G(U)$ is a complete graph.*
>
> *A sun G is* odd *if n is odd.*

For suns the following basic property holds.

Lemma 7.2.1 [603, 372, 180] *Each sun contains a complete sun.*

Theorem 7.2.1 [372, 180] *A graph is strongly chordal if and only if it is sun-free chordal.*

Odd-sun–free chordal graphs are an interesting class between strongly chordal and chordal graphs.

Theorem 7.2.2 (Lehel, Tuza [720]) *A chordal graph is neighborhood perfect if and only if it contains no induced odd sun.*

Thus strongly chordal graphs are neighborhood perfect. Chang and Nemhauser [184] observed that every sun S_k with $k \geq 5$ contains an induced $\overline{S_3}$. Thus $(S_3, \overline{S_3})$-free chordal graphs are odd-sun–free chordal. Recall that in the perfection chapter, after Definition 2.5.4, the notion of $\mathcal{T}_k$-perfection was given.

Theorem 7.2.3 (Chang, Nemhauser [184]) *Let G be a graph.*

(i) *G is $\mathcal{T}_k$-perfect for all $k \geq 2$ if and only if G is $(S_3, \overline{S_3})$-free chordal.*

(ii) *If G is $\mathcal{T}_k$-perfect for all even $k \geq 2$, then G is odd-sun-free chordal.*

(iii) *If the SPGC is true, then the converse direction of* (ii) *holds as well.*

The S_3-free chordal graphs are characterized in [183], and in [57] it is mentioned that S_3-free chordal graphs are pseudomodular.

Theorem 7.2.4 (Bandelt, Mulder [57]) *A graph is S_3-free chordal if and only if it contains no induced C_4 and all its induced subgraphs are pseudomodular.*

The well-known class of ptolemaic graphs (see Definition 10.2.2) is characterized as the gem-free chordal graphs; see Theorem 10.2.3. Note that every sun contains a gem and therefore gem-free graphs are sun-free.

7.2.3 Hole-free graphs and variants

There are many graph classes that are defined or characterized by forbidden induced cycles of length greater than k or similar conditions. Among these are

> chordal, strongly chordal, chordal bipartite, (k, l)-chordal, weakly chordal, Meyniel, Gallai, parity, distance-hereditary, and Berge graphs,

which are described in the chapter on cycles and chords (chapter 3).

The hole-free condition can be combined with other forbidden subgraphs such as the house, the domino, the gem, and the suns. Some examples are

> HHD-free graphs; see Proposition 3.1.1,
>
> HHDA-free (i.e., weak bipolarizable) graphs; see Theorem 5.2.3,
>
> HHP-free graphs; see Theorem 5.2.6,
>
> HH-free graphs.

For a characterization of distance-hereditary graphs by forbidden induced subgraphs, see Theorem 10.1.1: A graph is distance hereditary if and only if it is HHDG-free.

Definition 5.4.4 introduces the class of homogeneously orderable graphs as a common generalization of dually chordal and distance-hereditary graphs. The class of homogeneously orderable graphs is not hereditary—the hereditary homogeneously orderable graphs are characterized in the following theorem.

Theorem 7.2.5 [148] *A graph G is house-hole-domino-sun-free (HHDS-free) if and only if every induced subgraph of G is homogeneously orderable.*

Note that distance-hereditary graphs are HHDS-free since every sun S_k, $k \geq 3$, contains a gem.

Hertz [548] gives a characterization of bipolarizable graphs (see Definition 5.7.1) in terms of infinitely many forbidden subgraphs, including so-called k-wheels, the C_5, C_6, $\overline{P_5}$, A, domino, and a certain 8-vertex graph. This characterization implies that bipolarizable graphs are HHDA-free, i.e., weak bipolarizable.

Chvátal et al. [219] give a characterization of hereditary Matula perfect graphs (see Definition 5.7.8) in terms of infinitely many forbidden subgraphs, including so-called bicycles, the holes, the $\overline{P_5}$, and the domino. This characterization implies that hereditary Matula perfect graphs are HHD-free.

The next class is a common generalization of cographs and chordal graphs.

Definition 7.2.2 (Maire [764]) *A graph G is* slightly triangulated *if*

G is hole free and

for every induced subgraph H of G, there is a vertex v in H such that the neighborhood of v in H contains no P_4.

Obviously cographs, chordal graphs, and co-chordal graphs are slightly triangulated. Maire [764] showed that slightly triangulated graphs are perfect. Slightly triangulated graphs can be recognized in polynomial time as a consequence of the definition.

7.2.4 Asteroidal triples

This section deals with some classes of graphs with a linear structure. The first examples are the well-known interval graphs (see Definition 4.3.1) and some generalizations with the *linear intersection model* such as the d-trapezoid graphs (see Definition 4.7.6).

Lekkerkerker and Boland [723] gave a characterization of interval graphs by forbidden induced subgraphs. The list of forbidden induced subgraphs is infinite and there are several types of forbidden subgraphs. There is, however, a characterization that describes this in a more compact way.

Definition 7.2.3 [723] *Let G be a graph.*

Three vertices u, v, w of G form an asteroidal triple (AT) *of G if for every pair of them there is a path connecting the two vertices that avoids the neighborhood of the remaining vertex.*

G is AT-free *if no three vertices of G form an AT.*

Note that the vertices u, v, w of an AT are necessarily pairwise nonadjacent. Examples of graphs containing ATs are the suns S_k, $k \geq 3$, the $\overline{S_3}$, and the cycles C_k of length $k \geq 6$.

Theorem 7.2.6 [723] *G is an interval graph if and only if G is chordal and contains no AT.*

For a simpler proof of Theorem 7.2.6 see [506]. See Theorem 4.3.1 for other characterizations of interval graphs.

Since every sun contains an AT, interval graphs are strongly chordal, but this is also clear from the fact that interval graphs are directed-path graphs.

Theorem 7.2.7 (Golumbic, Monma, Trotter [461]) *Every co-comparability graph is AT-free.*

Thus, co-comparability graphs are net free, hole free, and sun free.

Recently Corneil, Olariu, and Stewart [250, 251, 252, 253] investigated the class of AT-free graphs, showing their linear structure.

Definition 7.2.4 [250] *Two vertices u, v of a graph G are a* dominating pair *of G if every path connecting u and v dominates G.*

For a general graph G, it can be tested in polynomial time whether G has a dominating pair, as indicated in [250].

Theorem 7.2.8 [250] *Every connected AT-free graph has a dominating pair.*

A dominating pair can be found in linear time for a graph known to be AT-free. The fastest known recognition algorithm for AT-free graphs takes $\mathcal{O}(n^3)$ time [251, 253]. Broersma et al. [160] show that the independent set problem can be solved in polynomial time on AT-free graphs, while the clique problem is $\mathbb{NP}$-complete. An important characterization of AT-free graphs is given in Theorem 11.1.11.

Deogun and Kratsch [301] study generalizations of AT-free graphs.

Definition 7.2.5 (Deogun, Kratsch [301]) *Let G be a graph.*

Two vertices u, v of a graph G are a diametral pair *of G if $d(u, v) = diam(G)$.*

A shortest path between a diametral pair u, v is a diametral path *of G.*

A graph G is a diametral path graph *if every connected induced subgraph of G has a dominating diametral pair.*

It turns out that the class of AT-free graphs is properly contained in the class of diametral path graphs [697]. For more generalizations of AT-free graphs see [698].

Kaplan and Shamir [643] and Parra and Scheffler [858] study AT-free claw-free graphs, which are an interesting special case of AT-free graphs; see Theorem 11.1.14.

Jackowski characterized proper interval graphs in terms of the following triples, which are similar to ATs.

Definition 7.2.6 (Jackowski [606]) *Three vertices u, v, w of a graph G are an* astral triple *of G if for every pair of them there is a path connecting the two vertices and not containing the third vertex and which does not have two consecutive vertices that are neighbors of the third.*

Theorem 7.2.9 [606] *A graph is a proper interval graph if and only if it contains no astral triple.*

Bipartite AT-free graphs are exactly the bipartite permutation graphs; see Theorem 6.2.1.

7.3 Forbidden minors—planarity and variants

Planarity of a graph is one of the most basic notions in graph theory from a purely graph-theoretic as well as from an algorithmic point of view. It goes back to the roots of graph theory; Euler's formula from 1736 may be the origin of graph theory. A good guide to the literature is the book of Nishizeki and Chiba [814].

Definition 7.3.1 *Let $G = (V, E)$ be a graph.*

> *A* crossing-free embedding of G in the plane *is given by drawing G in the plane with points representing vertices and curves representing edges such that no two curves for edges intersect except at common end vertices.*
>
> *G is* planar *if there is a crossing-free embedding of G in the plane.*
>
> *A graph G' is a* direct subdivision *of G if G' is obtained from G by subdividing an edge of G into two edges by inserting a new vertex:*
>
> > *There is an edge $uv \in E$ with $E' = (E \setminus \{uv\}) \cup \{ux, xv\}$ and $V' = V \cup \{x\}$, $x \notin V$.*
>
> *A graph G' is a* subdivision *of G if it is obtained from G by a sequence of direct subdivisions.*
>
> *Two graphs G_1, G_2 are* homeomorphic *if there is a graph G such that G_1 and G_2 can be obtained from G by a suitable sequence of subdivisions.*
>
> *G is a* minor *of a graph H if G can be obtained from H by a series of zero or more vertex deletions, edge deletions, and/or edge contractions (i.e., replacing two adjacent vertices v and w by a vertex that is adjacent to all neighbors of v or w).*

Theorem 7.3.1 (Kuratowski [702]) *A graph is planar if and only if it does not contain a subdivision of K_5 or $K_{3,3}$.*

Another statement of Kuratowski's theorem is that a graph is planar if and only if it does not contain the K_5 or $K_{3,3}$ as a minor. This is one of the most celebrated results in graph theory. An elegant proof can be found in [1022] (see [814]). Note that Kuratowski's criteria give no obvious polynomial-time algorithm for recognizing planar graphs. The following papers deal with the recognition of planar graphs. Note that linear time for planar graphs means $\mathcal{O}(n)$, since a planar graph with $n \geq 3$ vertices has at most $3n - 6$ edges.

- Auslander and Parter [36] give a planar embedding procedure;
- Goldstein [449] gives an implementation of the Auslander–Parter procedure that always halts;
- Shirey [976] gives an implementation of the Auslander–Parter procedure in $\mathcal{O}(n^3)$ steps;

- Lempel, Even, and Cederbaum [724] give a different planarity-recognition algorithm with $\mathcal{O}(n^2)$ implementation;
- Hopcroft and Tarjan [579, 580] give an $\mathcal{O}(n)$ implementation of the Auslander–Parter method using deep properties of DFS;
- Booth and Lueker [127] give an $\mathcal{O}(n)$ implementation for the Lempel–Even–Cederbaum method using PQ-trees;
- Chiba et al. [194] give a simplified $\mathcal{O}(n)$ algorithm using PQ-trees that constructs a planar embedding if one exists (see [814]).

Definition 7.3.2 *A graph is* outerplanar *if it has a crossing-free embedding in the plane such that all vertices are on the same face.*

Proposition 7.3.1 *A graph is outerplanar if and only if it contains no subgraph homeomorphic to K_4 or $K_{2,3}$ by a homeomorphism that contracts degree-2 vertices but does not add them.*

Series-parallel graphs, which are studied in the recursion chapter (chapter 11), are equivalent to the graphs that contain no subgraph homeomorphic to K_4; see Theorems 11.2.1 and 11.2.2. Thus, outerplanar graphs are series parallel.

Definition 7.3.3 (Baker [44]) *A graph G is k-*outerplanar *if for $k = 1$, G is outerplanar and for $k > 1$, G has a planar embedding such that if all vertices on the exterior face are deleted, the connected components of the remaining graph are all $(k-1)$-outerplanar.*

Proposition 7.3.2 *It can be determined in polynomial time whether G is*

(i) *an outerplanar graph,*

(ii) *a k-outerplanar graph.*

Halin [504] defined a graph class that was subsequently called *Halin graphs.*

Definition 7.3.4 *G is a* Halin graph *if G is formed by embedding a tree having no degree-2 vertices in the plane and connecting its leaves by a cycle that crosses none of its edges.*

These graphs were introduced as an example of a class of edge-minimal planar 3-connected graphs.

Proposition 7.3.3

(i) *The classes of Halin graphs and outerplanar graphs are incomparable with respect to set inclusion.*

(ii) *Halin graphs are 2-outerplanar.*

Halin graphs have some interesting properties mentioned in [258]. Bondy and Lovász have shown a Hamiltonicity property of them; see the "matching book" of Lovász and Plummer [742].

Theorem 7.3.2 *Halin graphs have a Hamiltonian circuit, and if any node is deleted, the remaining graph has a Hamiltonian circuit.*

Halin graphs can be recognized in linear time, as communicated by Sysło. The leaves of the tree must form a face consisting of more than half the vertices in an arbitrary planar embedding of the graph.

A very natural generalization of planarity is to admit embeddings into surfaces of higher genus (the plane has genus 0).

Definition 7.3.5 *Let G be a graph.*

> *G is* toroidal *if it is crossing-free embeddable in the torus (thus generalizing the crossing-free embedding of planar graphs in the plane).*
>
> *The* genus *of G is the smallest genus of a surface in which G can be embedded crossing free.*

The recognition of genus-k graphs for any fixed k is solvable in polynomial time [386] (see [419, 627]). Thomassen [1025] has shown that the problem of determining the genus of a given graph is $\mathbb{NP}$-complete.

Johnson [631] mentions two other generalizations of planar graphs. Another very fundamental generalization concerns all minor-closed graph classes, i.e., classes of graphs $\mathcal{G}$ for which minors from graphs $G \in \mathcal{G}$ are in $\mathcal{G}$.

Robertson and Seymour obtained in a series of seminal papers [909, 910, 913, 914, 915, 916, 917, 918, 919, 920, 921, 922, 923, 924, 925] the following deep theorems (the first has been known as the Wagner conjecture).

Theorem 7.3.3 *For every class of graphs $\mathcal{G}$ that is closed under minors there is a finite set $OB(\mathcal{G})$ of graphs (called the* obstruction set *of $\mathcal{G}$) such that for every graph G, $G \in \mathcal{G}$ if and only if there is no minor H of G in $OB(\mathcal{G})$.*

Theorem 7.3.4 *For every graph H there is an $\mathcal{O}(n^3)$ algorithm that tests whether H is a minor of an input graph G.*

It follows that every graph class closed under taking minors is recognizable in $\mathcal{O}(n^3)$ time. As Bodlaender [120] showed (see Theorem 11.1.6), every class of graphs that does not contain all planar graphs and that is closed under taking minors can be recognized in linear time. Since the proof of Theorem 7.3.3 is inherently nonconstructive and the hidden constants in the $\mathcal{O}$-notation may be quite large, the time-bound results say only that there are such algorithms without giving a construction of them.

Another consequence of Theorem 7.3.4 is that the Kuratowski theorem leads to the existence of a $\mathcal{O}(n^3)$-time planarity test—the planarity-testing algorithms listed in this section, however, are much faster. For a survey on minors, see Robertson and Seymour [911].

7.4 Forbidden induced ordered subgraphs

Many graph classes have been characterized by certain ordering properties, primarily of the following type: $G \in \mathcal{G}$ if and only if there is a vertex ordering $<$ of G with the property that $<$ contains no subordering from the set S. Examples of that type are the chordal, the comparability, the perfectly orderable, and the opposition graphs. For interval and proper interval graphs see Theorems 5.5.9 and 5.5.10.

In the following we give other examples of that type. Many small classes can be characterized in this way.

Definition 7.4.1 *The abbreviations ch, cp, cl denote the suborderings given in Figure 7.2.*

If L is an ordered graph, then L^C denotes the ordered complement of L, i.e., the ordering of the vertices remains unchanged and the edge set of L^C is the complement of the edge set of L.

L^* is obtained by reversing the ordering of L, and the edge set remains unchanged.

Let $p_1, \ldots, p_5$ be the ordered subgraphs of four vertices given in Figure 7.2.

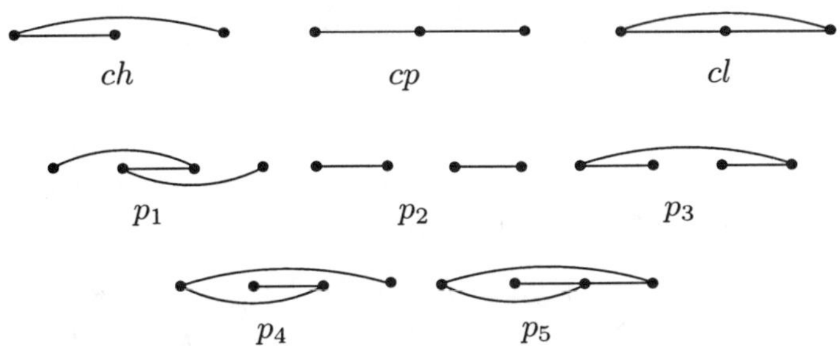

Figure 7.2: The suborders ch, cp, cl and the ordered subgraphs $p_1, \ldots, p_5$.

Theorem 7.4.1 (Damaschke [284]) *$G \in \mathcal{G}$ if and only if there is an order of G that contains no induced ordered subgraph from S.*

(1)	$\mathcal{G} =$ **forest**	$S = \{ch, cl\}$
(2)	$\mathcal{G} =$ **threshold**	$S = \{ch, ch^C\}$ *or* $S = \{ch^*, ch^C, cp\}$
(3)	$\mathcal{G} =$ **interval**	$S = \{ch, cp^C\}$
(4)	$\mathcal{G} =$ **proper interval**	$S = \{ch, ch^*\}$
(5)	$\mathcal{G} =$ **split**	$S = \{ch, ch^{C*}\}$
(6)	$\mathcal{G} =$ **bipartite**	$S = \{cp, cl\}$
(7)	$\mathcal{G} =$ **permutation**	$S = \{cp, cp^C\}$
(8)	$\mathcal{G} =$ **bipartite $\cap$ permutation**	$S = \{cp, cl, cp^C\}$
(9)	$\mathcal{G} =$ **cograph**	$S = \{cp, cp^C, p_1\}$
(10)	$\mathcal{G} =$ **chordal bipartite**	$S = \{cp, cl, p_2, p_3, p_4\}$
(11)	$\mathcal{G} =$ **strongly chordal**	$S = \{ch, p_5\}$

In [284] it is mentioned that circular-arc graphs can be characterized in this way.

For larger classes the work on ordered graphs has been influenced heavily by Chvátal's notion of admissible orderings (see Definition 5.6.1) and perfectly orderable graphs.

Olariu [822] studied all variations of Definition 5.6.1 which lead to perfect graphs. All orderings of a P_4 are shown in Figure 7.3.

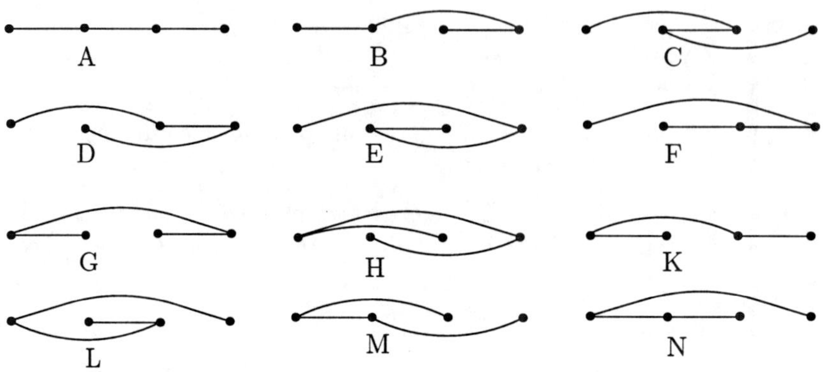

Figure 7.3: All orderings of a P_4.

Theorem 7.4.2 [822] *If the graph G admits an ordering that does not contain an induced ordered P_4 from the set S_i $(1 \le i \le 42)$, then G is perfect.*

S_1	$ACEFK$	S_{15}	$AEHK$	S_{29}	$CEHKN$
S_2	$ACEMN$	S_{16}	$ABCFL$	S_{30}	$BDGHL$
S_3	$CEGK$	S_{17}	$ACGHKL$	S_{31}	$BGHMN$
S_4	$ABHL$	S_{18}	$ABCDL$	S_{32}	$BGHLM$
S_5	$ACEKN$	S_{19}	$ABDGL$	S_{33}	$BGHLN$
S_6	$ABCEGH$	S_{20}	$CGMN$	S_{34}	$BCHLN$
S_7	$ACEKM$	S_{21}	$ABCLM$	S_{35}	$BFGHL$
S_8	$AEGKM$	S_{22}	$ABGLM$	S_{36}	$BCFHL$
S_9	$CDFG$	S_{23}	$EGHKM$	S_{37}	BDF
S_{10}	$ACDEK$	S_{24}	$DFGHK$	S_{38}	KMN
S_{11}	$ADEGK$	S_{25}	$DEGHK$	S_{39}	LMN
S_{12}	$ABCLN$	S_{26}	$EFGHK$	S_{40}	DEF
S_{13}	$ACDFL$	S_{27}	$CEFHK$	S_{41}	$ABCGHK$
S_{14}	$BCGL$	S_{28}	$EGHKN$	S_{42}	$ACEGHL$

Note that the case $S_{37} = \{B, D, F\}$ is the case of perfectly orderable graphs.

Skrien [979] proposed another concept: Let F be a set of digraphs.

Definition 7.4.2 *G is an F*-graph (*F^**-graph) *if it has an orientation (acyclic orientation) that has no induced subdigraphs isomorphic to any of the digraphs in F.*

Skrien [979] gives characterizations of chordal graphs, comparability graphs, proper-interval graphs, and proper circular-arc graphs in these terms.

Duffus, Ginn, and Rödl [342] study the complexity of recognizing graphs that are characterized by a forbidden induced ordered subgraph. They show that for a large number of classes defined by forbidding 2-connected induced ordered graphs, the recognition problem is $\mathbb{NP}$-complete. The same paper makes the conjecture that the forbidden induced ordered subgraph problem is $\mathbb{NP}$-complete for all 2-connected graphs that are not complete.

Chapter 8

Hypergraphs and Graphs

8.1 α-acyclicity of hypergraphs

This chapter is primarily based on the book of Berge [90] about hypergraphs. Furthermore, it has profited greatly from the survey of Golumbic [455] and refers to several papers about relational database schemes. For a recent survey on hypergraphs see Duchet [338].

For the basic notions and properties of hypergraphs see, e.g., the definitions in section 1.3 where some basic notions such as conformality, Helly hypergraph, hypertree, and dual hypertree are introduced. Recall also that by Theorem 1.3.1 a hypergraph H is a hypertree if and only if H has the Helly property and its line graph $L(H)$ is chordal.

Hypergraphs are a good model for expressing desirable properties of relational database schemes, such as the equivalence of local and global consistency of a relational database. It turns out that several of the desirable properties are equivalent to so-called α-acyclicity of a hypergraph. A hierarchy of hypergraph acyclicity types reflects a number of basic properties of databases; see Fagin [369, 370], Beeri et al. [74, 73], and Golumbic [455].

There are several natural definitions of cycles in hypergraphs. The following gives a basic one. Further types are given in Definition 8.2.1.

Definition 8.1.1 *Let H be a hypergraph.*

> *A sequence $(E_1, \ldots, E_k)$ of distinct hyperedges of H is a* hyperpath (*or* path) *of length $k-1$ if for all i, $1 \leq i < k$, $E_i \cap E_{i+1} \neq \emptyset$. If furthermore $E_1 \cap E_k \neq \emptyset$, then the hyperpath is a* hypercycle. *Its length is k.*
>
> *The path $(E_1, \ldots, E_k)$* connects *all vertex pairs $x \in E_1$, $y \in E_k$ and also* connects *the hyperedges E_1 and E_k. A set of hyperedges is* connected *if every pair of hyperedges in the set is connected.*
>
> *A* connected component *is a maximal connected set of hyperedges.*
>
> *A hyperedge F of H is a* chord *of the hypercycle $(E_1, E_2, \ldots, E_k)$ if there are indices i, j, l with $1 \leq i < j < l \leq k$ such that F contains the intersections $E_i \cap E_{i+1}$,*

$E_j \cap E_{j+1}$, $E_l \cap E_{l+1}$ (mod k). *Here and in subsequent definitions concerning cycles, all index arithmetic is done modulo k.*

A chordless hypercycle *is a hypercycle without any chord.*

There is a natural concept of an elimination ordering for hypergraphs that is called *Graham's reduction algorithm* [468] consisting of the repeated application of the following two rules.

Rule 1. If vertex v appears in exactly one hyperedge, then delete v from this hyperedge.

Rule 2. If one hyperedge is contained in another, then omit the smaller one from $\mathcal{E}$.

Definition 8.1.2 *Let $H = (V, \mathcal{E})$ be a hypergraph.*

Graham's reduction algorithm succeeds on $\mathcal{E}$ *if $\mathcal{E}$ can be reduced to the empty set by repeated application of rules* 1 *and* 2.

A hypergraph $\mathcal{E}$ is reduced *if rule* 2 *does not apply.*

The reduction *red($\mathcal{E}$) of $\mathcal{E}$ is obtained by applying rule* 2 *repeatedly until it no longer applies.*

Let $U \subseteq V$. The reduction $\mathcal{E}[U]$ of the induced subhypergraph $H(U)$ is the hypergraph red($\mathcal{E}(U)$), where $\mathcal{E}(U) = \{E_i \cap U : E_i \in \mathcal{E}\}$.

A set $A \subseteq V$ is an articulation set *for $\mathcal{E}$ if $A = E_1 \cap E_2$ for some pair of hyperedges $E_1, E_2 \in \mathcal{E}$ and $\mathcal{E}[V \setminus A]$ has more connected components than $\mathcal{E}$.*

A block *of $\mathcal{E}$ is a U-generated hypergraph of $\mathcal{E}$ with no articulation set for some $U \subseteq V$.*

A block is trivial *if it has only one hyperedge.*

A reduced hypergraph is α-acyclic *if all its blocks are trivial.*

H is α-acyclic *if its reduction red($\mathcal{E}$) has this property.*

Theorem 8.1.1 (Beeri et al. [74], Goodman, Shmueli [466]) *Let $H = (V, \mathcal{E})$ be a hypergraph. The following conditions are equivalent:*

(i) *H is α-acyclic;*

(ii) *H is a dual hypertree;*

(iii) *H has no chordless hypercycle of length at least* 3*;*

(iv) *Graham's reduction algorithm succeeds on $\mathcal{E}$.*

Note that α-acyclic hypergraphs are not cycle-free (see the remarks after Definition 1.3.6). Tarjan and Yannakakis [1021] give a linear-time recognition algorithm for testing α-acyclicity.

8.2 Further acyclicity types; clique and neighborhood hypergraphs

Several other types of cycles in hypergraphs appear under various names in the literature. We mention some of them.

Definition 8.2.1 (Berge [90], Fagin [369]) *Let $H = (V, \mathcal{E})$ be a hypergraph.*

A Berge cycle *in $\mathcal{E}$ is a sequence $C = (E_1, E_2, \dots, E_k)$ of $k \geq 2$ distinct hyperedges in $\mathcal{E}$ such that there exist k distinct vertices $v_1, v_2, \dots, v_k$ with the property that for every $i, 1 \leq i < k$, $v_i \in E_i \cap E_{i+1}$ and $v_k \in E_k \cap E_1$.*

A sequence $C = (v_1, E_1, v_2, E_2, \dots, v_k, E_k)$ of distinct vertices $v_1, v_2, \dots, v_k$ and distinct hyperedges $E_1, E_2, \dots, E_k$ is a special cycle *(or* chordless cycle *or* induced cycle *or, in* [742], unbalanced circuit*) if $k \geq 3$, for every i, v_i, $v_{i+1} \in E_i$ (recall that index arithmetic is done modulo k) and for all i, $1 \leq i \leq k$, $E_i \cap \{v_1, \dots, v_k\} = \{v_i, v_{i+1}\}$. The length of cycle C is k.*

A Berge cycle C is a pure cycle *if $k \geq 3$ and for every distinct i and j, $1 \leq i, j \leq k$, if $i \not\equiv j - 1 \pmod k$ and $i \not\equiv j + 1 \pmod k$, then $E_i \cap E_j = \emptyset$.*

C is a β-cycle *if for $E'_i = E_i \setminus \bigcap_{j=1}^{k} E_j$, the sequence $(E'_1, E'_2, \dots, E'_k)$ is a pure cycle.*

C is a γ-cycle *if C is a pure cycle or a cycle (E_1, E_2, E_3) such that $v_1 \notin E_3$ and $v_2 \notin E_1$.*

A hypergraph is θ-acyclic *if it has no θ-cycle ($\theta = \beta, \gamma$, Berge).*

The various acyclicity types form the following hierarchy:

Berge acyclic $\subset$ γ-acyclic $\subset$ β-acyclic $\subset$ α-acyclic [369].

Dahlhaus [276] discusses two other types of acyclicity called δ- and *ε-acyclic hypergraphs.* See also the survey of Duke [344] on types of cycles in hypergraphs.

Definition 8.2.2 [90] *A hypergraph is* balanced *if it contains no special cycles of odd length.*

Balanced hypergraphs are a natural generalization of bipartite graphs. Berge and Las Vergnas [100] have shown that they fulfill the König property (see Definition 1.3.14).

Definition 8.2.3 (Lovász [737]) *A hypergraph is* totally balanced *if it has no special cycle.*

Totally balanced hypergraphs are a natural generalization of trees.

Theorem 8.2.1 (D'Atri, Moscarini [289]) *A hypergraph $\mathcal{E}$ is totally balanced if and only if $\mathcal{E}$ is β-acyclic.*

There is also the following characterization.

Theorem 8.2.2 (Ryser [938], Lehel [718]) *A hypergraph $\mathcal{E}$ is totally balanced if and only if every subhypergraph of $\mathcal{E}$ is a hypertree.*

The various acyclicity types of hypergraphs have connections to graph classes as follows.

Theorem 8.2.3 *Let $G = (V, E)$ be a graph.*

(i) *G is chordal if and only if $\mathcal{C}(G)$ is α-acyclic (a corollary of Theorems* 1.3.1 *and* 8.1.1*).*

(ii) *G is odd-sun–free chordal if and only if $\mathcal{C}(G)$ is balanced* [720].

(iii) *G is strongly chordal if and only if $\mathcal{C}(G)$ is β-acyclic (or, equivalently, totally balanced)* [162, 372].

(iv) *G is ptolemaic if and only if $\mathcal{C}(G)$ is γ-acyclic* [289, 276].

(v) *G is a block graph if and only if $\mathcal{C}(G)$ is Berge acyclic* [90].

Theorem 8.2.3 demonstrates the correspondence between the hierarchy of acyclicity types of the clique hypergraph and the following hierarchy of graph classes:

block $\subset$ ptolemaic $\subset$ strongly chordal $\subset$ odd-sun–free chordal $\subset$ chordal.

A similar correspondence exists for acyclicity types of the neighborhood hypergraphs and a hierarchy of bipartite graphs (see Theorem 8.2.6).

Dahlhaus [276] mentions that δ-acyclicity corresponds to so-called *strongly ptolemaic* and ε-acyclicity corresponds to so-called *very strongly ptolemaic* graphs.

We now present theorems about neighborhood hypergraphs. Recall that in Definitions 1.3.8 and 1.3.9 various neighborhood hypergraphs of graphs and bipartite graphs are introduced.

Theorem 8.2.4 *Let $G = (V, E)$ be a graph.*

(i) *$\mathcal{N}(G)$ is totally balanced if and only if G is strongly chordal* [372].

(ii) *$\mathcal{N}(G)$ is balanced if and only if G is odd-sun–free chordal (i.e., chordal and neighborhood perfect; see Theorem* 7.2.2*)* [162].

Theorem 8.2.5 (Brouwer, Duchet, Schrijver [162]) *Let B be a bipartite graph.*

(i) *$\mathcal{N}_0(B)$ is totally balanced if and only if B is chordal bipartite.*

(ii) *$\mathcal{N}_0(B)$ is balanced if and only if B is bipartite without induced C_{4q+2} for $q \geq 1$.*

The next theorem describes the bipartite analogue of Theorem 8.2.3.

Theorem 8.2.6 (Ausiello, D'Atri, Moscarini [35]) *Let $B = (X, Y, E)$ be a bipartite graph.*

(i) *B is Y-chordal and Y-conformal if and only if $\mathcal{N}^X(B)$ is α-acyclic.*

(ii) *B is $(6,1)$-chordal (i.e., chordal bipartite) if and only if $\mathcal{N}^X(B)$ is β-acyclic (i.e., totally balanced).*

(iii) *B is $(6,2)$-chordal* (i.e., bipartite distance-hereditary) *if and only if $\mathcal{N}^X(B)$ is γ-acyclic.*

(iv) *B is $(4,1)$-chordal (i.e., a forest) if and only if $\mathcal{N}^X(B)$ is Berge acyclic.*

The same theorem holds when replacing X by Y and vice versa for reasons of symmetry.

8.3 Graphs with maximum neighborhood orderings and corresponding hypertrees

For a graph G, let $l(G)$ denote the number of edges of a longest induced cycle of G. Recall that maximum neighborhood orderings were introduced in Definition 5.4.1. There are several characterizations that give connections between graphs with such orderings and corresponding hypertrees based on the following lemma.

Lemma 8.3.1 (Dragan, Prisacaru, Chepoi [331]) *Let G be a graph. Then*

(i) $l(L(\mathcal{D}(G))) = l(L(\mathcal{N}(G)))$;

(ii) $l(L(\mathcal{N}(G))) \leq l(L(\mathcal{C}(G)))$.

Theorem 8.3.1 ([331], Brandstädt et al. [146]) *Let G be a graph. The following conditions are equivalent:*

(i) *G has a maximum neighborhood ordering (i.e., G is dually chordal);*

(ii) *$\mathcal{C}(G)$ is a hypertree;*

(iii) *$\mathcal{N}(G)$ is a hypertree;*

(iv) *$\mathcal{N}(G)$ is a dual hypertree;*

(v) *$\mathcal{D}(G)$ is a hypertree.*

Condition (iv) of Theorem 8.3.1 leads to a linear-time recognition of dually chordal graphs by using the linear-time α-acyclicity testing algorithm of Tarjan and Yannakakis [1021] for testing whether $\mathcal{N}(G)$ is a dual hypertree (i.e., α-acyclic). Note that $\mathcal{N}(G)$ has linear size with respect to G.

Dragan [323] shows that dually chordal graphs can be characterized as the Helly graphs, which do not contain certain generalized suns as isometric subgraphs.

Dragan, Prisacaru, and Chepoi [331] give the following characterizations of dually chordal graphs.

Theorem 8.3.2 *Let G be a graph. The following conditions are equivalent:*

(i) *G is dually chordal;*

(ii) *For all $k \geq 0$, G^{2k} is chordal and $\{N^k[v] : v \in V\}$ has the Helly property;*

(iii) *G^2 is chordal and neighborhood Helly;*

(iv) *G has a spanning tree T such that every clique C in G forms a subtree of T;*

(v) *G has a spanning tree T such that every disk of G forms a subtree of T;*

(vi) *G is the clique graph of some intersection graph of paths in a tree.*

Parts of Theorems 8.3.1 and 8.3.2 were found also by Szwarcfiter and Bornstein [1017] and later again by Gutierrez and Oubiña [485]. Thus, the equivalence of (i) and (vi) of Theorem 8.3.2 was shown independently in [1017].

Analogous properties hold for bipartite graphs. By using the standard constructions $B(G)$ and $B_C(G)$ (and others) one can switch between graphs and bipartite graphs; see [146] (see Definition 3.3.2 and Theorem 3.3.1).

Proposition 8.3.1 (Brandstädt et al. [146]) *Let G be a graph.*

(i) *The following conditions are equivalent:*

  (a) *$\mathcal{N}(G)$ is conformal;*
  (b) *$\mathcal{N}^Y(B(G))$ is conformal;*
  (c) *$B(G)$ is X-conformal;*
  (d) *$B(G)$ is X-conformal and Y-conformal.*

(ii) *$2SEC(\mathcal{N}(G))$ is chordal if and only if $2SEC(\mathcal{N}^Y(B(G)))$ is chordal.*

Proposition 8.3.2 [146] *Let $H = (V, \mathcal{E})$ be a hypergraph.*

(i) *$\mathcal{I}(\mathcal{E})$ is X-chordal if and only if $2SEC(\mathcal{N}^Y(\mathcal{I}(\mathcal{E})))$ is chordal. Analogously, $\mathcal{I}(\mathcal{E})$ is Y-chordal if and only if $2SEC(\mathcal{N}^X(\mathcal{I}(\mathcal{E})))$ is chordal.*

(ii) *$\mathcal{I}(\mathcal{E})$ is X-conformal if and only if $\mathcal{N}^Y(\mathcal{I}(\mathcal{E}))$ is conformal. Analogously, $\mathcal{I}(\mathcal{E})$ is Y-conformal if and only if $\mathcal{N}^X(\mathcal{I}(\mathcal{E}))$ is conformal.*

Corollary 8.3.1 *Let $H = (V, \mathcal{E})$ be a hypergraph.*

(i) *$\mathcal{E}$ is a hypertree if and only if $\mathcal{I}(\mathcal{E})$ has a maximum X-neighborhood ordering.*

(ii) *$\mathcal{E}^*$ is a hypertree if and only if $\mathcal{I}(\mathcal{E})$ has a maximum Y-neighborhood ordering.*

Proposition 8.3.3 [146] *Let $B = (X, Y, E)$ be a bipartite graph. Then*

(i) *$\mathcal{N}^X(B)$ has the Helly property if and only if $\mathcal{C}(split_X(B))$ has the Helly property (analogously for Y instead of X).*

(ii) *$L(\mathcal{N}^X(B))$ is chordal if and only if $L(\mathcal{C}(split_X(B)))$ is chordal.*

Corollary 8.3.2

(i) *Let $B = (X, Y, E)$ be a bipartite graph. Then B is X-chordal and X-conformal if and only if $split_X(B)$ is doubly chordal.*

(ii) *$\mathcal{E}$ is a hypertree if and only if $split_V(\mathcal{I}(\mathcal{E}))$ has a maximum neighborhood ordering.*

(iii) *B is chordal bipartite if and only if $split_X(B)$ is strongly chordal.*

A direct proof of equivalence (iii) of Corollary 8.3.2 was given by Dahlhaus [276].

Proposition 8.3.4 *Let G be a graph.*

(i) *G is dually chordal if and only if $B_C(G)$ has a maximum X-neighborhood ordering.*

(ii) *G is doubly chordal if and only if $B_C(G)$ has a maximum X-neighborhood ordering and a maximum Y-neighborhood ordering.*

By Theorem 5.4.1, a graph G is doubly chordal if and only if $\mathcal{C}(G)$ is a hypertree and a dual hypertree [799, 331]. This class was introduced in [798] under the name α-graphs.

It was shown in [35] that a hypergraph $\mathcal{E}$ is β-acyclic if and only if $\mathcal{E}^*$ is β-acyclic. Thus, by Theorem 8.2.3 (ii), for strongly chordal graphs G, the clique hypergraph $\mathcal{C}(G)$ and its dual $\mathcal{C}(G)^*$ are β-acyclic.

Dragan and Voloshin [332] studied the incidence graphs of biacyclic hypergraphs. A hypergraph H is *biacyclic* if H and H^* are a hypertree. For the following theorem, we need to define the 2-step graph of a graph G: Two vertices x and y are adjacent in the 2-step graph if and only if their distance in G is exactly two. Recall that we consider here only graphs without loops.

Theorem 8.3.3 [332] *Let G be a graph. The following conditions are equivalent:*

(i) *G is the incidence graph of a biacyclic hypergraph;*

(ii) *G is open-neighborhood Helly and the 2-step graph of G is chordal;*

(iii) *G is an absolute bipartite retract that has no isometric bipartite wheels with a cycle of length $2k$ and a central vertex of degree k, $k \geq 4$.*

The class of doubly chordal graphs was introduced primarily in order to refine the borderline for the Steiner tree problem, which is $\mathbb{NP}$-complete on chordal graphs and solvable in polynomial time on strongly chordal graphs. Moscarini [798] shows that the Steiner tree problem is solvable in polynomial time on doubly chordal graphs and $\mathbb{NP}$-complete on the class of chordal clique-Helly graphs, which will be defined below. The polynomial-time solutions can be generalized, since maximum neighborhood orderings are algorithmically very useful for domination-like problems, and the Steiner tree problem can be reformulated as a special connected r-domination problem (see [323, 143]). See also [148] and Definition 5.4.4 for the notion of *homogeneously orderable graphs*—a generalization of distance-hereditary and dually chordal graphs. Similarly to dually chordal graphs, these graphs have a characterization in terms of (dual) hypertrees given in Theorem 8.4.2, and the Steiner tree problem remains solvable in polynomial time on these graphs.

8.4 Further classes with dual hypertree characterizations

We first mention a dual hypertree characterization for distance-hereditary graphs. For a graph $G = (V, E)$, let $\mathcal{CC}(G)$ denote the hypergraph of all maximal subsets of V that in G induce a connected cograph.

Theorem 8.4.1 (Nicolai [810, 811]) *A graph G is distance hereditary if and only if its cograph hypergraph $\mathcal{CC}(G)$ is a dual hypertree.*

This characterization has nice consequences for bipartite graphs and for ptolemaic graphs. Recall the notion of the biclique hypergraph $\mathcal{BC}(B)$ of B (Definition 1.3.9).

Corollary 8.4.1 [810, 811] *Let $B = (X, Y, E)$ be a bipartite graph. The following conditions are equivalent:*

(i) *B is distance hereditary (i.e., $(6, 2)$-chordal bipartite);*

(ii) *There is an ordering $(v_1, \ldots, v_n)$ of $X \cup Y$ such that $D_2(v_i)$ induces a complete bipartite subgraph in G_i, $i \in \{1, \ldots, n\}$;*

(iii) *The biclique hypergraph $\mathcal{BC}(B)$ is a dual hypertree.*

Corollary 8.4.2 [810, 811] *Let G be a graph. The following conditions are equivalent:*

(i) *G is chordal and distance hereditary (i.e., ptolemaic);*

(ii) *There is a vertex ordering $(v_1, \ldots, v_n)$ of G such that $D_2(v_i)$ induces a trivially perfect subgraph in G_i, $i \in \{1, \ldots, n\}$;*

(iii) *The hypergraph of all maximal trivially perfect subgraphs of G is a dual hypertree.*

The analogue of the biclique hypergraph for bipartite graphs is the join-partitionable set hypergraph for arbitrary graphs (see Definition 1.3.8) (join-partitionable sets were called *join-splitted sets* in [148]).

Theorem 8.4.2 (Brandstädt, Dragan, Nicolai [148]) *Let G be a graph. The following conditions are equivalent:*

(i) *G is homogeneously orderable;*

(ii) *G^2 is chordal and every maximal set of vertices of pairwise distance at most two in G is join partitionable;*

(iii) *The hypergraph $\mathcal{JP}(G)$ of all maximal join-partitionable sets of G is a dual hypertree.*

In [148], yet another dual hypertree characterization of homogeneously orderable graphs is given, which is close to the homogeneous elmination ordering (see Definition 5.4.4).

8.5 Disk-Helly, clique-Helly, and neighborhood-Helly graphs

The Helly property (see Definition 1.3.4) is a hypergraph property of basic importance and with many structural consequences. Here we collect some properties of graphs having a Helly disk, clique, or neighborhood hypergraph.

Definition 8.5.1 *A graph G is a* disk-Helly (clique-Helly, neighborhood-Helly) graph *if the disk hypergraph $\mathcal{D}(G)$ (clique hypergraph $\mathcal{C}(G)$, neighborhood hypergraph $\mathcal{N}(G)$) has the Helly property.*

In what follows, disk-Helly graphs are called *Helly graphs.* Clearly every disk-Helly graph is neighborhood-Helly. It is easy to show that every neighborhood Helly graph is clique-Helly. Thus

Helly $\subset$ neighborhood Helly $\subset$ clique Helly.

Bandelt and Pesch [63] give a polynomial-time algorithm for recognizing Helly graphs. Dragan [321] gives polynomial-time algorithms for recognizing Helly, neighborhood-Helly, and clique-Helly graphs. For the case of clique-Helly graphs see also [1016].

Recall that every dually chordal graph is Helly and every doubly chordal graph is chordal and Helly.

The clique graph of a graph is a construction that is closely related to the Helly property.

Definition 8.5.2

> *The* clique graph *$K(G)$ of a graph G is the line graph $L(\mathcal{C}(G))$ of the set $\mathcal{C}(G)$ of maximal cliques of G.*
>
> *A graph G is a* clique graph *if there is a graph G' such that $G = K(G')$ is the clique graph of G'.*

The clique graphs are characterized in [908].

Theorem 8.5.1 (Roberts, Spencer [908]) *The graph G is a clique graph if and only if some family of complete subgraphs of G covers all edges of G and has the Helly property.*

A consequence of this is the following result (which was found earlier).

Theorem 8.5.2 (Hamelink [508]) *Every clique-Helly graph is a clique graph.*

Theorem 8.5.3 (Escalante [356]) *The clique graph of a clique-Helly graph is a clique-Helly graph, and every clique-Helly graph is the clique graph of some clique-Helly graph.*

The results of [331, 146, 1017] imply that the dually chordal graphs are the clique graphs of chordal graphs. In order to collect the corresponding results we use the following notions.

Definition 8.5.3 (Brandstädt et al. [146]) *Let G be a graph.*

G is a Helly-chordal graph *if G is chordal and clique Helly;*

G is a clique-chordal graph *if the line graph $L(\mathcal{C}(G))$ is chordal;*

G is a power-chordal graph *if all its powers are chordal.*

Theorem 8.5.4 [146] *Let G be a graph.*

(i) *G is Helly chordal if and only if G is the clique graph of some dually chordal graph.*

(ii) *G is dually chordal if and only if G is the clique graph of some chordal graph.*

(iii) *G is doubly chordal if and only if G is the clique graph of some doubly chordal graph.*

The following combinations are of interest:

dually chordal	=	**clique-Helly**	$\bigcap$	**clique chordal;**		
doubly chordal	=	**clique-Helly**	$\bigcap$	**clique chordal**	$\bigcap$	**chordal;**
power chordal	=	**clique chordal**	$\bigcap$	**chordal.**		

Definition 8.5.4 *A graph G* converges to the one-vertex graph *if there exists a natural number n such that the iterated clique graph $K^n(G) = K(K^{n-1}(G))$ consists of only one vertex.*

Theorem 8.5.5 (Bandelt, Prisner [65]) *Let G be a graph. The following conditions are equivalent:*

(i) *G is a Helly graph;*

(ii) *G is a dismantlable clique-Helly graph;*

(iii) *G is a clique-Helly graph that converges to the one-vertex graph.*

Dragan [321, 322] gives other characterizations of Helly graphs in terms of eccentricity functions. A graph is Helly if and only if for every vertex-weight function, every local minimum of the corresponding eccentricity function is a global minimum.

For a connection to absolute reflexive retracts see Theorem 10.4.1; for the bipartite case see Theorem 10.4.2.

Since every Helly graph is the clique graph of a Helly graph, we say that the class of Helly graphs is closed under the clique-graph operator. [65] gives more classes that are closed under the clique-graph operator, including clique-Helly, strongly chordal, ptolemaic, and block graphs.

For more results on graphs converging to one vertex see [65, 131, 880].

An interesting case is the clique graph of a chordal graph. In [331] (see [146]) it is shown that if G is chordal, then $K(G)$ is chordal if and only if G^2 is chordal. Bandelt and Prisner [65] show that if G is chordal, then $K(K(G))$ is chordal.

Since bridged graphs are dismantlable (see Theorem 5.8.2), Theorem 8.5.5 implies the following theorem.

Theorem 8.5.6 (Bandelt, Prisner [65], Dragan [321]) *A bridged graph is a Helly graph if and only if it is a clique-Helly graph.*

Dragan [321] gives more conditions characterizing bridged Helly graphs.

Theorem 8.5.7 [321] *G is a bridged Helly graph if and only if it is a Helly graph containing no induced cycles C_4 and C_5.*

The hereditary Helly graphs are characterized in the following way.

Theorem 8.5.8 [321] *Every induced subgraph of G is a Helly graph if and only if G is S_3-free chordal.*

For chordal graphs a result similar to Theorem 8.5.6 holds.

Theorem 8.5.9 [321] *A chordal graph is a Helly graph if and only if it is a clique-Helly graph.*

Dragan [321] gives several other conditions characterizing these graphs.

For bipartite graphs there are similar connections between different types of hypergraphs having the Helly property. [53] collects some of them.

8.6 Perfect graphs and normal hypergraphs

Perfect graphs can be characterized by a property of their clique hypergraph.

Definition 8.6.1 (Lovász [735, 739]) *Let $H = (V, \mathcal{E})$ be a hypergraph.*

> *The* degree $\deg(v)$ *of $v \in V$ in H is the number of hyperedges containing v. The maximum degree of vertices in H is denoted by $\Delta(H)$.*
>
> *The* chromatic index $\chi_E(H)$ *of H is the least number of colors by which the hyperedges can be colored such that edges with the same color are disjoint.*
>
> *H is* normal *if for every partial hypergraph $H' = (V, \mathcal{E}')$ of H, $\chi_E(H') = \Delta(H')$ holds.*

Obviously, $\chi_E(H) \geq \Delta(H)$ holds for every hypergraph H. The τ-normal hypergraphs are introduced in a similar way. These correspond to perfection of the complement graphs, and it is shown that a hypergraph is τ-normal if and only if it is normal.

Definition 8.6.2 [735, 739] *Let $H = (V, \mathcal{E})$ be a hypergraph.*

> *A hypergraph H is* τ-normal *if for every partial hypergraph $H' = (V, \mathcal{E}')$ of H, $\nu(H') = \tau(H')$ holds.*

Theorem 8.6.1 [735] *Let H be a hypergraph. The following conditions are equivalent:*

(i) *H is τ-normal;*

(ii) *H is normal;*

(iii) *H has the Helly property and its line graph $L(H)$ is perfect.*

Balanced hypergraphs are normal: If H is balanced then $\chi_E(H) = \Delta(H)$, which follows from Theorem 8.6.1.

For a connection between game theory and normal hypergraphs see Boros and Gurvich [133, 134] and Boros, Gurvich, and Vasin [135]. Using this connection Boros and Gurvich were able to show that every perfect graph is kernel solvable, which is one direction of a conjecture of Berge and Duchet; see Theorem 2.4.2.

8.7 Interval hypergraphs

In correspondence to the notion of interval graphs, there is an interesting notion of *interval hypergraphs.*

Definition 8.7.1 *A hypergraph H is an* interval hypergraph *if for some vertex ordering $(v_1, \ldots, v_n)$ of H, every hyperedge forms an interval.*

Duchet [335, 338] describes properties of interval hypergraphs. They form a subclass of hypertrees corresponding to interval graphs.

Theorem 8.7.1 *A hypergraph H is an interval hypergraph if and only if it has the Helly property and the line graph of the augmented hypergraph H^+, resulting from H by adding all singletons as new edges, is an interval graph.*

Duchet [335] mentions a list of sources for the theory of interval hypergraphs. Tucker [1039] gives a characterization by forbidden induced subhypergraphs.

Theorem 8.7.2 [334, 335] *Let G be a graph. $\mathcal{N}(G)$ is an interval hypergraph if and only if G is a proper-interval graph.*

Bandelt and Prisner [65] mention that the clique hypergraph of a proper interval graph (i.e., indifference graph) is an interval hypergraph: Proper interval graphs are exactly the 2-section graphs of hypergraphs H for which both H and the dual H^* are interval hypergraphs. Lehel [717, 718] mentions that interval hypergraphs are totally balanced. Another paper studying interval hypergraphs is [929].

Quilliot [887] studies the case of hypergraphs where the vertices have a circular arrangement such that all hyperedges are arcs of the circle (so-called *circular-representable* hypergraphs).

Chapter 9

Matrices and Polyhedra

Many graph and hypergraph properties can be expressed in terms of incidence matrices of different kinds. Thus, there is a close correspondence between many properties of graphs, bipartite graphs, hypergraphs, and matrices.

9.1 The consecutive 1s and the circular 1s properties

A variety of well-known graph classes can be characterized in terms of consecutive 1s and circular 1s in matrices derived from the corresponding graphs. This is particularly important for recognition algorithms, since Booth and Lueker [127] showed that these properties can be tested in time proportional to the number of 1 entries in the matrix.

Definition 9.1.1

If M is an $n \times n$ matrix and the entries of the main diagonal are all 0, then the augmented matrix *M^* to M is obtained from M by adding 1s along the main diagonal.*

The adjacency *(or* neighborhood*)* matrix *$A(G)$ of a graph $G = (V, E)$ and a vertex ordering of V is the $(0,1)$-matrix (a_{ij}) defined by*

$$a_{ij} = \begin{cases} 1 & \text{if } v_i v_j \in E, \\ 0 & \text{otherwise.} \end{cases}$$

The closed neighborhood matrix *$N(G) = (m_{ij})$ given by*

$$m_{ij} = \begin{cases} 1 & \text{if } v_i \in N[v_j], \\ 0 & \text{otherwise} \end{cases}$$

is then the same as the augmented adjacency matrix $A^*(G)$: $N(G) = A^*(G)$.

Let $\mathcal{C}(G) = \{C_1, \ldots, C_k\}$ be the maximal cliques of a graph G. The clique matrix *$C(G)$ of G is the $(0,1)$-matrix (c_{ij}) with entries*

$$c_{ij} = \begin{cases} 1 & \text{if } v_i \in C_j, \\ 0 & \text{otherwise.} \end{cases}$$

The (vertex-edge) incidence matrix *$I(G)$ of $G = (V, E)$ with vertex ordering $(v_1, \ldots, v_n)$ and edge ordering $(e_1, \ldots, e_m)$ is the $n \times m$ $(0,1)$-matrix i_{kl} with entries*

$$i_{kl} = \begin{cases} 1 & \text{if } v_k \in e_l, \\ 0 & \text{otherwise.} \end{cases}$$

Definition 9.1.2 *Let M be a $(0,1)$-matrix.*

M has the consecutive 1s property for columns *if its rows can be permuted in such a way that the 1s in each column occur consecutively.*

M has the circular 1s property for columns *if its rows can be permuted in such a way that the 1s in each column occur in a circular consecutive order—imagine the matrix as wrapped around a cylinder.*

Remark 9.1.1 *Let M be a $(0,1)$-matrix.*

(i) *M has the circular 1s property if and only if M has the circular 0s property.*

(ii) *Consecutive 1s implies circular 1s for columns but not conversely.*

Theorem 9.1.1 (Fulkerson, Gross [409]) *A graph G is an interval graph if and only if its clique matrix $C(G)$ has the consecutive 1s property for columns.*

Theorem 9.1.2 (Roberts [906]) *G is a proper interval graph if and only if the augmented adjacency matrix $A^*(G)$ has the consecutive 1s property for columns.*

Note that there is a simple relationship between the consecutive 1s property for columns of $C(G)$ and the convexity (in the sense of Definition 6.2.4) of the bipartite vertex-clique graph $B_C(G)$:

> $C(G)$ has the consecutive 1s property for columns if and only if in $B_C(G) = (V, \mathcal{C}, E')$, the vertices V can be ordered such that for all $c \in \mathcal{C}$, $N(c)$ is an interval in V.

Definition 9.1.3 *A circular-arc graph G is a* Helly circular-arc graph *if there is a circular-arc representation of G that satisfies the Helly property.*

Theorem 9.1.3 (Gavril [425]) *A graph G is a Helly circular-arc graph if and only if its clique matrix $C(G)$ has the circular 1s property for columns.*

The characterization above implies a polynomial time recognition algorithm for Helly circular-arc graphs.

It is natural to consider what happens if the augmented adjacency matrix $A^*(G)$ has the circular 1s property for columns. In this case, G must be a circular-arc graph, but not all circular-arc graphs have this property.

Theorem 9.1.4 (Tucker [1037, 1038]) *If the augmented adjacency matrix $A^*(G)$ of G has the circular* 1*s property for columns, then G is a circular-arc graph.*

Note that there is also a characterization of proper circular-arc graphs in terms of the circular 1s property for the augmented adjacency matrix and some additional conditions, which leads to a linear-time recognition algorithm for proper circular-arc graphs [1038, 994].

9.2 Balanced and totally balanced matrices; doubly lexical orderings

Golumbic describes in [452, chapter 12] the connections between Gaussian elimination and pivot sequences on one side and perfect elimination orderings of chordal and perfect edge elimination orderings of perfect elimination bipartite graphs (see Definition 5.9.2) on the other side.

Definition 1.3.10 describes a connection between matrices and hypergraphs.

Definition 9.2.1 *Let $H = (V, \mathcal{E})$ with $V = \{v_1, \dots, v_n\}$ and $\mathcal{E} = \{e_1, \dots, e_m\}$ be a hypergraph. Then $\mathcal{I}(V, \mathcal{E})$ denotes the following* incidence matrix *of the hypergraph $\mathcal{E}$:*

$$\mathcal{I}(V, \mathcal{E})(i, j) = 1 \text{ if } v_i \in e_j \text{ and } 0 \text{ otherwise.}$$

It is clear that the transposed matrix $\mathcal{I}(\mathcal{E}, V) = \mathcal{I}(V, \mathcal{E})^T$ is the incidence matrix of the dual hypergraph $\mathcal{E}^*$.

The many connections between hypergraphs and graphs with chordality properties are also reflected in corresponding connections between such graphs and matrix properties.

Definition 9.2.2 *A $(0,1)$-matrix is* balanced *if it does not contain as a submatrix the (vertex-edge) incidence matrix of an odd cycle.*

An important property of balanced matrices is the following.

Theorem 9.2.1 (Fulkerson, Hoffman, Oppenheim [410])
Let $e_n = \underbrace{(1, 1, \dots, 1)}_{n \text{ times}}$. If A is a balanced matrix, then the polyhedra

$$Q(A) = \{x : x \in \mathbb{R}^n \text{ and } Ax \geq e_n \text{ and } x \geq 0\} \text{ and}$$

$$P(A) = \{x : x \in \mathbb{R}^n \text{ and } Ax \leq e_n \text{ and } x \geq 0\}$$

have only 0, 1*-extreme points.*

This leads to polynomial-time optimization results in many cases where the matrix of the linear-programming problem is known to be balanced. One such case is given in the following theorem.

Theorem 9.2.2 (Chang [180]) *A graph is odd-sun–free chordal if and only if its closed neighborhood matrix is balanced.*

Thus, due to Theorem 7.2.2, a graph is odd-sun–free chordal if and only if it is chordal and neighborhood perfect (see also Theorem 8.2.4).

Definition 9.2.3 *A* $(0,1)$*-matrix is* totally balanced *if it does not contain as a submatrix the (vertex-edge) incidence matrix of a cycle of length at least three.*

By definition, a graph is a forest if and only if its (vertex-edge) incidence matrix is totally balanced.

Theorem 9.2.3 (Farber [372]) *A bipartite graph* B *is chordal bipartite if and only if its adjacency matrix* $A(B)$ *is totally balanced.*

Furthermore, one has the following result.

Theorem 9.2.4 [372] *A graph is strongly chordal if and only if its closed neighborhood matrix is totally balanced.*

Theorem 9.2.4 is based on the fact that the ordering of the vertices of G is a strong elimination ordering if and only if the matrix $\binom{11}{10}$ (the Γ) is not a submatrix of $N(G)$ obtained using this ordering of the vertices.

Theorem 9.2.5 [372] *A graph is strongly chordal if and only if its clique matrix is totally balanced.*

Note that for every graph G, $N(G)$ is totally balanced if and only if $C(G)$ is totally balanced [372].

Definition 9.2.4 (Hoffman, Kolen, Sakarovitch [577]) *A* doubly lexical ordering *of a matrix is an ordering of the rows and columns of the matrix such that the rows, as vectors, are lexically increasing and the columns, as vectors, are lexically increasing (called* lexical ordering *in* [577]*); here row vectors are read from right to left and column vectors from bottom to top.*

Note that every matrix has a doubly lexical ordering [744].

Definition 9.2.5 (Lubiw [744]) *A (doubly lexically ordered) matrix is* Γ-free *(or* greedy*) if it does not contain a submatrix of the form* $\binom{11}{10}$.

The name "greedy" matrix stems from the fact that for such matrices certain linear-programming problems can be solved by a greedy algorithm.

Theorem 9.2.6 [744] *Every doubly lexical ordering of a totally balanced matrix is Γ-free.*

Theorem 9.2.7 ([577, 744], Anstee, Farber [24]) *A $(0,1)$-matrix has a Γ-free ordering if and only if it is totally balanced.*

There are several algorithms for producing doubly lexical orderings. Let i be the number of columns and j be the number of rows of the matrix. [577] gives an $\mathcal{O}(i^2 j)$-time algorithm, [744] gives a $\mathcal{O}(m \log^2(i+j)+i+j)$-time algorithm, which was improved by Paige and Tarjan [850] to $\mathcal{O}(m \log(i+j)+i+j)$ time; the time bound for dense matrices is improved in [991]. This yields a $\mathcal{O}(n^2)$ recognition algorithm for recognizing strongly chordal graphs and chordal bipartite graphs.

Definition 9.2.6 [744]

An ordered $(0,1)$-matrix M is supported Γ *if and only if for every pair $r_1 < r_2$ of rows and pair $c_1 < c_2$ of columns that form a Γ, there is a row $r_3 > r_2$ with $M(r_3, c_1) = M(r_3, c_2) = 1$ (r_3* supports Γ*).*

A subtree matrix *is the incidence matrix of a collection of subtrees of a tree T.*

Theorem 9.2.8 [744] *A $(0,1)$-matrix is a subtree matrix if and only if it has a supported Γ ordering.*

This yields a matrix characterization of chordal graphs; due to the duality between chordal and dually chordal graphs it also gives a matrix characterization of graphs with maximum neighborhood ordering by transposing the incidence matrix.

In a series of papers [239], Conforti, Cornuéjols, and Rao design a polynomial-time recognition algorithm for balanced (0,1) matrices. These results are extended in [238] to $(0, \pm 1)$-matrices. The algorithm works by modeling the matrix as a bipartite graph, and decomposing the graph using "k-joins"; these are a generalization of the join decomposition in Definition 12.3.1 to the case in which edges that cross the partition must be in one of k complete bipartite graphs. A recognition algorithm for the special case of linear matrices (no 2×2 submatrix has all 1 entries) is contained in [241].

9.3 Perfect and totally unimodular matrices

Perfect graphs have a characterization in terms of polyhedra.

Definition 9.3.1 *Let A be an $m \times n$ matrix and let*

$$P(A) = \{x : x \in \mathbb{R}^n \text{ and } Ax \leq e_n \text{ and } x \geq 0\} ,$$

$$P_I(A) = \text{convex hull of } \{x : x \in P(A) \text{ and } x \text{ integral}\}.$$

Theorem 9.3.1 (Chvátal [201]) *A graph G is perfect if and only if $P_I(C(G)) = P(C(G))$.*

Definition 9.3.2 *A matrix A is* perfect *if $P(A)$ has only integer extrema: $P_I(A) = P(A)$.*

Thus, a graph is perfect if and only if its clique matrix is perfect. Note that because of Theorem 9.2.1, balanced matrices are perfect matrices. Shepherd [969] studies a generalization of perfect matrices, called *near-perfect matrices*, in which the integer hull can be obtained by adding one extra rank constraint. Some results on perfect $(0, \pm 1)$-matrices are contained in [132].

There is another interesting subclass of perfect matrices.

Definition 9.3.3

> *A matrix A is* totally unimodular *if the determinant of every induced square submatrix of A is equal to $0, +1$, or -1.*
>
> *A graph G is* totally unimodular *if its clique matrix is totally unimodular.*

In [452] it is mentioned in an exercise that total unimodularity is a hereditary property, bipartite graphs are totally unimodular, and totally unimodular graphs are perfect. Every interval graph is totally unimodular, since every $(0, 1)$-valued matrix satisfying the consecutive 1s property is totally unimodular.

Padberg [848] gives a characterization of perfect (0,1)-matrices in terms of forbidden submatrices. Totally unimodular matrices have received a great deal of attention, due to connections with matroid theory and because one can solve integer linear programs on totally unimodular matrices; relaxing the integrality constraints does not alter the solution. An efficient recognition algorithm for this class was known long before such an algorithm was designed for balanced matrices.

Yannakakis [1095] studies a restricted class of totally unimodular matrices with no odd cycles; for this class of matrices, integer linear programs can be solved using maximum-flow techniques. He also gives a linear-time recognition algorithm for the class.

Crama, Hammer, and Ibaraki [268] extend Yannakakis's results on solving linear programs using flow techniques to a larger class of matrices, which they call *strongly unimodular.* A $(0, 1)$-matrix is strongly unimodular if all nonsingular square submatrices are triangular. The paper [267] gives a recognition algorithm for strongly uniform matrices and hypergraphs, and shows that a hypergraph is strongly unimodular if and only if it is strongly balanced, i.e., every odd cycle uses two edges containing at least three vertices of the cycle.

Subclasses of totally unimodular matrices have been studied by requiring the matrix to remain totally unimodular if certain nonzero entries are replaced by 0. The classes defined by allowing any single nonzero entry to be replaced by 0 and any subset of entries to be replaced by 0 are studied by Conforti and Rao [240]; the latter class is called *restricted unimodular.* The class of *k-totally unimodular matrices* defined by setting at most k entries to 0 for fixed k is studied by Loebl and Poljak [733]. They show that 3-totally unimodular matrices are equal to restricted unimodular matrices, i.e., that 3-totally unimodular = k-totally unimodular for all $k > 3$ and that 1-unimodular matrices are the same as strongly unimodular matrices, and give a characterization of 2-totally unimodular matrices.

De Werra [309] gives several characterizations of totally unimodular, balanced, and perfect matrices; Padberg [849] gives a survey of totally unimodular matrices.

Totally unimodular matrices are also closely connected to matroid theory; a matroid is *regular* if it can be represented by a totally unimodular matrix. A short proof of Tutte's theorem saying that a matroid is regular if and only if it has no minor equal to the Fano matroid or its dual can be found in [433].

A cubic algorithm for recognizing totally unimodular matrices based on Seymour's decomposition of regular matroids [967] is given by Truemper [1035].

9.4 Birkhoff graphs and doubly stochastic matrices

By the definition of isomorphism of graphs the following holds:

> If a graph G_A is described by its adjacency matrix A, then a graph G_B is isomorphic to G_A if and only if there is a permutation of the vertices of G_A such that for the corresponding permutation matrix π,

$$\pi A \pi^{-1} = B. \tag{9.1}$$

Definition 9.4.1 *Let S_n be the set of all doubly stochastic matrices X of order n. Then*

$$S_n = \{X : X \in \mathbb{R}^{n \times n} : X \geq 0 \text{ and } Xe_n = X^T e_n = e_n\}$$

(i.e., the sum of each row and column is 1*).*

Definition 9.4.2 *Let $A \equiv_{ds} B$ (A is* ds-isomorphic *to B) if there is an $X \in S_n$ such that*

$$XAX^{-1} = B. \tag{9.2}$$

Clearly (9.2) is weaker than (9.1).

Definition 9.4.3 (Tinhofer [1026]) *A graph G_A is a* Birkhoff graph *if every doubly stochastic matrix X for which $XAX^{-1} = A$ is a convex sum of automorphisms of A.*

This notion can also be expressed in terms of extremal points of a polytope. Each real-valued matrix of order n can be identified with a point in $\mathbb{R}^{n \times n}$. Then S_n is a polytope (the so-called assignment polytope).

To determine whether two matrices A, B satisfy $A \equiv_{ds} B$, one has to solve the following linear-programming problem:

$$XAX^{-1} = B, \quad Xe_n = e_n, \quad X^T e_n = e_n, \quad X \geq 0. \tag{9.3}$$

The solution set of (9.3) is a (possibly empty) subpolytope of S_n, which in general has nonintegral extremal points.

The polytope S_n itself, however, has only integral extremal points (the set of permutation matrices of order n) because of a famous theorem of Birkhoff that every doubly stochastic matrix is a convex sum of permutation matrices.

Let $S_n(A)$ be the set of all doubly stochastic matrices that commute with A. Thus, G_A is a Birkhoff graph if and only if $S_n(A)$ is the convex hull of automorphisms of A. The isomorphism problem for such graphs can be solved in polynomial time.

A natural task is to find interesting subclasses of Birkhoff graphs. In [1026] strong tree-cographs (tree-cographs are a generalization of trees and cographs; see Definition 11.5.1) are shown to be Birkhoff graphs.

9.5 Forbidden sets of submatrices

The material in this section is taken primarily from [664], where the topic is treated at greater depth. A number of graph classes in the literature can be characterized in terms of sets of forbidden submatrices. The first class arose in the context of a problem called H-coloring: G is H-colorable if there is a mapping f from V_G to V_H such that for all edges xy in G, $f(x)f(y)$ is an edge of H.

Definition 9.5.1 *G is an* X-graph *if the vertices can be ordered $v_1, \ldots, v_n$ so that whenever v_iv_j and v_kv_l are edges, $v_{\min(i,k)}v_{\min(j,l)}$ is an edge.*

The H-coloring problem is solvable in polynomial time if G is an X-graph [486]; the problem of recognizing X-graphs in polynomial time is open. The definition translates directly into the following matrix characterization.

Remark 9.5.1 *G is an X-graph if and only if the vertices can be ordered so that its adjacency matrix does not contain either of the induced submatrices*

$$X_1 = \begin{pmatrix} 0 & 1 \\ 1 & 1 \end{pmatrix} \qquad \textit{and} \qquad X_2 = \begin{pmatrix} 0 & 1 \\ 1 & 0 \end{pmatrix}.$$

Grid-intersection graphs (mentioned in the model chapter (chapter 4)) are the bipartite intersection graphs defined by a set of horizontal and vertical line segments on a two-dimensional grid.

Theorem 9.5.1 (Hartman, Newman, Ziv [523]) *A bipartite graph is a grid-intersection graph if and only if there is an ordering of the bipartite adjacency matrix that contains no induced*

$$\begin{pmatrix} w & 1 & x \\ 1 & 0 & 1 \\ y & 1 & z \end{pmatrix}, \qquad \textit{with} \quad w, x, y, z \in \{0, 1\}.$$

Definition 9.5.2 *Given a matrix M with cost c_{ij} on entry $M[i,j]$, an ordering of the entries of M is a* Monge sequence *for M if for all $i < p$, $j < q$, if (i,j) precedes both (i,q) and (p,j), $c_{ij} + c_{pq} \leq c_{iq} + c_{pj}$.*

Monge sequences have been studied extensively, primarily because the assignment problem is easy if a Monge sequence is given. Although graph classes have not been explicitly defined for these classes of matrices, there have been papers that study graph problems if the adjacency matrix or bipartite adjacency matrix has one of these forms.

Theorem 9.5.2 (Hoffman [576]) *A $(0,1)$-matrix has a Monge sequence if and only if the complement matrix formed by switching all* 0*s and* 1*s has a* Γ*-free ordering.*

A number of variants on Monge sequences have been studied, motivated by variants on the assignment problem.

Definition 9.5.3 *A matrix $C = (c_{ij})$ is a* Monge matrix *if for all $i < p$, $j < q$, $c_{ij} + c_{pq} \leq c_{iq} + c_{pj}$.*

A $(0,1)$-matrix is a Monge matrix if and only if it can be permuted to avoid the five induced submatrices below:

$$P_1 = \begin{pmatrix} 1 & 0 \\ 0 & 1 \end{pmatrix}, P_2 = \begin{pmatrix} 1 & 1 \\ 0 & 1 \end{pmatrix}, P_3 = \begin{pmatrix} 1 & 0 \\ 1 & 1 \end{pmatrix}, P_4 = \begin{pmatrix} 1 & 0 \\ 0 & 0 \end{pmatrix}, P_5 = \begin{pmatrix} 0 & 0 \\ 0 & 1 \end{pmatrix}.$$

Klinz, Rudolf, and Woeginger [664] show that $(0,1)$ Monge matrices can be recognized very simply; after repeatedly deleting identical rows and columns, one must obtain one of four induced submatrices if the input is a Monge matrix.

Definition 9.5.4 *A matrix $C = (c_{ij})$ is a* bottleneck Monge matrix *if for all $i < p$, $j < q$, $\max\{c_{ij}, c_{pq}\} \leq \max\{c_{iq}, c_{pj}\}$.*

Transportation problems with the bottleneck objective function can be solved efficiently on bottleneck Monge matrices. Restricting attention to $(0,1)$-matrices leads to Definition 9.5.5.

Remark 9.5.2 *A $(0,1)$-matrix is bottleneck Monge if and only if it contains no induced*

$$B_1 = \begin{pmatrix} 1 & 0 \\ 0 & 1 \end{pmatrix}, \quad B_2 = \begin{pmatrix} 1 & 0 \\ 0 & 0 \end{pmatrix}, \quad B_3 = \begin{pmatrix} 0 & 0 \\ 0 & 1 \end{pmatrix}.$$

Definition 9.5.5 *A $(0,1)$-matrix is a* double-staircase matrix *if it has the consecutive* 0*s property for rows, and rows are sorted in increasing order of both the first and last* 0*s in the row.*

Theorem 9.5.3 (Klinz, Rudolf, Woeginger [665]) *A $(0,1)$-matrix can be permuted to be bottleneck Monge if and only if it can be permuted to be a double staircase matrix.*

Theorem 9.5.3 is used in [665] to design a polynomial-time algorithm for recognizing general (i.e., not necessarily (0,1)) matrices that can be permuted to be bottleneck Monge.

9.6 Eigenvalues and graphs

Since the adjacency matrix $A(G)$ of a graph G is real symmetric, its $|V(G)|$ eigenvalues are all real. In the following we denote by $\lambda_{\min}(G)$ and $\lambda_{\max}(G)$ the smallest and the largest eigenvalues of $A(G)$, respectively. In this section, assume that all graphs are connected.

There are a number of relationships between eigenvalues of the adjacency matrix and graph theoretic properties; see [272, 273, 199] for discussions of many of them. Two well-known examples of such relationships are given in the following.

Theorem 9.6.1

(i) *G is bipartite if and only if* $\lambda_{\min}(G) = -\lambda_{\max}(G)$;

(ii) *G is regular if and only if* $\lambda_{\max}(G) = \frac{2|E(G)|}{|V(G)|}$.

Eigenvalues give much information about regular graphs. In general, graphs with small maximum eigenvalue of the adjacency matrix have small diameter. Alon [15] shows that if the degree of a regular graph is fixed at k and the number of vertices n becomes large, the largest eigenvalue not equal to $\pm k$ must have absolute value at least $2\sqrt{k-1}$. Graphs that reach this bound are called *Ramanujan graphs.*

Definition 9.6.1 *A k-regular graph is a* Ramanujan graph *if the absolute value of its largest eigenvalue excluding k is* $2\sqrt{k-1}$.

Lubotzky, Phillips, and Sarnak [747] and Margulis [770] give methods for constructing Ramanujan graphs for various values of k and n. Ramanujan graphs are especially important for their expansion properties, as defined below.

Definition 9.6.2 *A k-regular graph on n vertices is an* (α, β, n, k)-expander *if for every subset X of vertices of size at most $\alpha|V|$, the set of neighbors of X has size at least $\beta|X|$.*

Expander graphs play an important role in parallel-algorithm design [11], cryptography [82], and other areas; for a more complete look at expanders see [746]. Although it was known that random graphs of sufficiently large size had good expansion properties, explicit expander graphs were difficult to construct. Ramanujan graphs have good expansion properties, as shown in the following theorem.

Theorem 9.6.2 (Kahale [636]) *A k-regular Ramanujan graph is a* $(k^{-1/\varepsilon}, (1-\mathcal{O}(\varepsilon))k/2, n, k)$*-expander, where the constant in the $\mathcal{O}$ term is a small absolute constant.*

Another class of graphs is defined by taking extremal graphs with respect to the number of different eigenvalues rather than the size of the largest eigenvalue. Let $cb(G)$ be the number of complete bipartite graphs needed to partition the edges of G. It can be shown that $cb(G)$ is at least as large as the number of different (positive or negative) eigenvalues of G. When $cb(G)$ is exactly equal to the larger of these two values, the graph is called eigensharp.

Definition 9.6.3 *G is* eigensharp *if $cb(G)$ is equal to the number of different positive eigenvalues or the number of different negative eigenvalues of the adjacency matrix of G.*

Kratzke, Reznick, and West [700] determine which graphs are eigensharp for a number of families of graphs, and study some closure properties for classes of eigensharp graphs.

The following classical eigenvalue bounds on the chromatic number, due to Wilf [1085] (the upper bound) and Hoffman [578] (the lower bound) are related to the characterization of bipartite graphs in Theorem 9.6.1 (i).

Theorem 9.6.3 [1085, 578] *For every graph G,*

$$1 + \frac{\lambda_{\max}(G)}{|\lambda_{\min}(G)|} \le \chi(G) \le 1 + \lambda_{\max}(G).$$

Mohar and Poljak [792] pointed out that, concerning the upper bound in Theorem 9.6.3, every graph G can be colored with $1 + \lambda_{\max}(G)$ colors in polynomial time. A very important use of eigenvalues is Lovász's ϑ-function [738]. A matrix F is called *feasible* for the graph G if F is real and symmetric, and $(F)_{i,j} = 1$ whenever $i = j$ or $v_i v_j \in E(G)$ (the other elements of F can be any real number). Let $\lambda_{\max}(F)$ be the largest eigenvalue of F. Then a characterization of Lovász's ϑ-function via eigenvalues is (see [740, 677])

$$\vartheta(G) = \min\{\lambda_{\max}(F) : F \text{ is a feasible matrix for } G\}.$$

Theorem 9.6.4 [476, 477]

(i) (The sandwich theorem) *For every graph G, $\omega(G) \le \vartheta(G) \le \chi(G)$.*

(ii) *$\vartheta(G)$ can be computed in polynomial time: For every $\varepsilon > 0$, one can compute in polynomial time a rational number r such that $|r - \vartheta(G)| < \varepsilon$.*

The polynomial-time algorithm claimed in Theorem 9.6.4 (ii) is based on the ellipsoid method. As a consequence, the basic algorithmic graph problems *clique number* and *chromatic number* are solvable in polynomial time for perfect graphs. We shall emphasize the role of eigenvalues here by remarking that this is the only known way to compute these parameters for perfect graphs in polynomial time.

Applications of eigenvalues and of the ϑ-function in combinatorial optimization have been intensively studied recently. See, e.g., the papers [446, 16, 645, 17]. The survey by Mohar and Poljak [792] gives an abundance of eigenvalue bounds for various combinatorial optimization problems.

Chapter 10

Distance Properties

Distance in graphs is one of the most important topics of algorithmic graph theory and has led to many important results on graphs and connections to other fields such as metric spaces. The first section gives further properties of distance-hereditary and parity graphs.

10.1 Distance-hereditary and parity graphs

We first recall some definitions and properties that come from previous chapters. Let G be a connected graph and $d_G(u,v)$ (or simply $d(u,v)$ if the graph G is fixed) denote its distance function. For the notion of an isometric path (subgraph) in G see Definition 1.1.4. For the notion of a distance-hereditary graph see Definition 3.1.2. Theorem 3.1.1 gives a characterization of distance-hereditary graphs in terms of crossing chords and Theorem 3.5.3 shows that the larger class of parity graphs represents a distance property. In connection with distance properties of graphs the intervals between vertices play an important role.

Definition 10.1.1 *For a graph $G = (V, E)$ and vertices $u, v \in V$, $u \neq v$, let*

$$I(u,v) = \{x : x \in V \text{ and } x \text{ is on a shortest path between } u \text{ and } v\}$$

denote the interval between u and v.

It is important to distinguish between *chordless paths* and *shortest paths*, i.e., *paths of minimum length*, between two vertices —a path of minimum length between two vertices is chordless but not vice versa.

Theorem 10.1.1 (Bandelt, Mulder [56]) *Let G be a connected graph with distance function d and interval function I. The following conditions are equivalent:*

(i) *G is distance hereditary;*

(ii) *For every two vertices u and v with $d(u,v) = 2$, there is no induced path between u and v of length greater than 2;*

(iii) *The house, holes, domino, and gem are not induced subgraphs of G, i.e., G is HHDG-free;*

(iv) *The house, holes, domino, and gem are not isometric subgraphs of G;*

(v) *The house, gem, and domino are not induced (or isometric) subgraphs of G, and $I(u,v) \cap I(v,w) = \{v\}$ implies $d(u,w) \geq d(u,v) + d(v,w) - 1$;*

(vi) *The gem is not an induced subgraph of G, and for every three vertices u, v, w, at least two of the following inclusions hold:*

$$\begin{array}{lcl} I(u,v) & \subseteq & I(u,w) \cup I(v,w), \\ I(u,w) & \subseteq & I(u,v) \cup I(v,w), \\ I(v,w) & \subseteq & I(u,v) \cup I(u,w); \end{array}$$

(vii) *For every four vertices u, v, w, x, at least two of the following distance sums are equal: $d(u,v) + d(w,x)$, $d(u,w) + d(v,x)$, $d(u,x) + d(v,w)$;*

(viii) *G satisfies condition* (vii), *and if in* (vii) *the smaller distance sums are equal, then the largest one exceeds the smaller ones by at most* 2.

The paper [56] contains another characterization of distance-hereditary graphs in terms of iterated neighborhood and module properties, which is used by other authors in algorithmic applications.

Hammer and Maffray [511] independently discovered the equivalence of conditions (i) and (iii) of Theorem 10.1.1 and gave a linear-time recognition algorithm for distance-hereditary graphs.

In [288] the notion of a *hanging of G by v* is used.

Definition 10.1.2 *Let $G = (V, E)$ be a connected graph.*

The hanging of G by v *is the function h_v that assigns to every vertex $u \in V$ the value $d_G(u,v)$.*

The level $h_v(u)$ *of the vertex u in the hanging h_v is the set of vertices that have the same distance to v as u. Let L_i denote the vertices with distance i from v.*

Given a hanging h_v of G, the horizontal part *of G with respect to h_v, denoted by $H(G,v)$, is the subgraph (V, E') of G in which E' is the set of edges in E between pairs of vertices having the same level in h_v, and the* vertical part *of G with respect to h_v, denoted $V(G,v)$, is the complement of $H(G,v)$ in G.*

Theorem 10.1.2 (D'Atri, Moscarini [288]) *G is a distance-hereditary graph if and only if for every hanging h of G and every pair of vertices $u, v \in L_i$, $i \leq 1$, that are connected in $G(V \setminus L_{i-1})$, $L_{i-1} \cap N(u) = L_{i-1} \cap N(v)$.*

This property is of importance for a polynomial-time solution of the Steiner tree problem on distance-hereditary graphs.

A recursive characterization of distance-hereditary graphs is given in Theorem 11.6.7.

Due to condition (iii) of Theorem 10.1.1, distance-hereditary graphs are HHDS-free and thus brittle and perfectly orderable graphs.

Cicerone and Di Stefano [226] generalize distance-hereditary graphs to *k-distance-hereditary graphs* by requiring that the lengths of every pair of induced paths between x and y differ by a factor of at most k.

For a generalization of distance-hereditary graphs in terms of the number of edges in Steiner trees (*Steiner distance-hereditary graphs*), see [291].

A metric characterization of parity graphs is given in [59], which immediately leads to a polynomial-time recognition algorithm.

Theorem 10.1.3 (Bandelt, Mulder [59]) *For a connected graph G the following conditions are equivalent:*

(i) *G is a parity graph;*

(ii) *For all vertices u, v, w, x, either all three sums* $d(u,v) + d(w,x)$, $d(u,w) + d(v,x)$, $d(u,x) + d(v,w)$ *have the same parity or two of them are equal;*

(iii) *If $d(u,v) + d(w,x) \leq d(u,w) + d(v,x) \leq d(u,x) + d(v,w)$ such that $d(u,w) + d(v,x)$ is odd, then either $d(u,v) + d(w,x)$ or $d(u,x) + d(v,w)$ is odd;*

(iv) *For every edge wx and vertices u, v such that* $d(u,w) = d(u,x) = d(u,v) + 1$, *$v$ is adjacent to w if and only if v is adjacent to x.*

The paper [59] contains a characterization of this class by forbidden isometric subgraphs.

10.2 Subclasses of distance-hereditary graphs

10.2.1 Cographs

Cographs are clearly distance hereditary since they are HHGD-free. Cographs are the 2-parity graphs [170]—chordless paths between nonadjacent vertices have length two.

Theorem 10.2.1 (Bandelt, Mulder [56]) *G is a cograph if and only if G is the disjoint union of distance-hereditary graphs with diameter at most 2.*

For further characterizations of cographs see Theorem 11.3.3.

10.2.2 Bipartite distance-hereditary graphs

These are the bipartite graphs that have at least two crossing chords in every cycle of length at least 6 [288]. In Definition 3.1.1 (k, l)-chordal graphs were introduced.

Definition 10.2.1 (Ausiello, D'Atri, Moscarini [35]) *A bipartite graph B is* (6,2)-chordal bipartite *if B is bipartite and every cycle in B of length at least* 6 *has at least two chords.*

Thus, the bipartite distance-hereditary graphs are exactly the (6,2)-chordal bipartite graphs.

Theorem 10.2.2 [56, 288] *Let G be a distance-hereditary graph. For every hanging h of G, the horizontal part of G with respect to h is a cograph and the vertical part of G with respect to h is $(6,2)$-chordal bipartite.*

10.2.3 Chordal distance-hereditary graphs

Definition 10.2.2 (Kay, Chartrand [647]) *A connected graph G is* ptolemaic *if for any four vertices u, v, w, x of G,*

$(*)$ $d(u,v)d(w,x) \leq d(u,w)d(v,x) + d(u,x)d(v,w)$.

In metric space, this inequality is known as the *ptolemaic inequality* (see [113]).

Theorem 10.2.3 (Howorka [588]) *Let G be a graph. The following conditions are equivalent:*

(i) *G is ptolemaic;*

(ii) *G is distance hereditary and chordal;*

(iii) *G is chordal and does not contain an induced gem.*

The original version of this theorem in [588] contains two more equivalent conditions. There is also a characterization in terms of separator properties.

Theorem 10.2.4 [588] *A connected graph G is ptolemaic if and only if for all distinct nondisjoint maximal cliques P, Q of G, $P \cap Q$ separates $P \setminus Q$ and $Q \setminus P$.*

The four-point condition $(*)$ of Definition 10.2.2 can be varied in several ways, e.g., as follows.

Theorem 10.2.5 [56] *The graph G is ptolemaic if and only if for every four vertices u, v, w, x, at least two of the distance sums*

$$d(u,v) + d(w,x), \quad d(u,w) + d(v,x), \quad d(u,x) + d(v,w)$$

are equal, and if the smaller sums are equal, then the largest one exceeds the smaller ones by at most 1.

10.2.4 Block graphs and related classes

Definition 10.2.3 [647] *G is a* block graph *if G is connected and every maximal 2-connected subgraph (i.e., block) is complete.*

Equivalently, block graphs can be seen as graphs constructed from trees by replacing each edge by a clique of arbitrary size; the cliques have at most one vertex in common.

Block graphs are ptolemaic, and [647, 587] contain some characterizations of them. A purely metric characterization is given in [587], showing the equivalence of conditions (i) and (ii) of the next theorem.

Theorem 10.2.6 (Howorka [587], Bandelt, Mulder [56]) *Let G be a graph. The following conditions are equivalent:*

(i) *G is a block graph;*

(ii) *For every four vertices u, v, w, x, the larger two of the distance sums $d(u,v) + d(w,x)$, $d(u,w) + d(v,x)$, $d(u,x) + d(v,w)$ are equal;*

(iii) *Neither $K_4 - e$ nor any chordless cycle C_n with $n \geq 4$ is an isometric subgraph of G.*

Block graphs are also known as *completed Husimi trees* (see [647]), and there is a close relationship to geodetic and weakly geodetic graphs.

Definition 10.2.4 (Ore [841]) *G is a* geodetic graph *if for every pair of vertices there is a unique path of minimum length between them.*

Definition 10.2.5 [647] *G is a* weakly geodetic graph *if for every pair of vertices of distance 2 there is a unique common neighbor of them.*

Definition 10.2.6 [587] *Let G be a graph.*

A cycle C of G is a b-cycle *of G if C is not contained in a clique of G.*

Let $l(C)$ denote the length of the cycle C. The bulge *of G is*

$$b(G) = \begin{cases} \min\{l(C) : C \text{ is a b-cycle in } G\} & \text{if there is a b-cycle in } G, \\ \infty & \text{otherwise.} \end{cases}$$

Theorem 10.2.7 [647, 587] *G is weakly geodetic if and only if $b(G) \geq 5$.*

Theorem 10.2.8 [647, 587] *Let G be a graph. The following conditions are equivalent:*

(i) *G is a block graph;*

(ii) *$b(G) = \infty$;*

(iii) *G is weakly geodetic and ptolemaic.*

As a consequence, **weakly geodetic $\cap$ ptolemaic $\subset$ geodetic.**
The similarity of block graphs with trees is also evident in the following theorem.

Theorem 10.2.9 *Let $G = (V, E)$ be a graph.*

(i) [166] *G is a tree if and only if G contains no triangles and G satisfies the following four-point condition: For every $u, v, x, y \in V$,*

$$(*)\ \ d(u,v) + d(x,y) \leq \max\{d(u,x) + d(v,y), d(u,y) + d(v,x)\};$$

(ii) [587] *G is a block graph if and only if G satisfies condition $(*)$.*

There are two other modifications of geodetic graphs.

Definition 10.2.7 (Mulder [805], Bandelt, Mulder [55]) *A graph is* interval regular of diameter 2 *if every pair of nonadjacent vertices has exactly two paths of length* 2 *between them.*

Definition 10.2.8 [998] *A graph is* bigeodetic *if every pair of vertices has at most two paths of minimum length between them.*

10.3 Interval conditions

For the definition of some of the subsequent classes, the following conditions are of importance. Recall that d is the distance function of the graph.

Triangle Condition For every three vertices u, v, w with $1 = d(v,w) < d(u,v) = d(u,w)$, there is a common neighbor x of v and w such that $d(u,x) = d(u,v) - 1$.

Quadrangle Condition For every four vertices u, v, w, z with $d(v,z) = d(w,z) = 1$ and $d(u,v) = d(u,w) = d(u,z) - 1$, there is a common neighbor x of v and w such that $d(u,x) = d(u,v) - 1$.

Bandelt, Chepoi, and Mulder [51, 61, 189] define weakly modular graphs as follows.

Definition 10.3.1 *A graph is* weakly modular *if its distance metric fulfills the triangle and quadrangle conditions.*

Definition 10.3.2 *Three vertices u, v, w form a* metric triangle *if the intervals $I(u,v)$, $I(v,w)$, and $I(w,u)$ pairwise intersect only in the common end vertices.*

Theorem 10.3.1 (Chepoi [189]) *The graph G is weakly modular if and only if for every metric triangle $\{u, v, w\}$ in G, all vertices of the interval $I(v,w)$ are at the same distance $k = d(u,v)$ from u.*

The number $k = d(u,v)$ is called the *size* of the metric triangle.

Distance-hereditary graphs and (6,2)-chordal bipartite graphs can be characterized in terms of conditions on the intervals $I(u,v)$, $u, v \in V$ [56]. In [58] the following interval conditions are introduced.

Definition 10.3.3 (Bandelt, Mulder [58]) *A graph satisfies the condition (I_k) for $k \in \{1, 2, 3\}$ if for every three vertices u, v, w, at least k of the following three inclusions hold:*

$$I(u, v) \subseteq I(u, w) \cup I(v, w);$$

$$I(u, w) \subseteq I(u, v) \cup I(v, w);$$

$$I(v, w) \subseteq I(u, v) \cup I(u, w).$$

Theorem 10.3.2 [58] *G is a block graph if and only if G satisfies (I_3).*

For condition (I_2), see Theorem 10.1.1.

Theorem 10.3.3 [58] *A chordal graph G satisfies (I_1) if and only if G is S_3-free.*

This follows from a more general theorem, which gives a connection between bridged graphs and (I_1). Note that bridged graphs are defined in terms of distances: A graph G is bridged if and only if G does not contain an isometric subgraph C_n, $n \geq 4$ (see Definition 3.6.1).

There are some classes that are directly defined by interval conditions.

Definition 10.3.4 (Howorka [589], Bandelt [48]) *Let G be a graph.*

> *G is* modular *if for every three vertices x, y, z there exists a vertex w that lies on a shortest path between every two of x, y, and z: $I(x, y) \cap I(x, z) \cap I(y, z) \neq \emptyset$ for all $x, y, z \in V$. (This means that in modular graphs every metric triangle has size 0, i.e., is degenerate.)*
>
> *G is* pseudomodular *if every metric triangle has size at most* 1.
>
> *G is* median *if $|I(x, y) \cap I(x, z) \cap I(y, z)| = 1$ for all triples x, y, z.*
>
> *G is* hereditary modular (pseudomodular, median, weakly modular) *if every isometric subgraph of G is modular (pseudomodular, median, weakly modular).*

Note that here the hereditary classes are defined in terms of isometric instead of induced subgraphs.

Examples of median graphs are trees, the hypercube, and the grid graphs. Median graphs have many characterizations, some of them in terms of certain subgraphs of the hypercube; a survey is given by Klavžar and Mulder in [661]. The polynomial-time recognition of median graphs was improved several times, and the current best time bounds are due to Hagauer, Imrich, and Klavžar [500], who have designed both $\mathcal{O}(m\sqrt{n})$ and $\mathcal{O}(n^{1.5} \log n)$ algorithms. Note that median graphs are a subclass of the so-called binary Hamming graphs, which have $\mathcal{O}(n \log n)$ edges; see [500].

Theorem 10.3.4 (Bandelt [48, 47], Mulder [804]) *A graph is hereditary median if and only if it is hereditary modular and does not contain an induced $K_{2,3}$.*

Theorem 10.3.5 [48]

(i) *Modular graphs are bipartite.*

(ii) *The hereditary modular graphs are the graphs in which all isometric cycles have length* 4 (*i.e., they contain no isometric holes*).

Theorem 10.3.6 (Bandelt, Mulder [57]) *For a connected graph G the following conditions are equivalent:*

(i) *G is hereditary pseudomodular;*

(ii) *G is pseudomodular and does not contain the house, the* 3*-sun* S_3*, the* C_5*, or the* C_6 *as an isometric subgraph;*

(iii) *G does not contain the house, the* 3*-sun* S_3*, or any hole* C_n*,* $n \geq 5$*, as an isometric subgraph.*

An $\mathcal{O}(n^4)$ recognition algorithm for hereditary pseudomodular graphs is given in [57]. Note that the following inclusions hold:

HHDS-free ⊂ hereditary pseudomodular;

S_3-free chordal ⊂ hereditary pseudomodular.

Theorem 10.3.7 (Chepoi [189]) *A graph is hereditary weakly modular if and only if it contains no house and no hole* C_n*,* $n \geq 5$*, as an isometric subgraph.*

For a characterization of the same graph class in terms of distance-preserving elimination orderings see Corollary 5.3.1.

The following inclusions are discussed in [192]:

bridged ⊂ hereditary weakly modular;

HH-free ⊂ hereditary weakly modular.

Theorem 10.3.8 (Bandelt [48]) *Let G be a graph. The following conditions are equivalent:*

(i) *G is bridged and bipartite;*

(ii) *G is modular and has no induced* C_6*;*

(iii) *G is hereditary modular;*

(iv) *G is bipartite and every isometric cycle in G has length* 4.

Condition (iv) of Theorem 10.3.8 implies

chordal bipartite ⊂ hereditary modular.

More precisely, a hereditary modular graph is chordal bipartite if and only if it does not contain certain wheels and double-wheels as isometric subgraphs (see [48]).

In [57] the pseudomodular graphs are defined in a different way and the following characterizations of them are given.

Lemma 10.3.1 [57] *For a connected graph G the following conditions are equivalent:*

(i) *G is pseudomodular;*

(ii) *$I(u,v) \cap I(v,w) = \{v\}$ implies $d(u,w) \geq d(u,v) + d(v,w) - 1$ for all vertices u, v, w;*

(iii) *If three intervals $I(u,v)$, $I(u,w)$, $I(v,w)$ have pairwise one vertex in common, then u, v, w are either identical or pairwise adjacent.*

Theorem 10.3.9 [57] *For a connected graph G the following conditions are equivalent:*

(i) *G is pseudomodular;*

(ii) *Every three pairwise intersecting disks of G have a nonempty intersection (i.e., G is a 3-Helly graph—a relaxation of the Helly property);*

(iii) *If $1 \leq d(u,w) \leq 2$ and $d(u,v) = d(v,w) = k \geq 2$ for vertices u, v, w of G, then there exists a vertex x such that $d(u,x) = d(w,x) = 1$ and $d(v,x) = k-1$.*

Conditions (ii) and (iii) of Theorem 10.3.9 and conditions (ii) and (iii) of Lemma 10.3.1 give polynomial-time recognition algorithms for pseudomodular graphs.

10.4 Absolute retracts of reflexive and bipartite graphs

Recall that a graph G is *reflexive* if every vertex of G has a loop. *Retracts* were introduced and studied by Hell [535]. Many other papers deal with retracts; see, e.g., [52]; for a survey see Pesch [869].

Definition 10.4.1

> *A* retraction *f from a graph $H = (V_H, E_H)$ to a subgraph $G = (V_G, E_G)$ is a mapping $f : V_H \to V_G$ such that for every edge $uv \in E_H$, $f(u)f(v) \in E_G$ and $f(w) = w$ for all $w \in V_G$ (i.e., an idempotent mapping that preserves or collapses edges). Then G is a* retract *of H.*
>
> *G is an* absolute retract *if G is a retract of every graph H containing G as an isometric subgraph, provided that $\chi(G) = \chi(H)$.*

Note that a retract G of H is necessarily an isometric subgraph of H. A retraction of a bipartite graph is always color preserving.

In [50, 52, 53, 63, 65, 321, 331, 537, 625, 817, 869, 886], characterizations were found for absolute reflexive and absolute bipartite retracts from which we mention the following equivalences.

Theorem 10.4.1 *Let G be a reflexive graph. The following conditions are equivalent:*

(i) *G is an absolute reflexive retract;*

(ii) *G is Helly;*

(iii) *G is dismantlable and clique Helly;*

(iv) *G is pseudomodular and neighborhood Helly.*

Note the close connection to Theorem 8.5.5. For the bipartite case there are similar properties.

Definition 10.4.2 *Let $B = (X, Y, E)$ be a bipartite graph with $V = X \cup Y$.*

A half-disk *(with center u and radius k) is*

$$\{v : v \in V \text{ and } d(u, v) \leq k \text{ and } d(u, v) \text{ is even}\} \text{ or}$$
$$\{v : v \in V \text{ and } d(u, v) \leq k \text{ and } d(u, v) \text{ is odd}\}.$$

The half-disk hypergraph *of a bipartite graph $B = (X, Y, E)$ has vertex set $X \cup Y$ and its half-disks in B as hyperedges.*

Theorem 10.4.2 *Let B be a bipartite graph. The following conditions are equivalent:*

(i) *B is an absolute bipartite retract;*

(ii) *The half-disk hypergraph of B has the Helly property;*

(iii) *B is dismantlable and the open neighborhood hypergraph $\mathcal{N}_O(B) = \mathcal{N}_X(B) \cup \mathcal{N}_Y(B)$ of B has the Helly property;*

(iv) *B is modular and the open neighborhood hypergraph $\mathcal{N}_O(B)$ has the Helly property.*

Bandelt, Farber, and Hell [53] used standard constructions such as $B(G)$ and $B_C(G)$ in order to derive connections between results about absolute reflexive retracts and absolute bipartite retracts.

Thus, G is dismantlable if and only if $B(G)$ is dismantlable, and G is clique Helly if and only if $B(G)$ is open-neighborhood Helly.

There are close connections to hypergraph properties, since $B(G)$ is the bipartite vertex-neighborhood incidence graph of the neighborhood hypergraph of G and $B_C(G)$ is the bipartite vertex-clique incidence graph of the clique hypergraph of G; see Definition 1.3.11 and the remarks after it.

The paper [660] characterizes the absolute retracts of split graphs. [64] characterizes n-chromatic absolute retracts in terms of convexity and Helly-type properties as well as in terms of elimination properties.

Modularity is preserved under retraction (see [52]): Every retract of a modular graph is modular.

Theorem 10.4.3 (Bandelt, Mulder [57])

(i) *Each hereditary modular graph is an absolute retract of a bipartite graph.*

(ii) *Each absolute retract of a bipartite graph is a modular graph.*

10.5 Convexity

Convexity is a very general notion in (not necessarily metric) spaces, which can be defined as follows.

Definition 10.5.1 *A* convex structure *consists of a set X together with a collection $\mathcal{C}$ of subsets of X (the* convex sets) *such that*

the empty set and the set X are convex, and

the intersection of convex sets is convex

(called alignment *of X in* [375]). *The following notions are defined in the standard fashion:*

The convex hull *of a set $S \subseteq X$ is the smallest convex set containing S,*

An element $x \in S$ is an extreme point *of the convex set S if $S \setminus \{x\}$ is also convex;*

A convex geometry *is a convex structure fulfilling the following additional property—the Minkowski–Krein–Milman property: Every convex set is the convex hull of its extreme points.*

See [329] for discussion of the Minkowski–Krein–Milman property. There is another way to describe convex geometries.

Definition 10.5.2 *Let E be a finite set and σ a* hull operator *on E, i.e.,*

$X \subseteq \sigma(X)$ *for all* $X \subseteq E$,

$\sigma(X) \subseteq \sigma(Y)$ *for all* $X \subseteq Y \subseteq E$,

$\sigma(\sigma(X)) = \sigma(X)$ *for all* $X \subseteq E$.

σ is the hull operator of a matroid *if in addition the following* exchange property *is fulfilled:*

$(*)$ *If $y \notin \sigma(X)$ and $y \in \sigma(X \cup \{x\})$, then $x \in \sigma(X \cup \{y\})$.*

σ is the hull operator of an antimatroid *if instead of condition $(*)$ the following* antiexchange property *is fulfilled:*

If $y, z \notin \sigma(X)$ and $z \in \sigma(X \cup \{y\})$, then $y \notin \sigma(X \cup \{z\})$.

It is known that convex geometries are exactly the alignments fulfilling the antiexchange property.

For convex geometries the following fundamental result holds.

Theorem 10.5.1 (Farber, Jamison [375]) *If $(X, \mathcal{M})$ is a convex geometry, then $S \in \mathcal{M}$ if and only if there is an ordering $(x_1, \ldots, x_k)$ of $X \setminus S$ such that x_i is an extreme point of $S \cup \{x_i, \ldots, x_k\}$ for each $i \in \{1, \ldots, k\}$.*

This is similar to vertex elimination orderings of graphs, and it is natural to ask which classes of graphs having characterizations in terms of elimination orderings have a characterization in terms of a convex geometry and vice versa. We will see that chordal graphs are one such class.

For papers studying the connection between graph properties and convexity properties in metric spaces see, e.g., [612, 188, 983, 375, 1065].

There are several interesting types of convexity in graphs. The g-convexity and the m-convexity seem to be studied most extensively; see, e.g., [188, 337, 339, 374, 375, 376, 611, 612, 614, 804, 982, 984, 985, 56].

Definition 10.5.3 *Let $G = (V, E)$ be a graph.*

> *A subset $S \subseteq V$ is* geodesically convex (g-convex) *if for every two vertices $u, v \in S$, all vertices on shortest paths between u and v are also contained in S: If $u, v \in S$ then $I(u, v) \subseteq S$.*
>
> *A subset $S \subseteq V$ is* monophonically convex (m-convex) *if S contains every vertex on every chordless path between vertices of S.*

Ptolemaic graphs can be characterized by convexity as Jamison [613, 614] and Soltan [982, 983] have shown.

Theorem 10.5.2 *A graph is ptolemaic if and only if its associated g-convexity space is an antimatroid.*

Some results of Soltan [982] were extended in [375] in the following way.

Theorem 10.5.3 [982, 375] *Let G be a graph. The following conditions are equivalent:*

(i) *G is a disjoint union of ptolemaic graphs;*

(ii) *G is chordal and every 5-cycle has at least 3 chords;*

(iii) *G is chordal and contains no induced 3-fan (i.e., gem);*

(iv) *G is chordal and all chordless paths are shortest paths;*

(v) *The geodesic alignment of G is a convex geometry;*

(vi) *G is chordal and the monophonic and geodesic alignments of G are identical.*

Farber and Jamison [375] give a characterization of strongly chordal graphs in terms of convexity. The next theorem characterizes chordal graphs in terms of m-convexity.

Theorem 10.5.4 [375] *The following conditions are equivalent:*

(i) *G is chordal;*

(ii) *For all $v \in V$, the disk $D_1(v) = N[v]$ is m-convex;*

(iii) *For all $v \in V$ and all $j > 0$, the disk $D_j(v)$ with radius j is m-convex;*

(iv) *For all $K \subseteq V$ that are m-convex, $N[K]$ is m-convex;*

(v) *For all $K \subseteq V$ with K m-convex and all $j > 0$, the disks $D_j(K)$ are m-convex.*

For a slightly different convexity characterization of chordal graphs in terms of g-convexity and m-convexity see [974] (where these convexity types are called c-convexity and s-convexity).

A generalization in terms of *peakless functions* on graphs is studied by Chepoi [190], which leads to a characterization of *totally convex sets* (a notion known from the geometry of geodesics; see [171, 172]) as level sets of peakless functions. Chordal and ptolemaic graphs have a characterization in terms of peakless functions.

For g-convexity the conditions (ii)–(v) of Theorem 10.5.4 are not equivalent. [984] and [376] study the different cases, which are generalizations of chordal graphs since g-convexity is weaker than m-convexity.

Theorem 10.5.5 (Soltan, Chepoi [984], Farber, Jamison [376]) *Let G be a graph. The following conditions are equivalent:*

(i) *G is bridged;*

(ii) *The disk $D_j(K)$ is g-convex for every g-convex set K and every $j > 0$;*

(iii) *$D_j(v)$ is g-convex for every $v \in V$ and $j > 0$ and every 5-cycle in G has a chord.*

This result leads to a polynomial-time recognition algorithm for bridged graphs, when combined with the following result.

Theorem 10.5.6 [376] *Let G be connected and $S \subseteq V$. Then one can determine in $\mathcal{O}(n^3)$ steps whether S is g-convex.*

Bridged graphs can be recognized in $\mathcal{O}(n^4)$ steps, as shown in [376]. Recall that fast recognition of chordal graphs is connected with the construction of perfect elimination orderings. Similar schemes also exist for bridged graphs; see Theorem 5.8.2 and the remarks after this theorem [191].

Bridged graphs have a characterization in terms of the following type of convexity.

Definition 10.5.4 [374] *In a graph $G = (V, E)$, a vertex set $K \subseteq V$ is m_3-*convex *if K contains every vertex on every chordless path of length at most 3 joining vertices in K.*

As usual, let $N_i(v_i)$ denote the open neighborhood of v_i in G_i.

Theorem 10.5.7 (Farber [374]) *Let G be a graph. The following conditions are equivalent:*

(i) *G is bridged;*

(ii) *G has an ordering $(v_1, \ldots, v_n)$ such that for each i, $1 \le i \le n$, $N_i(v_i)$ is connected and m_3-convex in G_{i+1}.*

The following modification of m-convexity turns out to be useful for HHD-free graphs and is different from the notion of m_3-convexity of Definition 10.5.4.

Definition 10.5.5 (Dragan, Nicolai, Brandstädt [329]) *Let $G = (V, E)$ be a graph.*

A subset $S \subseteq V$ is m^3-convex *if for every pair of vertices x, y of S, each induced path of length at least 3 connecting x and y is completely contained in S.*

The following theorem is another example of the close relationship between chordal and HHD-free graphs.

Theorem 10.5.8 [329] *Let $G = (V, E)$ be a graph. The following conditions are equivalent:*

(i) *G is HHD-free;*

(ii) *For all $v \in V$, the disk $D_1(v) = N[v]$ is m^3-convex;*

(iii) *For all $v \in V$ and all $j > 0$, $D_j(v)$ is m^3-convex;*

(iv) *For all connected sets $S \subseteq V$, the disk $D_1(S)$ is m^3-convex;*

(v) *For all connected sets $S \subseteq V$ and all $j > 0$, the disks $D_j(S)$ are m^3-convex.*

There is also a characterization of weak bipolarizable graphs, i.e., HHDA-free graphs, in terms of m^3-convexity. Recall that semisimplicial vertices (Definition 5.2.1) are vertices that are not midpoints of any P_4.

Theorem 10.5.9 [329] *Let $G = (V, E)$ be a graph. The following conditions are equivalent:*

(i) *G is HHDA-free;*

(ii) *In every induced subgraph F of G, each nonsemisimplicial vertex lies on an induced path of length at least 3 between semisimplicial vertices of F;*

(iii) *Each m^3-convex set of G is the hull of its semisimplicial vertices, i.e., the family of the m^3-convex sets of G is a convex geometry;*

(iv) *A set S of G is m^3-convex if and only if there is an ordering $(v_1, \ldots, v_k)$ of $V \backslash S$ such that for each $i \in \{1, \ldots, k\}$, the vertex v_i is semisimplicial in $G(\{v_i, \ldots, v_k\} \cup S)$.*

Bipartite distance-hereditary graphs have the following nice convexity property.

Theorem 10.5.10 (Bandelt, Mulder [56]) *The system of g-convex sets in a bipartite distance-hereditary graph has the Helly property.*

The median graphs can be characterized in terms of convexity. We will assume that singletons are convex.

Definition 10.5.6 *Let X together with the family of subsets $\mathcal{C}$ be a convex structure.*

A half-space *is a convex set with a convex complement.*

A convex structure fulfills the separation property S_4 *if disjoint convex sets extend to complementary half-spaces.*

A convex structure has Helly number $\leq n$ *if each finite collection of convex sets meeting* n *by* n *has a nonempty intersection.*

A convex structure is median *if it has the separation property* S_4 *and Helly number* ≤ 2.

In [66] median graphs are described as the graphs whose geodesic convex structure is median. See [62] for characterizations of *quasi-median graphs*, which are a generalization of median graphs in terms of convexity or algebraic properties. In [60] another interesting generalization of median graphs called *pseudomedian graphs* is given.

Definition 10.5.7 (Bandelt, Mulder [60]) *A graph* G *is* pseudomedian *if for every triple* u, v, w *of vertices of* G *either there exists a unique vertex on shortest paths between each pair of them (if their mutual distances sum up to an even number) or there exists a unique triangle whose edges lie on shortest paths between the three pairs of* u, v, w, *respectively (if the distance sum is odd).*

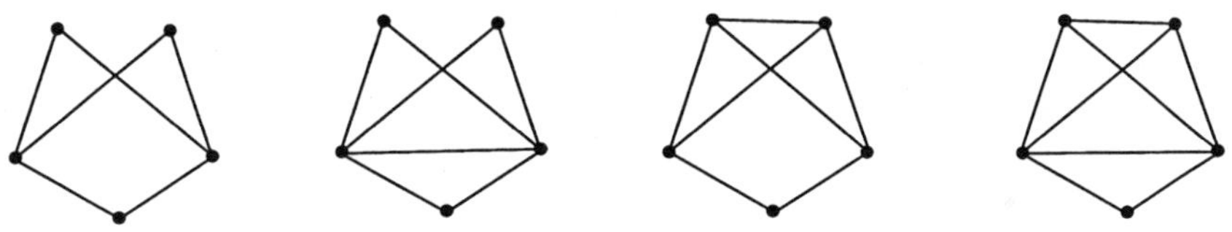

Figure 10.1: The graphs $K_{2,3}$, $K_2 * 3K_1$, $\overline{P_2 \cup P_3}$, and $K_2 * \overline{P_3}$.

Theorem 10.5.11 [60] *Let* G *be a connected graph. The following conditions are equivalent:*

(i) G *is a pseudomedian graph;*

(ii) G *is a pseudomodular graph and none of the graphs* $K_2 * 3K_1$, $K_{2,3}$, $\overline{P_2 \cup P_3}$, $K_2 * \overline{P_3}$ *is an induced subgraph of* G *(see Figure* 10.1*);*

(iii) *If* $1 \leq d(v, w) \leq 2$ *and* $d(u, v) = d(u, w) = k \geq 2$ *for vertices* u, v, w *of* G, *then there is a unique vertex* x *adjacent to* v *and* w *with* $d(u, x) = k - 1$.

10.6 Powers of graphs

Recall that the kth power of a graph $G = (V, E)$ is the graph $G^k = (V, E^k)$ with $uv \in E^k$ if and only if $d(u, v) \leq k$.

The main purpose of this section is to collect some closure properties of graph classes under the power operation. A graph class $\mathcal{C}$ is *closed under powers* if for every $G \in \mathcal{C}$ and every $k \in \mathbb{N}$, $G^k \in \mathcal{C}$. $\mathcal{C}$ is *strongly closed under powers* if for every $k \in \mathbb{N}$, if $G^k \in \mathcal{C}$, then $G^{k+1} \in \mathcal{C}$.

It should be mentioned that there are many other results on powers of graphs that are not of this form. Among the first ones was the characterization of the class of graphs that are the squares of trees [930]; see also [803]; more recently, Kearney and Corneil [648] give a polynomial-time recognition algorithm for graphs that are powers of trees. Obviously, a graph G' is the square of a graph G if and only if G' is the line graph $L(\mathcal{N}(G))$ of the closed neighborhoods $\mathcal{N}(G)$ of G. Harary and McKee [522] characterize the squares of chordal multigraphs.

If $H^2 = G$, then H is called a *square root of* G. Note that in the case of chordal multigraphs the square roots are unique [522]. A characterization of graphs that are squares of another graph is contained in [803]. The problem of recognizing such square graphs was shown to be $\mathbb{NP}$-complete by Motwani and Sudan [800]. For special cases the problem is easier; tree square roots and square roots of planar graphs can be found in linear time [729].

Our next topic is closure properties with respect to powers. The following theorem is of crucial importance.

Theorem 10.6.1 (Duchet [335]) *If G^k is chordal, then G^{k+2} is chordal.*

Thus, odd powers of chordal graphs are chordal. Since the square G^2 of homogeneously orderable graphs (and thus also the square of dually chordal graphs and of distance-hereditary graphs) is chordal (see Theorem 8.4.2) we have the following corollary.

Corollary 10.6.1

(i) *Even powers of homogeneously orderable graphs (and thus of dually chordal graphs and of distance-hereditary graphs) are chordal.*

(ii) *All powers of graphs that are chordal and homogeneously orderable (and thus also of doubly chordal graphs, including trees) are chordal.*

In particular, all powers of doubly chordal graphs are doubly chordal [146].

For trees, the chordality of all powers has been rediscovered several times (see, e.g., [729]). The fact that the class of chordal graphs is closed under odd powers was shown in [46]; Theorem 10.6.1, however, is stronger.

There is a characterization of the chordal graphs whose squares are chordal. A sun S_r with independent vertex set $\{w_1, \dots, w_r\}$ and inner vertices $\{u_1, \dots, u_r\}$ is *suspended* if there is a vertex $x \notin \{u_1, \dots, u_r, w_1, \dots, w_r\}$ such that x is adjacent to at least one pair of vertices w_i, w_j with $i \notin \{j+1, j-1\} \bmod r$.

Theorem 10.6.2 (Laskar, Shier [704]) *For a chordal graph G, the graph G^2 is chordal if and only if each sun S_r of G with $r > 3$ is suspended.*

There is also the following characterization using the line graph of the clique hypergraph of G.

Theorem 10.6.3 [331, 146] *For a chordal graph G, the graph G^2 is chordal if and only if the clique graph $K(G) = L(\mathcal{C}(G))$ is chordal.*

Note that this result appears later in [1076]. For strongly chordal graphs, i.e., sun-free chordal graphs, the following stronger theorem holds.

Theorem 10.6.4 (Lubiw [743]) *All powers of strongly chordal graphs are strongly chordal.*

See also the paper [279] by Dahlhaus and Duchet. This result has been improved as follows.

Theorem 10.6.5 (Raychaudhuri [899]) *The class of strongly chordal graphs is strongly closed under powers.*

There is an extension using special suns.

Theorem 10.6.6 (Wallis, Wu [1076]) *Let G be a chordal graph.*

(i) *If G is S_3-free and G^2 is chordal, then for all $k \geq 1$, G^k is S_3-free.*

(ii) *If G is odd-sun free and G^2 is chordal, then for all $k \geq 1$, G^k is odd-sun free.*

Bandelt and Mulder [57] show that powers of S_3-free chordal graphs are pseudomodular. This follows from Theorem 7.2.4 and the following.

Theorem 10.6.7 [57] *Every power of a pseudomodular graph is pseudomodular.*

A systematic investigation of some classes closed and strongly closed under powers has been made by Flotow [391], which gives a generalization of Theorem 10.6.1. A set family $\mathcal{S}$ is called *connected* if for every two properly intersecting sets $A, B \in \mathcal{S}$ (i.e., $A \cap B \neq \emptyset$), $A \cup B \in \mathcal{S}$ holds.

Theorem 10.6.8 [391] *Let $\mathcal{S}$ be a connected family of sets. If G^k is an intersection graph with respect to $\mathcal{S}$, then G^{k+2} is also an intersection graph with respect to $\mathcal{S}$.*

Examples of such set families are $\mathcal{S}$ = subtrees of a tree and $\mathcal{S}$ = arcs of a circle. Flotow [391] sharpens a result on cocomparability graphs.

Theorem 10.6.9 [391] *The class of cocomparability graphs is strongly closed under powers.*

This improves the result of [286], where it was shown that the same class is closed under powers.

In [391, 392] the powers of d-trapezoid graphs (see Definition 4.7.6) are considered.

Theorem 10.6.10 (Flotow [391, 392]) *If G^k is a d-trapezoid graph, then G^{k+1} is a d-trapezoid graph.*

Corollary 10.6.2 *The classes of interval graphs and trapezoid graphs are strongly closed under powers.*

Theorem 10.6.11 (Raychaudhuri [898, 899])

(i) *The class of proper interval graphs is strongly closed under powers.*

(ii) *The class of circular-arc graphs is closed under powers.*

It is open whether circular-arc graphs are also strongly closed under powers. A partial answer to this question is given by the following theorem.

Theorem 10.6.12 (Flotow [391, 393])

(i) *If $diam(G^k) \geq 4$ and G^k is a circular-arc graph, then G^{k+1} is a circular-arc graph.*

(ii) *If G^k is a circular-arc graph, then G^{k+2} is a circular-arc graph (see Theorem 10.6.8).*

It is shown in [391, 393] that the class of proper circular-arc graphs is closed under powers. An open question is whether this class is strongly closed under powers.

There is an interesting fact about ATs in powers of graphs.

Theorem 10.6.13 [898] *If G^k contains no AT then G^{k+1} contains no such triple.*

This means that the class of AT-free graphs is strongly closed under powers.

Some examples of classes not closed under powers are given in [391]—including the permutation graphs and cointerval and tolerance graphs, which are not closed under square.

Powers of distance-hereditary graphs are investigated in [54]. Recall that distance-hereditary graphs are exactly the HHDG-free graphs.

Theorem 10.6.14 (Bandelt, Henkmann, Nicolai [54]) *Let G be a distance-hereditary graph. Then the following properties hold:*

(i) *Every power of G is HHD-free;*

(ii) *Even powers of G are chordal.*

Claim (ii) is part of Corollary 10.6.1. Furthermore, the following results hold.

Theorem 10.6.15 [54]

(i) *If G is distance-hereditary, then for even k, G^k contains no induced $(2k+2)$-fan.*

(ii) *If G is ptolemaic (i.e., chordal and distance hereditary) or if k is odd then G^k contains no induced $(k+2)$-fan.*

It is mentioned in [54] that powers of distance-hereditary graphs are cop-win and pseudomodular graphs.

In [330] it is shown that odd powers of HHD-free graphs are HHD-free; in fact, a stronger condition is fulfilled.

Theorem 10.6.16 (Brandstädt, Le, Szymczak [154]) *Let G be a graph and $k \geq 1$ an integer. If G^k is HHD-free (HHDA-free), then G^{k+2} is HHD-free (HHDA-free).*

Dragan, Nicolai, and Brandstädt [330] show that odd powers of weak bipolarizable graphs are chordal.

There are many results on common elimination orderings for powers of graphs. If the ordering is a perfect elimination ordering, then see, e.g., Theorem 5.1.4 for odd powers of chordal graphs and Theorem 5.1.8 for even powers of distance-hereditary graphs. If the ordering is a semiperfect elimination ordering, then see Theorem 5.2.4 for all powers of distance-hereditary graphs and Theorem 5.2.5 for odd powers of HHD-free graphs.

Several of the closure properties are related to hypergraph properties. In [49] the following connection between the Helly property and powers is given.

Theorem 10.6.17 (Bandelt [49])

(i) *A reflexive graph G (i.e., with loops on all vertices) is a Helly graph if and only if all powers G^k, $k \geq 1$, of G are neighborhood Helly.*

(ii) *A bipartite (irreflexive) graph B is Helly if and only if all powers of B are neighborhood Helly.*

The fact that every power of a Helly graph must be Helly and every power of a dually chordal graph (doubly chordal graph) is dually chordal (doubly chordal) was mentioned in [321].

Chapter 11

Algebraic Compositions and Recursive Definitions

In many cases we can reverse an elimination process and think of using the same property as a method for constructing a graph. However, recursive definitions are not completely described in terms of reversing elimination orderings, and it is worthwhile to devote a chapter to algebraic compositions and recursive definitions of graphs.

11.1 Trees, k-trees, and partial k-trees

Trees have a recursive description.

Proposition 11.1.1

(i) *A one-vertex graph is a tree.*

(ii) *If* $T_1 = (V_1, E_1)$, $T_2 = (V_2, E_2)$ *are trees and* $V_1 \cap V_2 = \emptyset$, $x \in V_1$, $y \in V_2$, *then* $T = (V_1 \cup V_2, E_1 \cup E_2 \cup \{xy\})$ *is a tree.*

(iii) *There are no further trees.*

It is well known that this concept of trees coincides with that given in Definition 1.1.7. There are many generalizations of this recursive definition. One of them gives the notion of k-trees. For this purpose we recall that trees can be recursively defined in another way; see the last part of Proposition 1.1.1. The following definition of k-trees can be viewed as generalizing the notion of repeatedly adding degree-1 vertices.

Definition 11.1.1

*A clique with k vertices is a k-*tree.

If $T = (V, E)$ *is a k-tree and C is a clique of T with k vertices and* $x \notin V$, *then* $T' = (V \cup \{x\}, E \cup \{cx : c \in C\})$ *is a k-*tree.

There are no further k-trees.

Corollary 11.1.1

(i) *Every k-tree is chordal.*

(ii) *For a k-tree T, $\omega(T) = k$ if T is a k-clique and $\omega(T) = k+1$ otherwise.*

(iii) *The 1-trees are the trees.*

Properties of k-trees were studied by several authors, such as Beineke and Pippert [78, 79, 80, 872], Moon [797], and Rose [927].

The following notion of a partial k-tree turned out to be of crucial importance for algorithmic applications.

Definition 11.1.2 *A graph is a* partial k-tree *if it is a spanning subgraph of a k-tree.*

It is clear that every graph with n vertices is a partial n-tree. The problem of determining a smallest k such that G is a partial k-tree, however, is $\mathbb{NP}$-complete.

Theorem 11.1.1 (Arnborg, Corneil, Proskurowski [26]) *$\{(G,k) : G$ is a partial k-tree$\}$ is $\mathbb{NP}$-complete.*

The partial 2-trees correspond to series-parallel graphs as discussed in Theorems 11.2.1, 11.2.2, and 11.2.3. For this reason, we discuss the special cases of partial 2-trees and partial 3-trees in the next section.

It turns out that partial k-trees are closely connected to the concept of *treewidth* introduced by Robertson and Seymour.

Definition 11.1.3 *A* tree decomposition *of a graph $G = (V_G, E_G)$ is a pair $(T, \mathcal{X})$ consisting of a tree $T = (V_T, E_T)$ and $\mathcal{X} = \{X_t : X_t \subset V_G, t \in V_T\}$ for which*

$$\bigcup_{t \in V_T} X_t = V_G,$$

for each edge $uv \in E_G$ of the graph G there is a vertex t of T such that $\{u,v\} \subseteq X_t$, and

if $t' \in V$ is on the path between t and t'' in T then $X_t \cap X''_t \subseteq X'_t$.

The width *of a tree decomposition is*

$$tw(G,(T,\mathcal{X})) = \max\{|X_t| - 1 : t \in V_T\}.$$

The treewidth *of G is*

$$tw(G) = \min\{tw(G,(T,\mathcal{X})) : (T,\mathcal{X}) \text{ a tree decomposition of } G\}.$$

Theorem 11.1.2 (Scheffler [948], Wimer [1086]) *G has treewidth at most k if and only if G is a partial k-tree.*

For fixed k, partial k-trees can be recognized in polynomial time. The first algorithm is due to Arnborg, Corneil, and Proskurowski [25], and runs in $\mathcal{O}(n^{k+2})$ time. This time bound was improved to $\mathcal{O}(n^2)$ with an astronomically high constant due to results of Robertson and Seymour. This is done using nonconstructive methods, where only the existence of such an algorithm is shown. Finally, a linear-time algorithm was designed by Bodlaender [120] for all fixed k.

Theorem 11.1.3 [120] *For fixed k, there is a linear-time algorithm that tests whether a given graph G has treewidth at most k, and if so, outputs a tree-decomposition of G with treewidth at most k.*

Partial k-trees play a very important role in algorithmic graph theory, since many algorithmic graph problems are solvable in polynomial time (and quite often even in linear time) for partial k-trees with fixed k using dynamic programming by Arnborg and Proskurowski [29] or description methods from mathematical logic (monadic second-order logic; see, e.g., the survey given by Courcelle [261] and [964, 965, 28, 949, 950, 951, 115]) and algebra; see the paper of Arnborg et al. [27].

There is another important characterization of treewidth in terms of chordal graphs and clique size, as given below.

Definition 11.1.4 *A* triangulation *of a graph $G = (V, E)$ is a chordal graph $G' = (V, E')$ with $E \subseteq E'$.*

Theorem 11.1.4 *The treewidth of G is equal to the minimum value of $\omega(G') - 1$ over all triangulations G' of G.*

Due to the following result of Robertson and Seymour, Theorem 11.1.3 has important consequences.

Theorem 11.1.5 (Robertson, Seymour [909, 912, 914]) *A class of graphs that is closed under taking minors has bounded treewidth if and only if it does not contain all planar graphs.*

Theorem 11.1.6 [120] *Every class of graphs that is closed under taking minors and does not contain all planar graphs has a linear time recognition algorithm.*

Good surveys on treewidth are given by Bodlaender [119] and Kloks [668]. Due to the many applications of treewidth, especially for the case of small treewidth (see also the recent dissertation of de Fluiter [294] on this topic), the problem of determining the treewidth of a graph is important. Several interesting graph classes turn out to be partial k-trees for small k; see, e.g., Bodlaender [114, 116, 117]. We first collect some of them.

Definition 11.1.5 *G is a* cactus *if every edge is part of at most one cycle in G.*

Cactus graphs are outerplanar since they cannot contain K_4 or $K_{2,3}$ as a minor. A more general class is the class of almost k-trees.

Definition 11.1.6 (Coppersmith, Vishkin [242], Gurevich, Stockmeyer, and Vishkin [478]) *G is an* almost tree(k) *if every biconnected component has the property that there are at most k edges not in a spanning tree of this biconnected component.*

Thus, G is a cactus if and only if G is an almost tree(1).

Theorem 11.1.7 (Bodlaender [114]) *Almost trees(k) have treewidth $\leq k+1$ (and thus cactus graphs have treewidth ≤ 2).*

Since a graph can be divided into its biconnected components in linear time, almost trees(k) can be recognized in polynomial time for fixed k.

The fact that Halin graphs are partial 3-trees was observed by several authors, such as Wimer [1086], Wimer and Hedetniemi [1087], Elmallah and Colbourn [350], and Bodlaender [117].

Theorem 11.1.8 (Bodlaender [117]) *Every k-outerplanar graph is a partial $(3k-1)$-tree.*

For some classes for which treewidth is not constant bounded, the treewidth can be determined in polynomial time. A trivial case is the chordal graphs, where the treewidth is simply the maximum clique size minus 1. For other classes it is known that the problem of determining the treewidth is NP-hard.

Theorem 11.1.9 *Treewidth can be determined in polynomial time for the following classes:*

- *circle graphs* [666] (*hence, for cographs* [122], *for permutation graphs* [121], *and for distance-hereditary graphs* [891, 19, 159]),
- *circular-arc graphs* [1008],
- *co-comparability graphs of bounded dimension* [675],
- *chordal bipartite graphs* [670],
- *weakly chordal graphs* [140].

Treewidth is NP-hard for co-bipartite graphs [26], *dually chordal graphs* [143], *and graphs with maximum degree at most* 9 [123].

The *pathwidth* (the *proper pathwidth*) of a graph was defined in an analogous way to Theorem 11.1.4. Pathwidth (proper pathwidth) of a graph G can be characterized as the smallest maximum clique size minus 1 of an interval graph (a proper interval graph) G' that contains G. For some VLSI applications of embedding graphs into interval graphs with small clique size see the survey paper of Möhring [789].

There is an analogous result to Theorem 11.1.5 for pathwidth.

Theorem 11.1.10 [909, 912, 914] *A class of graphs that is closed under taking minors has bounded pathwidth if and only if it does not contain all forests.*

Note that the pathwidth of a graph coincides with other parameters. Kirousis and Papadimitriou [658] show that the so-called node-search number of a graph G is the same as the pathwidth of G. Thus, it follows from a result of Kirousis and Papadimitriou [659] that determining the pathwidth of a graph is $\mathbb{NP}$-complete (see Theorem 11.1.15 for more results in this direction).

There are several interesting classes for which treewidth coincides with pathwidth. This is a good indicator of linear structure. One interesting case is the class of AT-free graphs.

Theorem 11.1.11 (Möhring [790], Parra, Scheffler [858, 859]) *A graph G is AT-free if and only if every minimal triangulation of G is an interval graph.*

Thus, for AT-free graphs treewidth coincides with pathwidth. Theorem 11.1.11 extends results of Habib and Möhring [497].

There are also interesting classes for which the pathwidth is almost as small as the treewidth, such as multitolerance graphs [854] (see Theorem 4.8.5).

Theorem 11.1.12 *For multitolerance graphs the pathwidth is at most treewidth plus one.*

Kaplan and Shamir [643] and Parra and Scheffler [858] show that proper pathwidth is closely related to a classical parameter, namely *bandwidth*, of a graph $G = (V, E)$, which can be defined as the smallest integer k such that G is a subgraph of the kth power of the induced path P_n with $n = |V|$ vertices. The bandwidth minimization problem is $\mathbb{NP}$-complete even for special kinds of trees; see the results of Garey et al. [418] and Monien [793] on bandwidth for caterpillars (roughly speaking, a caterpillar is a tree with a dominating path).

Theorem 11.1.13 [643] *For every graph G, bandwidth and proper pathwidth of G coincide.*

Theorem 11.1.14 [858] *A graph G is AT-free claw-free if and only if every minimal triangulation of G is a proper interval graph.*

For pathwidth there are also some special classes where the problem is hard.

Definition 11.1.7 *A graph G is a* domino *graph if every vertex of G is contained in at most two maximal cliques of G.*

Note that this differs from the domino as a special 6-vertex graph.

Theorem 11.1.15 *The pathwidth problem is $\mathbb{NP}$-complete on*

- *co-comparability graphs* [497],
- *starlike graphs* [484] *(a subclass of chordal graphs), and*
- *chordal domino graphs* [673].

The class of domino graphs properly contains the line graphs of bipartite graphs. Every domino graph is claw free. A characterization of domino graphs is given below.

Theorem 11.1.16 (Kloks, Kratsch, Müller [673])

(i) *A graph is a domino graph if and only if it is claw-, gem-, and 4-wheel–free.*

(ii) *A chordal graph is a domino graph if and only if it is claw-free and gem-free.*

The following notion, which has similarities to k-trees, leads to a unified description of several graph classes.

Definition 11.1.8 (Wimer, Hedetniemi [1087]) *Let $G = (V, E)$ be a graph.*

(G, T) *is a* k-terminal graph *if $T \subseteq V$, $1 \leq |T| \leq k$, denotes a set of terminals.*

A k-terminal recursive family $\mathcal{F}$ *is a family of k-terminal graphs for which there exists a finite collection of k-terminal basis graphs $\{B_i\}_{i\in\{1,\ldots,m\}}$ and a finite collection of k-terminal composition operations $\{O_j\}_{j\in\{1,\ldots,l\}}$ such that*

for each index $i \in \{1, \ldots, m\}$, $B_i \in \mathcal{F}$;

if $G_1, G_2 \in F$, then $G_1 O_j G_2 \in \mathcal{F}$ for each applicable operation O_j, and

there are no further elements of $\mathcal{F}$.

With appropriate choices of a basis and operations the following graph classes can be represented in this way: trees, series-parallel graphs, outerplanar graphs, Halin graphs, cacti, k-trees, and partial k-trees.

The algorithmic advantage of such representations is the polynomial-time or linear-time solvability of many algorithmic problems on such graphs (see [1018, 101, 29, 1012, 128] etc.).

11.2 Series-parallel graphs

Series-parallel graphs are an important example of recursively defined graphs with a nice tree decomposition. Several variants of these graphs were studied by many authors, such as MacMahon [753], Riordan and Shannon [905], Dirac [317], Elgot and Wright [349], Duffin [341], Nishizeki [813], Lawler [705], Nishizeki and Saito [815, 816], Shinoda et al. [975], Takamizawa, Nishizeki, and Saito [1018], Kikuno, Yoshida, and Kakuda [657], Valdes, Tarjan, and Lawler [1064], and Wald and Colbourn [1075] and in many other papers.

We discuss here only undirected variants of series-parallel graphs. Directed variants are defined in a completely analogous way. For a systematic discussion of such concepts see [1064].

Note that, in general, series-parallel graphs can have multiple edges and loops (so-called multigraphs).

Series-parallel graphs are obtained recursively by applying two operations, the *series* and the *parallel compositions*, which closely resemble operations used to define the series-parallel posets given in Definition 6.4.1.

Definition 11.2.1

A two-terminal labeled graph *is a triple* (G, s, t)*, where* $G = (V, E)$ *is a multigraph and* $s, t \in V$ *(*s *is called the* source *of* G *and* t *is called the* sink *of* G*).*

The series composition *of two-terminal labeled graphs* $((V_1, E_1), s_1, t_1)$ *and* $((V_2, E_2), s_2, t_2)$ *with* $t_1 = s_2$ *is the two-terminal labeled graph* $((V_1 \cup V_2, E_1 \cup E_2), s_1, t_2)$ *(assuming that* $V_1 \cap V_2 = \{t_1\}$*).*

The parallel composition *of two-terminal labeled graphs* $((V_1, E_1), s_1, t_1)$ *and* $((V_2, E_2), s_2, t_2)$ *with* $s_1 = s_2$ *and* $t_1 = t_2$ *is the two-terminal labeled graph* $((V_1 \cup V_2, E_1 \cup E_2), s_1, t_1)$ *(assuming that* $V_1 \cap V_2 = \{s_1, t_1\}$*).*

The two-terminal labeled graph (G, s, t) *is a* two-terminal series-parallel graph *if it consists of only one edge with its end vertices as source and sink (a so-called* basic two-terminal series-parallel graph*) or it results from the applications of the series or the parallel composition to two-terminal labeled graphs.*

The series and the parallel compositions can easily be generalized to more than two operands.

The following *decomposition tree* describes how a two-terminal series-parallel graph (G, s, t) is composed (see, e.g., de Fluiter [294]).

An *sp*-tree $T_{(G,s,t)}$ of (G, s, t) is a rooted tree in which each node has one of the types *p-node, s-node,* and *leaf-node* and has an ordered pair (u, v) of vertices as label. Every node of an *sp*-tree corresponds to a unique two-terminal series-parallel graph (G', a, b), where G' is a subgraph of G and (a, b) is the label of the node. The root of the tree has label (s, t) and corresponds to the graph (G, s, t). The leaves of the tree are leaf-nodes and correspond to the edges of G. The children of an s-node are ordered, while the children of a p-node are not ordered. The two-terminal series-parallel graph defined by an s-node is the result of the series composition applied to its children in their given order. The two-terminal series-parallel graph defined by a p-node is the result of the parallel composition applied to its children.

Lemma 11.2.1 *Two-terminal series-parallel graphs have treewidth at most* 2.

The *sp*-trees of two-terminal series-parallel graphs were used in the design of linear-time algorithms (using dynamic programming on the *sp*-tree); see, e.g., Bern, Lawler, and Wong [101], Borie, Parker, and Tovey [129], Kikuno, Yoshida, and Kakuda [657], and Takamizawa, Nishizeki, and Saito [1018]. These results for two-terminal series-parallel graphs were generalized to graphs of bounded treewidth, where for many problems a similar dynamic-programming technique using the underlying tree structure leads to efficient algorithms (see, e.g., [29]).

The converse direction of Lemma 11.2.1 does not hold—the $K_{1,3}$ (i.e., the claw) is an example of a graph that cannot be generated from an edge using the series and parallel compositions. This is easy to see but is also a consequence of the following.

Lemma 11.2.2 (He [533], Eppstein [353]) *If a two-terminal series-parallel graph is not* 2*-connected, then its blocks form a path: every cut vertex of the graph is in exactly* *two blocks, all blocks have at most two cut vertices, and there are exactly two blocks that contain only one cut vertex.*

Thus, trees (and, moreover, outerplanar graphs) are in general not two-terminal series parallel.

Duffin defined in [341] the *series-parallel graphs* as follows.

Definition 11.2.2 *Let G be a multigraph. Two edges e, e' in G are* confluent *if there are no cycles C_1, C_2 in G such that C_1 meets e and e' in the same direction (assuming that both edges have an orientation) but C_2 meets e and e' in the opposite direction. G is* confluent *if every pair of its edges is confluent.*

Trees are confluent since they contain no cycles. The K_4 is an example of a graph that is not confluent.

Theorem 11.2.1 [341] *A multigraph is confluent if and only if it contains no subgraph homeomorphic to K_4.*

Moreover, Duffin defines the following operations:

- S*: Replace an edge by two edges *in series* (i.e., subdivide an existing edge by a new vertex).
- P*: Replace an edge by two edges *in parallel.*

Definition 11.2.3 *A multigraph is* series parallel *if its 2-connected components can be generated by repeatedly applying operations S* and P*, starting with a loop.*

Theorem 11.2.2 [341] *A multigraph is series parallel if and only if it is confluent.*

Reversing the operations S* and P* in the obvious way leads to reduction schemes. The reduction steps are the following:

- S: Delete a node between two series edges.
- P: Delete an edge parallel to another edge.

Corollary 11.2.1 *A multigraph is series parallel if and only if its 2-connected components can be reduced to a loop by a suitable sequence of reduction steps S and P.*

Note that series-parallel graphs are exactly the partial 2-trees, as the following theorem shows.

Theorem 11.2.3 (Wald, Colbourn [1075]) *A graph is a partial 2-tree if and only if it contains no subgraph homeomorphic to K_4.*

Since outerplanar graphs cannot contain a homeomorphic K_4, it follows that they are series parallel. Since K_5 and $K_{3,3}$ contain a homeomorphic K_4, it follows from Kuratowski's theorem (Theorem 7.3.1) that series-parallel graphs are planar.

Partial 3-trees, which generalize series-parallel graphs, also have a forbidden minor characterization [30, 31].

11.3 Cographs and domination

11.3.1 Cograph characterizations

Cographs were introduced in Definition 1.5.3 as the graphs whose modular decomposition tree contains only series and parallel modules. The series and parallel operations for posets can also be expressed by the union and complement operation for the comparability graph. Therefore, the following recursive characterization describes the cographs as well.

Theorem 11.3.1 *Let $G = (V, E)$ be a graph.*

(i) *If $|V| = 1$, then G is a cograph.*

(ii) *If $G_1 = (V_1, E_1)$ and $G_2 = (V_2, E_2)$ are vertex disjoint cographs, then $G = (V_1 \cup V_2, E_1 \cup E_2)$ is a cograph.*

(iii) *If $G = (V, E)$ is a cograph, then $\overline{G} = (V, \overline{E})$ is a cograph.*

(iv) *There are no further cographs.*

Threshold graphs, i.e., the $(2K_2, C_4, P_4)$-free graphs, have the following recursive characterization.

Theorem 11.3.2 *Let $G = (V, E)$ be a graph.*

(i) *If $|V| = 1$, then G is a threshold graph.*

(ii) *If G is a threshold graph and $x \notin V$, then $G' = (V \cup \{x\}, E)$ is a threshold graph (adding an isolated vertex).*

(iii) *If G is a threshold graph and $x \notin V$, then $G' = (V \cup \{x\}, E \cup \{xy : y \in V\})$ is a threshold graph (adding a universal vertex).*

(iv) *There are no further threshold graphs.*

As shown by Lerchs [725], cographs are exactly the graphs that contain no induced P_4, and cographs have a unique tree representation reflecting the operations generating a cograph G. This tree representation is the basis for fast solutions of some algorithmic problems on cographs, which are $\mathbb{NP}$-hard in general.

Cographs were rediscovered several times under different names (*D^*-graphs* by Jung [635], *hereditary-Dacey graphs* by Sumner [1005], *2-parity graphs* by Burlet and Uhry [170]) and in different fields such as logic and learning theory by Gurvich [479, 480]. Several earlier papers such as Kelmans [651, 652] study the operations (disjoint) union and join on graphs.

Definition 11.3.1 *Let $G = (V, E)$ be a graph.*

G has the clique-kernel intersection property *if for every maximal clique C in G and for every maximal stable set K in G, $|C \cap K| = 1$.*

G is a Dacey graph *if for every pair of distinct vertices u, v and every maximal clique C in G, $C \subseteq N(u) \cup N(v)$ implies that $uv \in E$.*

G is a hereditary-Dacey graph (HD-graph) *if every induced subgraph is a Dacey graph* [1005].

The fundamental theorem on cographs contains the results of several authors, such as Lerchs [725, 726], Seinsche [966], Sumner [1005], Jung [635], and Corneil, Lerchs, and Stewart-Burlingham [249].

Theorem 11.3.3 *The following conditions are equivalent:*

(i) *G is a cograph;*

(ii) *Every nontrivial subgraph of G has at least one pair of twins;*

(iii) *Every subgraph of G has the clique-kernel intersection property;*

(iv) *G is P_4-free;*

(v) *The complement of every nontrivial connected subgraph of G is disconnected;*

(vi) *G is an HD-graph;*

(vii) *Every connected subgraph of G has diameter ≤ 2.*

Another characterization of cographs is given in Theorem 11.3.5. Corneil, Perl, and Stewart [256] give a linear-time recognition algorithm for cographs. For a characterization of cographs that leads to an NC recognition algorithm for cographs see the paper of Lin and Olariu [728].

11.3.2 Domination properties

There are many generalizations of cographs in different directions. Bacsó and Tuza [42] give a characterization of P_k-free graphs, $k \geq 4$, in terms of graph centers and domination; the characterizing conditions depend on the parity of k. Fouquet et al. [402] give another characterization of P_k-free graphs, give structural properties of $(P_5, \overline{P_5})$-free graphs, and characterize (P_5, C_4)-free graphs.

C_4-free cographs (i.e., trivially perfect graphs) have a characterization in terms of the following domination property.

Theorem 11.3.4 (Wolk [1090, 1091]) *All connected induced subgraphs of a graph G have a dominating vertex if and only if G is P_4-free and C_4-free.*

Theorem 11.3.5 (Cozzens, Kelleher [264]) *A graph G is P_4-free if and only if in every connected induced subgraph G' of G, all maximal cliques dominate G'.*

In [43] the (P_5, C_5)-free graphs are characterized in the same spirit.

Theorem 11.3.6 (Bacsó, Tuza [43]) *A graph G is (P_5, C_5)-free if and only if every connected induced subgraph of G contains a dominating clique.*

In [732] this is extended to triangle-free graphs.

Theorem 11.3.7 (Liu, Zhou [732]) *Let G be a triangle-free graph.*

(i) *G is (P_6, C_6)-free if and only if every connected induced subgraph of G has a dominating complete bipartite subgraph.*

(ii) *G is P_6-free if and only if every connected induced subgraph has a dominating complete bipartite subgraph or a dominating induced C_6.*

For the non–triangle-free case the following holds.

Theorem 11.3.8 [732] *Every connected (P_6, C_6)-free graph has a dominating subgraph H containing an edge that dominates H.*

Theorem 11.3.9 [732] *A connected graph is P_5-free if and only if each of its connected induced subgraphs has a dominating clique or a dominating induced C_5.*

See [42] for characterizations of graphs without long induced paths in terms of dominating centers.

11.4 Bounding the number of P_4s

This section is closely related to the notion of homogeneous decomposition, which is studied in the decomposition chapter (chapter 12). Theorem 12.2.3 is of particular importance.

11.4.1 P_4-reducible and P_4-sparse graphs and variants

Several generalizations of cographs are obtained by bounding the number of P_4s in different ways. A first example is the following definition.

Definition 11.4.1 (Jamison, Olariu [615]) *A graph is* P_4-reducible *if every vertex belongs to at most one P_4 of the graph.*

This generalization of cographs has properties similar to cographs: Jamison and Olariu give in [617] a unique tree representation of the P_4-reducible graphs, which is based on the subsequent structure theorem and leads to a linear-time recognition of this class.

Theorem 11.4.1 [615] *A graph G is P_4-reducible if and only if for every induced subgraph H of G, exactly one of the following conditions is satisfied:*

(i) *H is disconnected;*

(ii) $\overline{H}$ *is disconnected;*

(iii) *There is a unique* P_4 *abcd in* H *such that every vertex in* H *outside* $\{a, b, c, d\}$ *is adjacent to both b and c and nonadjacent to both a and d.*

As Damaschke has noted, every P_4-reducible graph is a permutation graph.
As a subclass of the alternately orientable graphs, Hoàng [553] has introduced P_4-sparse graphs.

Definition 11.4.2 [553] *A graph* G *is* P_4-sparse *if no set of* 5 *vertices in* G *induces at least two distinct* P_4*s.*

In [553] and [620] various characterizations of P_4-sparse graphs are given. One of them uses the notion of a spider.

Definition 11.4.3 (Jamison, Olariu [620]) *A graph* $G = (V, E)$ *is a* spider *if there is a partition of* V *into sets* S, K, R *such that*

> S *is a stable set,* K *is a clique, and* $|S| = |K| \geq 2$;
>
> *every vertex in* R *is adjacent to all vertices in* K *and nonadjacent to all vertices in* S;
>
> *there is a bijection* f *between* S *and* K *such that either* $N(x) = \{f(x)\}$ *for all vertices* $x \in S$ *(a* thin spider*) or* $N(x) = K \setminus \{f(x)\}$ *for all vertices* $x \in S$ *(a* thick spider*).*

R *is called the* head of the spider. *A spider is* headless *if* R *is empty. A headless spider is* proper *if it has more than four vertices.*

It is easy to see that the smallest spider is the P_4 and a spider is prime if and only if its head has at most one vertex. Furthermore, the complement of a thin spider is a thick one and vice versa. The basic structure theorem for P_4-sparse graphs is the following.

Theorem 11.4.2 [620] *A graph* G *is* P_4*-sparse if and only if for every induced subgraph* H *of* G *exactly one of the following conditions is satisfied:*

(i) H *is disconnected;*

(ii) $\overline{H}$ *is disconnected;*

(iii) H *is isomorphic to a spider.*

A consequence of this is that a P_4-sparse graph is prime if and only if it is a prime spider.
From the structural characterizations given in Theorems 11.4.1 and 11.4.2 one obtains generalizations based on the characterization of P_4-reducible and P_4-sparse graphs in terms of forbidden subgraphs given in Theorem 7.1.6. Admitting the C_5 leads to the notion of extended P_4-reducible and extended P_4-sparse graphs (see Definition 7.1.3).

A modification, the *P_4-extendible graphs*, is introduced in [619], where it is shown that these graphs, which contain the extended P_4-reducible graphs, also have a unique tree structure. A further generalization is the class of *P_4-tidy graphs* proposed by Rusu; these contain the class of extended P_4-sparse graphs and are used as part of a characterization of P_4-lite graphs in [439].

Definition 11.4.4 (Jamison, Olariu [618]) *A graph G is P_4-*lite *if every induced subgraph H of G with at most six vertices satisfies exactly one of the following conditions:*

H contains at most two induced P_4s;

H or $\overline{H}$ is isomorphic to the graph S_3.

Jamison and Olariu [618] show that P_4-lite graphs are brittle. This is further extended by generalizing P_4-lite graphs as follows.

Definition 11.4.5 (Giakoumakis [437]) *A graph G is P_4-*laden *if every induced subgraph H of G with at most six vertices satisfies exactly one of the following conditions: H contains at most two induced P_4s, or else H is a split graph.*

The *extended P_4-laden graphs* are defined by generalizing split graphs to $(2K_2, C_4)$-free graphs [437].

Clearly, P_4-lite graphs are P_4-laden. In [437] it is shown that P_4-laden graphs are brittle. Thus,

$$\textbf{cograph} \subset \textbf{P}_4\textbf{-reducible} \subset \textbf{P}_4\textbf{-sparse} \subset \textbf{P}_4\textbf{-lite} \subset \textbf{P}_4\textbf{-laden} \subset \textbf{brittle}.$$

In Definition 7.1.4 the class of semi–P_4-sparse graphs is introduced. In [401] it is shown that the techniques used in [219] for finding optimal colorings and solving similar problems for $(P_5, \overline{P_5}, C_5)$-free graphs can be carried over to semi–P_4-sparse graphs.

For more characterizations of the classes mentioned above and a containment diagram see [439]. Using the modular decomposition tree, which can be obtained in linear time (see [775, 262, 280]), a unified approach leads to linear-time recognition for the following graph classes:

- extended P_4-reducible and P_4-sparse graphs [440] (similar techniques also imply linear recognition of P_4-reducible and P_4-sparse graphs);
- P_4-tidy graphs [439];
- P_4-laden graphs and extended P_4-laden graphs [437];
- semi–P_4-sparse graphs [401].

11.4.2 p-trees

Definition 12.2.1 introduces the notion of p-connectedness: A graph $G = (V, E)$ is *p-connected* if for every partition V_1, V_2 of V, there is a *crossing* P_4 between V_1 and V_2, i.e., a P_4 containing vertices from V_1 and V_2.

Babel and Olariu [39, 37] define a generalization of cographs with similarities to trees.

Definition 11.4.6 [39, 37] *Let G be a graph.*

G is a p-cycle *if every vertex of G is contained in at least two P_4s and G is minimal with this property, i.e., every proper induced subgraph of G has a vertex occurring in at most one P_4.*

G is a p-tree *if G is p-connected and contains no p-cycle.*

G is a p-forest *if G contains no p-cycle.*

Note that there are simple trees that are p-cycles. However, it is worth noting that a p-cycle with at least eight vertices is P_4-isomorphic to a chordless cycle [37].

Definition 11.4.7 (Babel [37]) *Let G be a p-connected graph.*

A vertex v is a p-articulation vertex *in G if $G - v$ is not p-connected.*

v is a p-end-vertex *if v is contained in exactly one P_4.*

For the notion of p-chains see Definition 12.2.3.

Theorem 11.4.3 [37] *Let G be a graph. The following conditions are equivalent:*

(i) *G is a p-tree;*

(ii) *G is p-connected and every p-connected induced subgraph H of G contains at least one p-end-vertex;*

(iii) *G is p-connected, contains no proper induced headless spider, and has exactly $n-3$ P_4s;*

(iv) *G contains no induced p-cycle and has exactly $n-3$ P_4s;*

(v) *G is p-connected and contains no proper induced headless spider, and each vertex of a p-connected induced subgraph H of G is either a p-end-vertex or a p-articulation-vertex in H;*

(vi) *G contains no proper induced headless spider and each pair of vertices is connected either by a unique nontrivial p-chain or by trivial p-chains only.*

Babel [37] shows that recognition and isomorphism of p-trees can be done in linear time.

11.4.3 (q,t)-graphs

A generalization of some classes in terms of (q,t)-graphs is given by Babel and Olariu [39, 37]: A graph G is a *(q,t)-graph* if no set of at most q vertices induces more than t distinct P_4s. Of special interest are the $(q,q-3)$- and $(q,q-4)$-graphs.

Babel [37] investigates the structure of $(q,q-4)$-graphs and $(q,q-3)$-graphs. From the definition it is clear that every graph is (4,1), every C_5-free graph is (5,2), P_4-sparse graphs are exactly the (5,1)-graphs, and the C_5-free P_4-extendible graphs are the (6,2)-graphs. Furthermore, Babel [37] shows the following inclusions:

P_4-lite $\subset$ **(7,3);**

P_4-lite $\subset$ **(7,4);**

P_4-extendible $\subset$ **(9,6);**

P_4-lite $\subset$ **(9,6);**

p-forest $\subset$ **(9,6);**

$(q, q-4) \subset$ **brittle** for $q \in \{6, 7, 8\}$.

Some of these inclusions follow from the subsequent characterization of $(q, q-3)$-graphs for $q \geq 7$.

Theorem 11.4.4 [37] *Let G be a p-connected graph with n vertices.*

(i) *If $n \geq 7$, then G is an $(n, n-3)$-graph if and only if G is a p-tree.*

(ii) *If $n > q$, $q \in \{7, 9\}$, then G is a $(q, q-3)$-graph if and only if precisely one of the following conditions holds*

  (a) *G is a p-tree;*

  (b) *G is a hole or antihole;*

  (c) *G is a headless spider.*

(iii) *If $n > q$, $q = 8$, or $q \geq 10$, then G is a $(q, q-3)$-graph if and only if precisely one of the following conditions holds*

  (a) *G is a p-tree;*

  (b) *G is a hole or antihole.*

This also leads to a linear-time recognition of $(q, q-3)$-graphs for every $q \geq 7$.

11.5 Tree-cographs and hookup classes

Let A be the adjacency matrix of a graph G and let M_1 denote the matrix of 1*s* and I the unit matrix.

Definition 11.5.1 (Tinhofer [1026]) *The class TC of* tree-cographs *is recursively defined as follows:*

(i) *Every tree belongs to TC;*

(ii) *If $A \in TC$, then $\overline{A} = M_1 - I - A \in TC$ ($\overline{A}$ is the adjacency matrix of the complement graph);*

(iii) *If $A_1, \ldots, A_k \in TC$, $k \geq 2$, are connected, then the disjoint union*

$$A_1 \cup \ldots \cup A_k = \begin{bmatrix} A_1 & & & \\ & A_2 & & \\ & & \ddots & \\ & & & A_k \end{bmatrix} \in TC.$$

Thus, cographs are the special case of tree-cographs where the rule (i) is replaced by the corresponding one-vertex graph rule. For tree-cographs the recursion starts with any tree. Since the class of perfect graphs is closed under union and complement the tree-cographs are perfect. Tree-cographs have nice parse trees similar to those of cographs and can be recognized in polynomial time (see [1026]).

If the components in (iii) are pairwise nonisomorphic (so-called *strong tree-cographs*) then the recursively defined graph is a Birkhoff graph (see Definition 9.4.3 and [1026]).

A similar extension is made in defining hookup classes. The recursion does not start with a single-vertex graph, a tree, or a clique but with an arbitrary graph A.

Definition 11.5.2 (Klawe, Corneil, Proskurowski [662]) *The* hookup class $[A, B]$ *denotes the set of graphs formed by starting with graph A and recursively adding a vertex adjacent to all vertices in an induced subgraph isomorphic to B.*

Thus k-trees are the hookup class $[K_k, K_k]$. In [662], it is shown that the isomorphism problem for hookup classes is polynomial in some cases, based on the unique tree representation for such graphs.

11.6 Recursively defined perfect graphs

There are some well-known operations that preserve perfection. This section deals with classes defined by such operations.

The *clique-bonding* operation is introduced in Definition 12.7.1. Clique bonding preserves perfection (see Theorem 12.7.1). It is interesting that this property is restricted to bonding of cliques.

Theorem 11.6.1 (Corneil, Kirkpatrick [248]) *For any perfect noncomplete graph G there exist perfect graphs H_1, H_2, both containing an induced copy of G, such that H_1 and H_2 bonded at G is not perfect.*

In the following "bonding" means clique bonding. It is known that replacing a vertex v in a perfect graph G by another perfect graph G' preserves perfection; see Lemma 2.4.1. Replacement means creating a module G' in the new graph where the adjacencies of G' are all the vertices adjacent to v in G.

Definition 11.6.1 [248] *The following classes $\mathcal{F}$ are defined recursively. The parameter k restricts the size of the bonding clique.*

The single-vertex graph is in $\mathcal{F}$;

Class $\mathcal{F}$	*Operations*
CU	*complement, union (cographs)*
CUR	*complement, union, replacement*
k-CUB	*complement, union, and bonding*
CUB	$\bigcup_k$ *k-CUB*
k-CURB	*as with k-CUBs and allowing replacements*
CURB	$\bigcup_k$ *k-CURB*

If $G_1, G_2 \in \mathcal{F}$, then $G_1 \circ G_2$ is in $\mathcal{F}$ for an operation $\circ$ admitted to $\mathcal{F}$;

There are no further graphs in $\mathcal{F}$.

Theorem 11.6.2 [248]

(i) *CUR = CU.*

(ii) *CUB graphs can be recognized in polynomial time.*

(iii) *CU $\subset$ 1-CUB. For $k \geq 2$, the k-CUB isomorphism problem is isomorphism complete.*

The recognition algorithm for CUB graphs in the above theorem is based on Whitesides's clique-separator decomposition algorithm [1081].

Strict inclusion (iii) in Theorem 11.6.2 holds since the P_4 is in 1-CUB. It is also noted in [248] that CURB and permutation graphs are incomparable classes.

There are also other more complicated operations that preserve perfection.

Definition 11.6.2 (Burlet, Fonlupt [169]) *Let $G_1 = (V_1, E_1), G_2 = (V_2, E_2)$ be disjoint graphs and $x_i \in V_i$, $K_i \subseteq N(x_i)$, $i \in \{1, 2\}$, such that*

K_1, K_2 are cliques with $|K_1| = |K_2|$;

if $z \in K_i$ and $t \in N(x_i) \setminus K_i$, then $zt \in E_i$, $i \in \{1, 2\}$;

$N(x_1) \setminus K_1 = \emptyset$ if and only if $N(x_2) \setminus K_2 = \emptyset$.

A graph G is the amalgam *of G_1 and G_2 if G is obtained from G_1, G_2 by*

one-to-one identification of the vertices of K_1 with the vertices of K_2,

creating an edge between every vertex of $N(x_1) \setminus K_1$ and every vertex of $N(x_2) \setminus K_2$,

deleting the vertices x_i, $i \in \{1, 2\}$.

The resulting graph will be denoted by $G = (G_1, x_1, K_1) \; \phi \; (G_2, x_2, K_2)$.

Conversely, if G has such a representation, then G can be decomposed by the amalgam operation into two graphs G_1, G_2.

Theorem 11.6.3 [169]

(i) *The amalgam operation preserves perfection.*

(ii) *If G_1 and G_2 are Meyniel graphs, then their amalgamation is also a Meyniel graph.*

Note that it also follows from the star-cutset lemma—see condition (P11) in the SPGC chapter (chapter 14)—that the amalgam operation preserves perfection.

Definition 11.6.3 [169] $G = (V, E)$ *is a* basic Meyniel graph *if*

G is connected;

V can be partitioned into A, K, S such that $G(A)$ is a two-connected bipartite graph, K is a clique, S is a stable set;

if $x \in A, y \in K$, then $xy \in E$;

for all $x \in S$, $|N(x) \cap A| \leq 1$.

Theorem 11.6.4 [169] *A connected graph G that is not a basic Meyniel graph is a Meyniel graph if and only if G can be ϕ-properly decomposed into two Meyniel graphs G_1 and G_2 by the amalgam operation.*

This theorem is the key for a polynomial-time recognition of Meyniel graphs given in [169]. The same paper describes a similar approach for Gallai graphs, and Burlet and Uhry [170] give a similar algorithm for recognizing parity graphs based on Theorem 11.6.6.

For the class of parity graphs, the notion of twins plays an important role. Obviously, the operation of adding a twin, true or false, preserves perfection. Recall the notion of a prime graph (see Definition 1.5.2) and the fact that a prime graph has no twins.

Theorem 11.6.5 (Burlet, Uhry [170]) *In a prime parity graph, every minimal separating set is bipartite.*

Definition 11.6.4 [170] *The* extension of a graph G by a bipartite graph $B = (X_1, X_2, A)$ *is the operation that generates a new graph by identifying certain vertices of X_1 with a (possibly one-element) set of false twins of G.*

Theorem 11.6.6 [170] *Every connected parity graph is obtained from a single vertex by repeatedly applying the following operations:*

- *creating twins,*
- *extension by a bipartite graph,*

applied successively and in any order.

Cicerone and Di Stefano [225] restrict the operation of extension by a bipartite graph to particular bipartite graphs and get an infinite hierarchy between distance-hereditary and parity graphs. In another paper [224], the same authors design a linear-time algorithm for recognizing parity graphs based on Theorem 11.6.6 and the linear-time split decomposition of Dahlhaus (see the decomposition chapter (chapter 12)).

Note that twins (which are modules of size 2) also play an important role for distance-hereditary graphs.

Theorem 11.6.7 (Bandelt, Mulder [56]) *Let $G = (V, E)$ be a graph with $|V| \geq 2$. Then G is distance hereditary if and only if G is obtained from an edge by a sequence of one-vertex extensions consisting of attaching pendant vertices and creating twins.*

An easy consequence of this recursive characterization is the fact that every distance-hereditary graph is a circle graph, as noticed by Damaschke.

The class of ptolemaic graphs can be described recursively by an operation called *subatomic identification* introduced by Howorka in [588]. Subatomic identification is a special case of clique bonding. He shows that a graph is ptolemaic if and only if it can be obtained from cliques using subatomic identification. For another recursive description of ptolemaic graphs see [56].

The amalgam operation was further generalized in [257] to a 2-amalgam operation that preserves perfection and also generalizes and unifies union, clique bonding, replacement, and the operation of join [108, 269, 271]. One motivation for generalizing perfection-preserving operations is to try to design a recognition algorithm for perfect graphs that is similar to the Meyniel graph-recognition algorithm due to Burlet and Fonlupt [169].

The approach of [169] is generalized in [591, 592] by using similar but more general operations in order to give decompositions of perfect and planar perfect graphs. In the case of planar perfect graphs this leads to a polynomial-time recognition algorithm.

Planar perfect graphs are decomposed into essentially two special classes of inseparable component graphs that are easy to recognize. They are

(1) planar comparability graphs and

(2) planar line graphs of those planar bipartite graphs whose maximum degrees are not greater than 3.

Chvátal and Sbihi [223] use a similar approach to recognize claw-free perfect graphs in polynomial time. Olariu [823] used this technique to recognize paw-free perfect graphs, using a theorem that every paw-free perfect graph is either bipartite or a complete multipartite graph.

Chmeiss and Jegou [195] define an infinite number of classes of recursively defined graphs that generalize the chordal graphs. Let CSG^0 be the class of complete graphs. The class CSG^i is defined by a vertex-addition scheme in which the neighborhood of each vertex added must induce a graph in CSG^{i-1}. Thus, CSG^1 is exactly the class of chordal graphs. Chmeiss and Jegou devise polynomial-time algorithms for recognizing graphs in CSG^k for all fixed k, and give bounds on the number of maximal cliques generalizing those known for chordal graphs.

Chapter 12

Decompositions and Cutsets

In this chapter we discuss methods of decomposing a graph into subgraphs. We divide the decompositions into two groups, depending on whether the decomposition splits a graph into connected components by removing a subset of vertices. Decompositions that reduce the size of the graph without removing specific cutsets are considered first.

12.1 Modular decomposition—the poset aspect

The modular, substitution, or lexicographic decomposition of graphs and partial orders arose in connection with Gallai's results investigating the structure and recognition of comparability graphs [416] (see also [939, 941, 494, 271, 270, 269, 650, 1084, 971, 552]).

The substitution-decomposition theory is by now a well-understood theory with many applications in discrete mathematics. There are, e.g., substitution decompositions for Boolean functions, set systems, and relations. For survey articles see the papers of Möhring [786, 788] and Möhring and Radermacher [791].

We describe here very briefly the substitution decomposition for partial orders and graphs following Möhring [788].

Definition 12.1.1 *Let $Q = (V', <_Q)$ be a partial order, let $a_1, \dots, a_h \in V'$ be pairwise-distinct elements from V', and let $P_1 = (V_1, <_1)$, ..., $P_h = (V_h, <_h)$ be partial orders with mutually disjoint ground sets $V', V_1, \dots, V_h$. Then*

> $P = Q^{P_1, \dots, P_h}_{a_1, \dots, a_h}$ *is the partial order $P = (V, <_p)$ resulting from substituting the elements a_i of Q with the associated partial order P_i, $i = 1, \dots, h$, such that*
>
> $V = (V' \setminus \{a_1, \dots, a_h\}) \cup V_1 \cup \dots \cup V_h$ *and*
>
> $a <_p b$ *if and only if there is an i with $a, b \in V_i$ and $a <_i b$ or there are $i, j, i \neq j$, with $a \in V_i, b \in V_j$ and $a_i <_Q a_j$.*
>
> *The substitution is* proper *if $1 < |V_i| < |V|$ for some i.*

A partial order is decomposable *if it can be obtained by proper substitution. Otherwise it is* indecomposable (*or* prime).

Every decomposable partial order P is obtained by a sequence of single substitutions

$$Q_1 = Q_{a_1}^{P_1},\ Q_2 = (Q_1)_{a_2}^{P_2},\ \ldots,\ P = Q_m = (Q_{m-1})_{a_m}^{P_m},$$

in which each P_i is prime. The reverse of such a sequence is also referred to as a *composition series.* Composition series fulfill the following theorem.

Theorem 12.1.1 *Every two composition series of a partial order have the same length and the same factors up to rearrangement* (Jordan–Hölder property) *and the same final factor Q_1 up to isomorphism* (Church–Rosser property).

Definition 12.1.2 *Let $P = (V, <)$ be a poset.*

A subset $B \subseteq V$ is autonomous *if for all $a \in V \setminus B$, the following condition is fulfilled:*

If there is an element $b_0 \in B$ with $a < b_0$ ($b_0 < a$), then for all $b \in B$ $a < b$ ($b < a$), i.e., for all $a \in V \setminus B$, either a is related to all $b \in B$ or to none of $b \in B$.

P is decomposable *if and only if it has a nontrivial autonomous set B(i.e., $1 < |B| < |V|$).*

The following theorem shows that the substitution composition is a generalization of the series-parallel composition (see Definition 6.4.1 and [788]).

Theorem 12.1.2 (Gallai [416]) *For each decomposable partial order $P = Q_{a_1 \ldots a_h}^{P_1 \ldots P_h}$ one of the following cases applies:*

(i) *Q is an antichain. Then P is obtained by* parallel composition *of $P_1, \ldots, P_h$.*

(ii) *Q is a linear order (chain). Then P is obtained by* series composition *of $P_1, \ldots, P_h$.*

(iii) *Q is a (uniquely determined) prime partial order. Then P is said to be of the* prime type *and Q is called the associate* prime quotient.

Thus, any partial order can be represented in a *decomposition tree,* which is unique if we require the quotient in (i) and (ii) to be as large as possible. This tree is called the *canonical decomposition tree.*

Corollary 12.1.1 *A partial order is series parallel if and only if its (canonical) decomposition tree contains no prime-type node.*

An early application of substitution decomposition comes from poset-dimension theory (see the poset chapter (chapter 6)).

Theorem 12.1.3 (Hiraguchi's formula, [552])

$$dim\ Q^{P_1 \dots P_h}_{a_1 \dots a_h} = \max\{dim\ Q, dim\ P_1, \dots, dim\ P_h\}.$$

For the notion of a *module* recall Definition 1.5.1.

Theorem 12.1.4 (McConnell, Spinrad [776]) *Comparability graph orientation and recognition can be done in $\mathcal{O}(n+m)$ and $\mathcal{O}(MM)$ time, respectively.*

The orientation and recognition algorithms are based on modular decomposition. The composition can also be used for recognition of two-dimensional posets and permutation graphs using the fact from [345] that a poset P is two-dimensional if and only if P has a nonseparating linear extension (and using Hiraguchi's formula).

Theorem 12.1.5 [776] *It is possible to recognize permutation graphs in $\mathcal{O}(n+m)$ time and construct two linear extensions of the corresponding two-dimensional poset in the same time bound if the graph is a permutation graph.*

The first polynomial-time solution for isomorphism of permutation graphs is due to Colbourn, and uses the modular decomposition tree (see [232]). There are many other combinatorical (optimization) problems that have a fast solution via the modular composition (see, e.g., the papers of Möhring [786, 787, 788]).

In particular, many $\mathbb{NP}$-complete problems can be solved in time that is exponential in the decomposition diameter, i.e., the maximum number of children of a prime module. For posets, Steiner [1000] and Habib, Morvan, and Rampon [498] show that some problems are solvable efficiently if the poset has bounded decomposition width, that is, if the width of every poset induced by one element from each child of a prime module is bounded by a constant.

12.2 Homogeneous decomposition

Jamison and Olariu [622] study another decomposition called *homogeneous decomposition*, which, like modular decomposition, yields a unique decomposition tree for arbitrary graphs. It can be seen as a natural extension of modular decomposition, since it goes further in decomposing graphs that are prime with respect to the modular decomposition in the following way.

Definition 12.2.1 [622] *Let $G = (V, E)$ be a graph.*

G is p-connected *if for every partition V_1, V_2 of V, there is a* crossing P_4 *between V_1 and V_2, i.e., a P_4 containing vertices from V_1 and V_2.*

The p-connected components *of G are the maximal-induced subgraphs that are p-connected.*

Obviously, the p-connected components are connected subgraphs of G and $\overline{G}$. Furthermore, each graph G has a unique partition into p-connected components and so-called weak vertices, which are not contained in any P_4.

Definition 12.2.2 [622] *A p-connected graph $G = (V, E)$ is* separable *if V can be partitioned into nonempty V_1, V_2 in such a way that each crossing P_4 has its midpoints in V_1 and its endpoints in V_2. Then V_1, V_2 is a* separation *of G.*

The complement of a separable p-connected graph is clearly also separable.

Theorem 12.2.1 [622] *Every separable p-connected graph has a unique separation. Furthermore, every vertex belongs to a crossing P_4 with respect to the separation.*

The graph obtained by shrinking every maximal module of a graph G to a single vertex is called the *characteristic graph* of G.

Theorem 12.2.2 [622] *A p-connected graph is separable if and only if its characteristic graph is a split graph.*

The p-connectedness structure theorem for graphs is similar to Theorem 1.5.1 on substitution decomposition, Theorem 11.4.1 on P_4-reducible graphs, and Theorem 11.4.2 on P_4-sparse graphs.

Theorem 12.2.3 [622] *For an arbitrary graph G exactly one of the following conditions holds:*

(i) *G is not connected;*

(ii) *$\overline{G}$ is not connected;*

(iii) *There is a unique proper separable p-connected component H of G with partition H_1, H_2 such that every vertex outside H is adjacent to all vertices in H_1 and to no vertex in H_2;*

(iv) *G is p-connected.*

Theorem 12.2.3 leads to the homogeneous decomposition tree as follows. The labels of the interior nodes of the decomposition tree correspond to the first three cases of Theorem 12.2.3, while the leaves are the p-connected components and weak vertices.

Recently, Baumann [72] showed that the homogeneous decomposition tree can be constructed in linear time.

A further extension and refinement of this decomposition called *separable-homogeneous decomposition* is given by Babel [37]. Babel and Olariu [37, 40] study the structure theory of p-connected graphs. It is interesting to note that unbreakable graphs are p-connected and p-connectedness of a graph can be checked in linear time [37] based on Theorems 12.2.2 and 12.2.3 and a linear-time modular-decomposition algorithm.

The following notion of a p-chain leads to a characterization of p-connected graphs.

Definition 12.2.3 **[37, 40]** *Let G be a graph. A* p-chain *connecting x and y is a sequence of distinct vertices $(v_1, \ldots, v_k)$ such that $x = v_1, y = v_k$ and for all $i \in \{1, \ldots, k-3\}$, the set $X_i = \{v_i, v_{i+1}, v_{i+2}, v_{i+3}\}$ induces a P_4 in G.*

The simplest examples for p-chains are given by the P_4s of a P_k, $k \geq 4$. There are, however, more interesting examples. The importance of the notion of p-chains becomes more clear in the following.

Theorem 12.2.4 **[37, 40]** *A graph is p-connected if and only if every pair of vertices in the graph is connected by a p-chain.*

Raschle and Simon [894] generalize the homogeneous decomposition by combining the modular decomposition with the substitution of p-connected components. They use this decomposition to recognize P_4-comparability graphs; see the remarks after Definition 5.7.1.

12.3 Split decomposition

The *split* (or *join*) *decomposition* of a graph generalizes the substitution decomposition and was introduced by Cunningham [270] (under the name *join decomposition*) for directed graphs.

There is a natural correspondence between cographs, which are "completely decomposable" by substitution decomposition, and distance-hereditary graphs (or "completely separable" graphs; see [511]), which are "completely decomposable" by split decomposition.

Ma and Spinrad [750] mention another correspondence of this kind between permutation graphs and circle graphs: a permutation graph has a unique representation if it is indecomposable with respect to the substitution decomposition, while a circle graph has a unique representation if it is indecomposable with respect to the split decomposition.

Definition 12.3.1 *Let $G = (V, E)$ be a graph. A* split *of G is a partition V_1, V_2 of V such that*

$|V_1|, |V_2| \geq 2$ and

there exist subsets $W_1 \subseteq V_1$, $W_2 \subseteq V_2$ such that
$\{uv : uv \in E \text{ and } u \in V_1 \text{ and } v \in V_2\} = \{xy : x \in W_1 \text{ and } y \in W_2\}$.

A graph is split decomposable *if it has a split, otherwise it is called* prime.

The split decomposition of G is formed by taking any split V_1, V_2 and recursively decomposing $V_1 \cup \{m\}$, $V_2 \cup \{m\}$, where m is a new vertex such that $N(m) = W_1 \cup W_2$.

Cunningham and Edmonds [270, 271] developed a theory of split decomposition and give unique decomposition theorems. The final factors of the decomposition are unique, up to the decomposition of certain highly decomposable graphs, such as complete graphs and stars. The split decomposition discussed here is the special symmetric case of the decomposition theory, which was designed to work on directed graphs.

Theorem 12.3.1 (Bouchet [138], Gabor, Hsu, and Supowit [411], Naji [808]) *A graph is a circle graph if and only if every component created by the split decomposition is a circle graph.*

Based on this theorem, Naji [808], Bouchet [138], and Gabor, Hsu, and Supowit [411] develop polynomial-time recognition algorithms for circle graphs whose time bound was improved to $\mathcal{O}(n^2)$ by Spinrad [990].

Theorem 12.3.2 (Dahlhaus [278]) *The split decomposition of a graph can be found in time* $\mathcal{O}(n+m)$.

Hsu [593] uses the split-decomposition technique and the transformation from circular-arc to circle graphs in order to give fast isomorphism test algorithms for both classes and fast recognition of circular-arc graphs.

Theorem 12.3.3 [593]

(i) *Isomorphism of circular-arc and circle graphs can be tested in* $\mathcal{O}(nm)$ *time.*

(ii) *Circular-arc graphs can be recognized in* $\mathcal{O}(nm)$ *time.*

Hsu [591] also defines a generalization of the split decomposition as follows.

Definition 12.3.2 *Let* $G = (V, E)$ *be a graph.*
A generalized split *of* G *is a partition* V_1, V_2 *of* V *such that*

the subgraphs induced by V_1, V_2 *have an edge, and*

if $Cross(V_1, V_2)$ *is the bipartite graph formed by edges that go between* V_1 *and* V_2, *then* $Cross(V_1, V_2)$ *does not contain* $2K_2$ *as an induced subgraph.*

This decomposition is shown by Hsu [591] to maintain the property that G is perfect if and only if each component of the generalized split decomposition of G is perfect. It is not known whether this generalized split decomposition can be found in polynomial time.

The split decomposition has also been applied to structures other than graphs; see, for example, D. Wagner [1073] for an application of split decomposition to partial orders and submodular functions.

12.4 Other decompositions

This section describes a number of other decomposition techniques, which do not work by separating a graph by removal of a cutset. The first decomposition is based on the lexicographic product of two graphs; see Definition 2.5.1.

Definition 12.4.1 (Harary [519], Sabidussi [939]) *The graph* G *is* reducible (*with respect to the lexicographic product*) *if there are* G_1, G_2 *with* $G = G_1 \circ G_2$; *otherwise it is* irreducible.

It is easy to see that for graphs, testing irreducibility is at least as difficult as testing graph isomorphism. Suppose that we want to decide whether the connected graphs G_1 and G_2 are isomorphic. The graph $G = (V_1 \cup V_2, E_1 \cup E_2)$ (for $V_1 \cap V_2 = \emptyset$) has connected components G_1, G_2 and is reducible if and only if G_1 is isomorphic to G_2. The problem is also hard for connected graphs.

Theorem 12.4.1 (Feigenbaum, Schäffer [381]) *Testing a connected graph for irreducibility is at least as difficult as graph isomorphism.*

On the other hand, [381] shows that the problem is no more difficult than solving a polynomial number of graph-isomorphism problems of the same size.

Definition 12.4.2 (Sabidussi [940])

The Cartesian product $G_1 \times G_2$ *of two graphs* G_1, G_2 *is the graph with vertex set* $V(G_1) \times V(G_2)$ *and edge set*

$$\{(g_1, g_2)(g_1', g_2') : (g_1 g_1' \in E(G_1) \wedge g_2 = g_2') \vee (g_2 g_2' \in E(G_2) \wedge g_1 = g_1')\}.$$

G *is* reducible *(with respect to the Cartesian product) if there are* G_1, G_2 *with* $G = G_1 \times G_2$*; otherwise it is* irreducible.

Note that $G_1 \times G_2 \sim G_2 \times G_1$, while in general $G_1 \circ G_2 \not\sim G_2 \circ G_1$. Sabidussi [940] and independently Vizing [1067] showed that there is a unique decomposition by Cartesian product into irreducible components. Although testing for reducibility in a disconnected graph is as hard as graph isomorphism using arguments similar to those for the previous decomposition, the irreducible factors of a connected graph with respect to Cartesian-product decomposition can be found in polynomial time. The first polynomial algorithms can be found in [380, 1088]; the best known algorithm has cost $\mathcal{O}(m \log n)$ [34]. A polynomial-time algorithm for the natural directed variant of this decomposition can be found in [379].

Parity graphs can be characterized in terms of perfect Cartesian products as follows.

Theorem 12.4.2 (Jansen [623], de Werra, Hertz [310]) G *is parity if and only if* $G \times K_2$ *is perfect.*

A *simplicial decomposition* is the recursively defined analogue to writing G as the union of two proper subgraphs overlapping in a complete graph ("a simplex"). Primes are graphs that cannot be decomposed further. Simplicial decompositions were used by K. Wagner [1072] for the characterization of the class of all graphs that are not contractible to a complete K_5. For the theory of simplicial decomposition, see also Halin [503, 505, 507] and Diestel [311].

Note that there are also several other types of decompositions of graphs. Two examples are as follows. In [1053, 1054, 1055, 1051], a decomposition for unigraphs is studied, which leads also to decomposition results for matroidal, matrogenic, and threshold graphs (chapter 13).

A number of other operations introduced for dealing with perfect graphs can be viewed as decompositions; these include the even, odd, and partner decompositions as well as the amalgam operation; these are discussed in the perfect graph chapter (chapter 2); see Theorem 2.2.2.

12.5 Minimal separators

Separation properties of graphs are algorithmically important. These properties allow a graph to be cut into subgraphs, and allow a problem to be solved on the graph by solving simpler instances of the problem on the subgraphs. For basic notions of a cutset see Definition 1.1.6; for cutset properties of chordal graphs see Theorem 1.2.1. The next section contains some general ideas that are used repeatedly in the following sections.

Definition 12.5.1 *Let $G = (V, E)$ be a connected graph.*

*For vertices $a, b \in V$, a subset $S \subset V$ is an a, b-*separator *if after the removal of S the vertices a and b are in different connected components of $G - S$.*

$S \subset V$ is a minimal a, b-separator *if no proper subset of S is an a, b-separator.*

$S \subset V$ is a minimal separator *if there are nonadjacent vertices $a, b \in V$ such that S is a minimal a, b-separator.*

$S \subset V$ is an inclusion-minimal separator *of G if no proper subset of S is a separator of G.*

Note that there are minimal separators that are not inclusion minimal. The following lemma is a technical tool for dealing with separators (see [697]).

Lemma 12.5.1 *Let $G = (V, E)$ be a connected graph and $S \subset V$ be a separator of G.*

(i) *S is a minimal separator of G if and only if there are two connected components of $G - S$ such that every vertex of S has a neighbor in both the components.*

(ii) *S is an inclusion-minimal separator of G if and only if every vertex of S has a neighbor in every connected component of $G - S$.*

There is another concept that is helpful in connection with minimal triangulations of graphs (see Definition 11.1.4) and finding the treewidth of graphs.

Definition 12.5.2 (Kloks, Kratsch [671], Kratsch [697]) *Two minimal separators S_1, S_2 are* noncrossing *if $S_1 \setminus S_2$ is contained in one connected component of $G - S_2$ and $S_2 \setminus S_1$ is contained in one connected component of $G - S_1$.*

Lemma 12.5.2 [671, 697] *Let G be a chordal graph. Then every pair of minimal separators in G is noncrossing.*

12.6 Classes with a polynomially bounded number of minimal separators

In general, a graph may have an exponential number of minimal separators. Many graph classes have a polynomially bounded number of minimal separators, which proves useful for solving a variety of algorithmic problems. A good example are chordal graphs: It is easy to see that a chordal graph $G = (V, E)$ has at most $n = |V|$ maximal cliques (see Theorem 1.2.2). Since chordal graphs are characterized by the existence of clique trees (Theorem 1.2.3), it follows that the set of minimal separators of G consists of all sets $C \cap C'$, where C and C' are maximal cliques of G assigned to adjacent nodes in the clique tree. Thus, G has at most $n - 1$ minimal separators.

For d-trapezoid graphs (see Definition 4.7.6) and subclasses such as the permutation graphs, the minimal separators can be described by *scanlines*; see [121] for permutation graphs and [857, 697] for d-trapezoid graphs.

It follows that the number of minimal separators of a d-trapezoid graph with n vertices is at most $(2n-3)^{d-1}$ [697].

Similar results hold for co-comparability graphs of bounded dimension, trapezoid graphs, and permutation graphs as well as for some subclasses of weakly chordal graphs [697]. For chordal bipartite graphs, analogously to Theorem 12.7.4, the minimal separators are closely related to complete bipartite subgraphs.

Lemma 12.6.1 (Kratsch [697]) *Let B be a chordal bipartite graph and S be a minimal separator of B. Then either there is a vertex v of B with $S = N(v)$ or S is the intersection of two vertex sets S_1, S_2, which induce complete bipartite subgraphs in B.*

It follows that a chordal bipartite graph has at most $n + \binom{m}{2}$ minimal separators. For weakly chordal graphs, Bouchitté and Todinca [140] have recently shown the following.

Theorem 12.6.1 *A weakly chordal graph $G = (V, E)$ has at most $|\overline{E}|$ minimal separators.*

For circle graphs and circular-arc graphs there are also polynomial bounds.

Lemma 12.6.2 [697] *A circle graph (circular-arc graph) on $n \geq 2$ vertices has at most $2n^2 - 3n$ minimal separators.*

Kloks and Kratsch [670] give an algorithm to enumerate all minimal separators in $\mathcal{O}(n^5 R)$ time, where R is the number of separators.

A similar result can also be achieved for edge separators. Tsukiyama et al. [1036] give an algorithm to enumerate all minimal edge separators for a pair of vertices in $\mathcal{O}((n+m)R)$ time, where R is the number of separators. Thus, all minimal edge separators of a graph can be enumerated in $\mathcal{O}(mn^2 R)$ time.

12.7 Clique, biclique, and stable cutsets

The reverse direction of separating a graph by its clique cutsets can also be understood as an operation on graphs if one defines the following operation.

Definition 12.7.1 *The* clique-bonding operation *is defined as follows: Let $G_1 = (V_1, E_1)$, $G_2 = (V_2, E_2)$ be two graphs with $V_1 \cap V_2 = \emptyset$, which both contain cliques C_1, C_2 of size k. Then the resulting graph G of the clique-bonding operation is obtained by identifying C_1 with C_2, i.e., G has* *vertex set* $(V_1 \cup V_2) \setminus C_2$ *and*

edge set $E_1 \cup E_2 \setminus (\{xy : x, y \in C_2\} \cup \{xy : x \in C_2 \text{ and } y \in V_2 \setminus C_2\})$.

This operation is perfection preserving.

Theorem 12.7.1 (Berge [90]) *If* G_1 *and* G_2 *are perfect graphs, then every clique-bonding result of* G_1, G_2 *is perfect.*

The recognition of clique cutsets plays an important role in algorithmic questions. Gavril [427] introduces clique cutset trees and Whitesides [1081] gives an $\mathcal{O}(nm)$-time algorithm for determining whether a given graph has a clique cutset. The complete decomposition, i.e., the computation of the clique cutset tree, takes $\mathcal{O}(n^3m)$ steps if one applies the Whitesides algorithm repeatedly. An improvement, which takes $\mathcal{O}(nm)$ steps and uses elimination scheme properties (see the elimination chapter (chapter 5)), is described by Tarjan [1020], where algorithmic applications (divide-and-conquer algorithms) are also given. In particular, problems such as maximum weighted independent set, minimum fill-in, maximum-weighted clique, and coloring are solvable in polynomial time on G if they are solvable on the final indecomposable graphs found by the clique-cutset decomposition. This is used to design polynomial problems for the maximum weighted clique and independent-set problems on EPT graphs (see Definition 4.4.3).

Gavril [427] defines the following graphs.

Definition 12.7.2 *Let* G *be a graph.*

G is a type-1 graph *if its vertex set can be partitioned into two disjoint sets* V_1, V_2 *such that* $G(V_1)$ *is a connected bipartite graph,* $G(V_2)$ *is a clique, and every vertex of* V_1 *is adjacent in* G *to every vertex of* V_2 *(a join* $V_1 * V_2$ *between* V_1 *and* V_2*). Here, one of* V_1, V_2 *may be empty.*

G is a type-2 graph *if it is a complete* k*-partite graph for some* k.

G is primitive *if* H *has no clique cutset.*

G is clique separable *if every induced primitive subgraph of* G *is of type* 1 *or type* 2.

Obviously, type-1 and type-2 graphs are perfect.

Theorem 12.7.2 (Gallai [415]) *A Gallai graph that is primitive must be of type* 1 *or type* 2.

Thus, Gallai graphs are clique separable. It is clear that clique-separable graphs are perfect, since their primitive subgraphs are perfect and clique bonding preserves perfectness. Moreover, clique-separable graphs are alternately orientable, preperfect, and strict quasi parity.

Theorem 12.7.3 (Gavril [427], Whitesides [1082]) *It can be tested in polynomial time whether a graph is clique separable.*

An edge-separation property of chordal bipartite graphs is the following.

Definition 12.7.3 *Let $B = (X, Y, E)$ be a bipartite graph. A pair of edges ab and cd of B is* separable *if there exists a set S of vertices such that after removal from B the edges ab and cd are in distinct connected components of $B((X \cup Y) \setminus S)$. In this case S is an* edge separator *for ab and cd.*

Theorem 12.7.4 (Golumbic, Goss [456]) *A bipartite graph is chordal bipartite if and only if every minimal (with respect to set inclusion) edge separator induces a complete bipartite subgraph.*

Weakly chordal graphs also have a nice separator property: If $x, y \in V$ form a two-pair in the graph G, then $N(x) \cap N(y)$ is a cutset of G.

It is natural to study cutsets that are stable instead of being a clique. Tucker [1045] discussed stable cutsets for coloring graphs. He showed that if a graph has a stable cutset such that no odd hole goes through it, then the coloring problem of the graph is reducible to that of its components.

Corneil and Fonlupt [245] used stable cutsets for generating new classes of perfect graphs. Perhaps surprisingly, it is harder to find stable cutsets of a graph than it is to find clique cutsets.

Theorem 12.7.5 (Chvátal [206]) *Determining whether a line graph has a stable cutset is $\mathbb{NP}$-complete.*

This problem has also been shown to be difficult for other classes. The *connectivity* of an incomplete graph is the minimum cardinality of a cutset in that graph; the connectivity of the complete graph K_n is $n - 1$.

Theorem 12.7.6 (Brandstädt et al. [147]) *Determining whether a K_4-free graph or a graph with connectivity 2 has a stable cutset is $\mathbb{NP}$-complete.*

This theorem is best possible in the sense that finding a stable cutset in triangle-free graphs and in separable graphs (graphs with connectivity 1) is easy. For hole-free graphs and AT-free graphs the problem can be solved in polynomial time [147].

12.8 Small and balanced separators

In this section we study separators of small size with the property that the size of connected components once the separator is removed is much smaller than the size of the entire graph. For some applications of such separators, see Chung [198] and Kratsch [697]. Some algorithms using this technique work efficiently because the natural recurrence that comes from splitting the graph on the basis of a separator has a polynomial bound if the components are roughly balanced.

Let $S : \mathbb{N} \to \mathbb{N}$ be a monotone increasing function.

Definition 12.8.1 (Galil, Kannan, Szemerédi [413])

> *A connected graph G has an S-*separator *C if there is a partition $V = A \cup B \cup C$, $|A|, |B| \geq n/3$, $|C| \leq S(n)$, and $G(V \setminus C)$ is disconnected.*
>
> *A class of graphs is S-*separable *if every graph in the class has an S-separator.*
>
> *A class is* separable *if it is S-separable for some function $S(n)$ that is $o(n)$.*

Theorem 12.8.1 (The planar separator theorem, Lipton, Tarjan [731]) *The class of planar graphs is $\mathcal{O}(\sqrt{n})$-separable.*

In [18] this result is extended to every class of graphs with an excluded minor. [441, 318] extend the result to graphs of genus g for any fixed g.

Other examples of classes with especially good separators are trees and outerplanar graphs [722].

There is also a separator theorem for chordal graphs: Each chordal graph can be cut in half by removing $\mathcal{O}(\sqrt{m})$ vertices [442]. This bound, however, can only be of restricted use, since the cliques K_n, $n \geq 1$, are chordal.

For interesting connections between separator properties of k-pushdown graphs—a special class of k-page graphs—and simulation properties in computational complexity, see [413].

For the Hamiltonian circuit problem, not only the size of separators is interesting, but also the number of connected components after deleting the separator. This is reflected in the following notion of *toughness* of a graph.

Definition 12.8.2 (Chvátal [200]) *Let $c(G)$ denote the number of connected components of a graph G. The* toughness *of a complete graph K_n is $t(K_n) = \infty$. If G is not complete, then the toughness $t(G)$ of G is*

$$t(G) = \min\{ \tfrac{|S|}{c(G(V \setminus S))} : S \text{ is a separator of } G\}.$$

Obviously from the existence of a Hamiltonian circuit in G it follows that $t(G) \geq 1$ but not vice versa. The toughness is related to some other interesting parameters. Unfortunately $t(G)$ is hard to compute.

Theorem 12.8.2 (Bauer, Hakimi, Schmeichel [71]) *The problem $\{(G, k) : t(G) \geq k\}$ is co-$\mathbb{NP}$-complete.*

For some special intersection graph classes the problem becomes easier [673]. There is a well-known open problem regarding toughness and Hamiltonicity.

Conjecture 12.8.1 [200] *There is a constant k such that all graphs with toughness $\geq k$ are Hamiltonian.*

Bauer, Broersma, and Veldman [70] show by example that k must be > 2. For chordal graphs, such a constant is known to exist.

Theorem 12.8.3 (Chen et al. [187]) *Every 18-tough chordal graph is Hamiltonian.*

Chapter 13

Threshold Graphs and Related Concepts

There are many results on threshold graphs and related concepts, such as degree sequences, unigraphs, split graphs, the Dilworth number, and matroidal and matrogenic graphs. The monograph of Mahadev and Peled [762] is an excellent source describing the world of threshold graphs. Note that there are also other survey papers such as those by Tyshkevich, Chernyak, and Chernyak [1058, 1059, 1060], which describe several of the papers on this topic that were written in Russian and are not widely known.

Threshold graphs are also closely related to Ferrer's digraphs, as discussed by Cogis in [230].

13.1 The threshold dimension

Threshold graphs were introduced in Definition 4.8.4 as a special case of threshold tolerance graphs, and in Theorem 6.6.3 threshold graphs are characterized as the comparability graphs of threshold orders. The same theorem gives a forbidden-subgraph characterization of this class: A graph is threshold if and only if it contains no induced $2K_2$, C_4 and no P_4.

Originally, however, threshold graphs were introduced by Chvátal and Hammer [216] as follows (see [452]):

Let $G = (V, E)$ be a graph with $V = \{v_1, \ldots, v_n\}$. Every subset $X \subseteq V$ can be represented by its characteristic vector $x = (x_1, \ldots, x_n)$, namely,

$$x_i = \begin{cases} 1 & \text{if } v_i \in X, \\ 0 & \text{otherwise,} \end{cases}$$

and can be interpreted as a corner of the unit hypercube in $\mathbb{R}^n$.

Definition 13.1.1 *The* threshold dimension $\theta(G)$ *of the graph G is the minimum num-*

ber k of linear inequalities

$$(1)\quad \begin{array}{l} a_{11}x_1 + a_{12}x_2 + \cdots + a_{1n}x_n \le t_1, \\ \vdots \\ a_{k1}x_1 + a_{k2}x_2 + \cdots + a_{kn}x_n \le t_k, \end{array}$$

such that X is an independent set if and only if its characteristic vector $x = (x_1, \ldots, x_n)$ satisfies the system (1) *of inequalities.*

Theorem 13.1.1 ([216], Henderson, Zalcstein [540]) *The graph G is a threshold graph if and only if $\theta(G) \le 1$.*

This means that G is a threshold graph if and only if there is a hyperplane that cuts the unit hypercube in half in such a way that on one side all corners of the hypercube correspond to stable sets of G and on the other side all corners correspond to nonstable sets.

Equivalently, the threshold dimension of a graph $G = (V, E)$ is the minimum number k of threshold graphs $G_i = (V, E_i)$, $i \in \{1, \ldots, k\}$, such that $\bigcup E_i = E$ (see [452]).

Note that determining the threshold dimension is hard even for threshold dimension at most 3.

Theorem 13.1.2 (Yannakakis [1094]) $\{G : \theta(G) \le 3\}$ *is* $\mathbb{NP}$*-complete.*

Definition 13.1.2 (Hammer, Mahadev, Peled [516]) *The graph G is a k-*threshold graph *if $\theta(G) \le k$.*

Of special interest is the case $k = 2$. Ibaraki and Peled [602] conjectured that a graph G is 2-threshold if and only if the *conflict graph* G^* defined below is bipartite.

Definition 13.1.3 (Peled [866]) *Let $G = (V, E)$ be a graph.*

> *An* alternating 4-cycle *of G is a subgraph with four vertices a, b, c, d such that $ab, cd \in E$ and $ac, bd \notin E$, i.e., the four vertices induce a C_4, P_4, or $2K_2$ in G.*
>
> *The vertices of G^* are the edges of G and two vertices e_1, e_2 of G^* are adjacent if and only if e_1, e_2 form an alternating 4-cycle in G.*

The conjecture of Ibaraki and Peled was proved for split graphs in [602] and independently in [229, 320]. The conjecture was proved in general by Raschle and Simon in [893], which immediately gives an $\mathcal{O}(|E|^2)$-time recognition algorithm for 2-threshold graphs. An earlier recognition algorithm is given by Ma in [749]. Sterbini and Raschle [1001] give an $\mathcal{O}(n^3)$ implementation of Ma's algorithm.

2-threshold graphs are alternately orientable [556], perfectly orderable [219], and weakly chordal [516]. An interesting subclass of 2-threshold graphs is given in the following definition.

Definition 13.1.4 [516] *The graph G is a* strict 2-threshold graph *if G is the union of two threshold graphs T_1, T_2 such that every triangle of G is a triangle in T_1 or T_2.*

Hammer, Mahadev, and Peled [516] show the following inclusion:

strict 2-threshold $\subset$ comparability.

The polynomial-time recognition algorithm for strict 2-threshold graphs given by Mahadev and Peled [761] was improved to linear-time recognition by Petreschi and Sterbini [871].

A concept that is closely related to 2-threshold graphs is given in the following.

Definition 13.1.5 (Hammer, Mahadev [514]) *G is a* bithreshold graph *if G is the edge intersection of two threshold graphs T_1, T_2 and if every independent set of G is independent in T_1 or in T_2.*

Hammer and Mahadev [514] show that the complements of bithreshold graphs are 2-threshold graphs and give an easy $\mathcal{O}(n^4)$ time-bounded recognition algorithm for these graphs. De Agostino, Petreschi, and Sterbini [297] show that bithreshold graphs can be recognized in $\mathcal{O}(n^3)$ time. Hammer, Mahadev, and Peled [517] study the structure of bithreshold and bipartite bithreshold graphs and give a forbidden subgraph characterization of bipartite bithreshold graphs.

Hammer and Mahadev [514] show that every bithreshold graph is perfectly orderable. This result is generalized by Hertz [544] by introducing a class called *bipartable graphs*:

bithreshold $\subset$ bipartable;

bipartable $\subset$ perfectly orderable.

13.2 Constant-bounded Dilworth number

The notion of the Dilworth number dilw(G) of a graph G is given in Definition 1.1.17. The Dilworth number of a graph G can be determined by constructing a graph $D(G) = (V, E')$ with $xy \in E'$ if and only if x dominates y or vice versa. Hoàng and Mahadev [570] show that $\alpha(D(G)) = \text{dilw}(G)$. $D(G)$ is a comparability graph, and for such graphs the bottleneck step for computing the independence number $\alpha(D(G))$ is computing the size of a maximum matching in a bipartite graph [400]. $D(G)$ can clearly be constructed in $\mathcal{O}(MM)$ time, which gives a polynomial-time algorithm for computing the Dilworth number. Felsner, Raghavan, and Spinrad give faster algorithms for determining whether $D(G) = k$ for a fixed constant k [384].

The constant-bounded Dilworth number gives rise to some interesting graph classes.

Theorem 13.2.1 (Földes, Hammer [397], Chvátal, Hammer [217]) *The graph G is a threshold graph if and only if $dilw(G) = 1$.*

Dilworth number 2 graphs also have an interesting threshold characterization.

Definition 13.2.1 (Benzaken, Hammer, de Werra [86]) *The graph $G = (V, E)$ is a* threshold signed graph (TS-graph) *if there are two positive real numbers s, t and real vertex weights $a_1, \ldots, a_n$ such that*

$a_i < \min\{s,t\}$ *for* $i \in \{1, \ldots, n\}$;

$v_i v_j \in E$ *for* $i \neq j$ *if and only if either* $|a_i + a_j| \geq s$ *or* $|a_i - a_j| \leq t$.

Theorem 13.2.2 [86] *Let G be a graph.*

(i) *G is a TS-graph if and only if dilw(G)* ≤ 2.

(ii) *G is a threshold graph if and only if there is a TS representation of G with* $s = t$.

The definition of TS-graphs leads to two subclasses called *type* I and *type* II:

Type I: distinct vertices v_i, v_j are adjacent if and only if $|a_i + a_j| \geq s$;

Type II: distinct vertices v_i, v_j are adjacent if and only if $|a_i - a_j| \leq t$.

For the next property see the book of Mahadev and Peled [762].

Proposition 13.2.1

(i) *The graphs of type* I *are exactly the unions of two vertex-disjoint threshold graphs.*

(ii) *The graphs of type* II *are bipartite.*

Type II graphs appear also in connection with Ferrer's digraphs under the name *difference graphs* (see [762]) and in a paper of Yannakakis [1094] under the name *chain graphs.* For more properties of these graphs and the use of these graphs for proving several NP-completeness results including order dimension, interval order dimension, boxicity, and threshold dimension given by Yannakakis [1094], see [762].

A forbidden induced subgraph characterization of TS-graphs is given in [86]. The class of TS-graphs is closed under complement. Benzaken, Hammer, and de Werra [87] give a linear-time recognition algorithm for TS-graphs.

Theorem 13.2.3 [86] *If G is a TS-graph, then G is a permutation graph and G or* $\overline{G}$ *is an interval graph. The boxicity of TS-graphs is at most* 2.

A characterization of the permutations corresponding to threshold graphs can be found in [452].

Theorem 13.2.4 [87] *G is a split graph with dilw(G)* ≤ 2 *if and only if G and* $\overline{G}$ *are interval graphs.*

A forbidden induced subgraph characterization of this class is given by Benzaken, Hammer, and de Werra [87].

In [875] it is shown that graphs with Dilworth number 3 are strongly perfect. A stronger statement follows from [570] (see Remark 5.7.1), which shows that for every graph G of Dilworth number 3, either G or $\overline{G}$ contains a simplicial vertex. Thus, graphs of Dilworth number at most 3 are good [570].

The best containment known for Dilworth number 4 graphs is much weaker. Payan [863] shows that graphs with Dilworth number 4 are perfect. Clearly, C_5 has Dilworth number 5; so graph classes of Dilworth number larger than 4 will not be perfect graph classes.

13.3 Degree sequences

For a graph G let $d_1 \geq d_2 \geq \cdots \geq d_n$ be the ordered sequence of the degrees of its vertices. Several graph classes can be characterized by properties of their degree sequences.

Definition 13.3.1 *A sequence of integers $d_1 \geq d_2 \geq \cdots \geq d_n$ with $n-1 \geq d_1$ is* graphic *if there is a graph having $d_1, \ldots, d_n$ as its degree sequence.*

There are several criteria for graphic sequences; see Sierksma and Hoogeveen [977] for seven of them and the monograph of Mahadev and Peled [762] for further discussion. We mention here only the following.

Theorem 13.3.1 (Erdős, Gallai [354]) *A sequence of integers $d_1 \geq d_2 \cdots \geq d_n$ with $n-1 \geq d_1$ is graphic if and only if*

(i) $\sum_{i=1}^{n} d_i$ *is even, and*

(ii) $\sum_{i=1}^{r} d_i \leq r(r-1) + \sum_{i=r+1}^{n} \min\{r, d_i\}$ *for* $r \in \{1, 2, \ldots, n-1\}$ (*the* rth Erdős–Gallai inequality).

Split graphs have a degree-sequence characterization.

Theorem 13.3.2 (Hammer, Simeone [518], Tyshkevich, Melnikow, and Kotov [1051, 1061]) *Let G be a graph with degree sequence $d_1 \geq d_2 \geq \cdots \geq d_n$ and $\omega = \max\{i : d_i \geq i-1\}$. Then G is a split graph if and only if*

$$\sum_{i=1}^{\omega} d_i = \omega(\omega - 1) + \sum_{i=\omega+1}^{n} d_i.$$

Thus, split graphs are those graphs for which equality holds in the ωth Erdős–Gallai inequality. Note that in this case $\omega(G) = \omega$. This degree sequence characterization implies a linear-time recognition for split graphs; in fact, the time complexity is $\mathcal{O}(n)$ if the degree sequence is given. Furthermore, if G is a split graph, then every graph with the same degree sequence as G is a split graph as well. Note, however, that the degree sequence of a split graph does not determine the graph up to isomorphism.

Definition 13.3.2 *G is a* unigraph *if G is determined by its degree sequence up to isomorphism, i.e., if a graph H has the same degree sequence as G, then H is isomorphic to G.*

The following papers deal with characterizations and recognition of unigraphs: [663, 727, 686, 687, 1053, 1054, 1055, 1056, 287]. Tyshkevich and Chernyak [1056] give a linear-time recognition algorithm for unigraphs. Note that unigraphs are not necessarily perfect since C_5 is determined up to isomorphism by its degree sequence (2,2,2,2,2). The next theorem shows that threshold graphs are unigraphs. Let

$0 < \delta_1 < \cdots < \delta_k < |V|$ be the degrees of the nonisolated vertices,

$\delta_0 = 0, \delta_{k+1} = |V| - 1$, and

$D_i = \{v : v \in V$ and $deg(v) = \delta_i\}$.

Then $V = D_0 \cup D_1 \cup \cdots \cup D_k$ is the *degree partition* of G.

Theorem 13.3.3 (Chvátal, Hammer [216]) *The following conditions are equivalent:*

(i) *G is a threshold graph;*

(ii) *For all $x \in D_i$, $y \in D_j$, $x \neq y$, $xy \in E$ if and only if $i + j > m$;*

(iii) *The recursions below are satisfied:*

$$\delta_{i+1} = \delta_i + |D_{m-i}| \text{ for all } i \in \{0, 1, \ldots, \lfloor m/2 - 1 \rfloor\};$$
$$\delta_i = \delta_{i+1} - |D_{m-i}| \text{ for all } i \in \{m, m-1, \ldots, \lfloor m/2 \rfloor + 1\}.$$

Thus, in this case the structure of the graph G is completely described by its degree sequence. This implies that threshold graphs are unigraphs.

Theorem 13.3.4 (Hammer, Ibaraki, Simeone [509]) *G is a threshold graph if and only if equality holds in each of the Erdős–Gallai inequalities.*

For more characterizations of threshold graphs in terms of degree sequences see [762].

Although testing whether a sequence is graphic is easy, the following generalization of graphic sequences to (open-neighborhood) hypergraphs proves to be much more difficult to test for. The *neighborhood list (NL) problem* is defined as follows:

A set V and a list (multiset) $\mathcal{N} = (N_1, \ldots, N_n)$ of subsets of V are given. Is there a graph G with $\mathcal{N} = \mathcal{N}(G)$?

Theorem 13.3.5 (Aigner, Triesch [8, 7]) *The decision problem NL is $\mathbb{NP}$-complete.*

It should be remarked that Aigner and Triesch [8, 7] also show that for the restriction to forests the problem is easy, and a bipartite variant is isomorphism complete.

13.4 Matroidal and matrogenic graphs

Matroidal graphs are a generalization of threshold graphs as can be seen from the forbidden-subgraph characterization of threshold graphs: G is a threshold graph if and only if G contains no induced $2K_2, P_4$, or C_4.

Definition 13.4.1 *Let $G = (V, E)$ be a graph.*

An edge set $E' \subseteq E$ is threshold independent *in G if the subgraph of G induced by E' is a threshold graph. Let $\mathcal{I}_E$ denote the family of threshold-independent edge sets.*

Analogously, a vertex set $V' \subseteq V$ is threshold independent *if $G(V)$ is a threshold graph. Let $\mathcal{I}_V$ denote the family of threshold-independent vertex sets.*

Definition 13.4.2 (Peled [866]) *A graph $G = (V, E)$ is* matroidal *if the threshold-independence system $(E, \mathcal{I}_E)$ is a matroid.*

Definition 13.4.3 (Földes, Hammer [396]) *A graph $G = (V, E)$ is* matrogenic *if the threshold-independence system $(V, \mathcal{I}_V)$ is a matroid.*

In [866] the matroidal graphs are defined in slightly different terms (alternating 4-cycles and couples). We have used here a description from the survey paper of Tyshkevich, Chernyak, and Chernyak [1059]. Matrogenic and matroidal graphs have very similar structure. Peled [866] characterizes matroidal graphs in terms of forbidden induced subgraphs on five vertices. For this purpose we need a configuration $\mathcal{F}$ on five vertices a, b, c, d, e for which ab, ac, de are edges and ad, be, ce are nonedges. The other pairs are not specified.

Theorem 13.4.1 [866] *A graph is matroidal if and only if it contains neither the configuration $\mathcal{F}$ nor an induced C_5.*

It follows that matroidal graphs can be recognized in polynomial time, and the complement of a matroidal graph is also matroidal. Peled [866] shows by structural investigations of matroidal graphs that matroidal graphs are chordal or cochordal.

Theorem 13.4.2 [396] *A graph is matrogenic if and only if it does not contain the configuration $\mathcal{F}$.*

Obviously this implies the following corollary.

Corollary 13.4.1 *A graph is matroidal if and only if it is matrogenic and contains no induced C_5.*

Furthermore, matrogenic graphs are closed under complementation.

Tyshkevich and Chernyak [1053, 1054, 1055, 1051] introduce a decomposition having the property that a graph is a unigraph (matrogenic, matroidal, threshold) if and only if all its indecomposable components are a unigraph (matrogenic, matroidal, threshold) (see [1052]). The indecomposable graphs of this type are described. Tyshkevich [1052] and Marchioro et al. [768] show that matrogenic graphs are unigraphs.

The degree sequences of matroidal and matrogenic graphs are described in [1052], which leads to linear-time recognition algorithms for these classes. The description of matrogenic graphs is independently obtained in [768].

For more informations on these graphs see [762], where further extensions of threshold graphs are described:

- *domishold graphs* [85, 767] and generalizations;
- *box-threshold graphs* [897, 867, 1057];
- *pseudothreshold graphs* [217];
- *equistable graphs* [862, 763];

- *generalized threshold graphs* [903];
- *universal threshold graphs* [510].

For some generalizations of threshold graphs in terms of tolerances, see the chapter on models and interactions (chapter 4).

Chapter 14

The Strong Perfect Graph Conjecture

Recall that Berge graphs are those graphs containing no odd holes and no odd antiholes as induced subgraphs. Since odd holes and odd antiholes are imperfect, perfect graphs are necessarily Berge. The SPGC, Conjecture 2.1.1, states that the perfect graphs are exactly the Berge graphs.

Berge proposed this conjecture around 1960; its truth is still unknown, although it has been proved for successively larger graph classes. In 1992, however, Prömel and Steger [884] showed that the SPGC is at least asymptotically true: Let $\mathcal{P}(n)$ and $\mathcal{B}(n)$ denote the the set of all perfect graphs and set of all Berge graphs on n vertices. While the SPGC states that $\mathcal{P}(n) = \mathcal{B}(n)$ for all n, Prömel and Steger proved that $\lim_{n\to\infty} \frac{|\mathcal{P}(n)|}{|\mathcal{B}(n)|} = 1$. Note that this fact does not exclude the possibility that the SPGC could be false.

The following concept plays an important role in all known attempts to prove the SPGC.

Definition 14.0.4 *A graph is* minimal imperfect *if it is not perfect, but all its proper induced subgraphs are perfect.*

Clearly, the SPGC can be reformulated as follows.

Conjecture 14.0.1 (The SPGC, second version) *The only minimal imperfect graphs are the odd holes and odd antiholes.*

Thus, it is conjectured that there exist no minimal imperfect Berge graphs, called *monsters*. Gurvich and Hougardy [483] and Gurvich and Udalov (cited in [213]) have shown that monsters must have at least 25 vertices, improving a previous result by Lam et al. [703].

In view of this conjecture, the structure of minimal imperfect graphs has been studied intensively. New properties of minimal imperfect graphs mean that new classes of perfect graphs can be generated.

14.1 Properties of minimal imperfect graphs

We shall only list those properties of minimal imperfect graphs that have been used, directly or indirectly, in proving the SPGC for particular cases or in generating classes of perfect graphs.

Consider a minimal imperfect graph G. Then it follows from the PGT (Theorem 2.1.2) that

(P1) $\overline{G}$ *is a minimal imperfect graph.*

Theorem 2.1.5 and the perfection of $G-v$ further imply that $G-v$ can be partitioned into $\omega(G)$ stable sets of cardinality $\alpha(G)$ and into $\alpha(G)$ cliques of cardinality $\omega(G)$. This motivated Bland, Huang, and Trotter [110] (see also [215]) to consider the following graphs.

Definition 14.1.1 *Let G be a graph and α, ω be two integers.*

> *G is* (α,ω)-partitionable *if G has $\alpha \cdot \omega + 1$ vertices and for each vertex v of G, $G-v$ admits a partition into α cliques of cardinality ω and also a partition into ω stable sets of cardinality α.*
>
> *G is* partitionable *if it is an (α,ω)-partitionable graph for some α, ω.*

As mentioned above, all minimal imperfect graphs G are (α,ω)-partitionable with $\alpha = \alpha(G)$ and $\omega = \omega(G)$. Thus, the class of partitionable graphs contains all potential counterexamples to the SPGC.

There are infinitely many partitionable graphs that are not minimal imperfect. Two constructions for certain classes of partitionable graphs are given in [215]. However, these constructions do not yield counterexamples to the SPGC, as shown by Sebő [961, 963] and Bacsó et al. [41].

There are many characterizations of partitionable graphs; for example, two of them are given in [164] and [969]. Recently, Gasparian [422] proved the following characterization for partitionable graphs; his short and simple proof made no use of Theorem 2.1.5, as well as no use of the substitution lemma for perfect graphs (Lemma 2.4.1).

Theorem 14.1.1 [422] *A graph G is (α,ω)-partitionable if and only if $\omega(G-S)=\omega$ for every stable set S of G, and G has a stable set S_0 of cardinality α such that for all $s \in S_0$, $\chi(G-s)=\omega=\omega(G-s)$.*

Since every minimal imperfect graph obviously satisfies the conditions of the theorem above, minimal imperfect graphs are partitionable. Thus Lovász's characterization of perfect graphs (Theorem 2.1.5) follows from Gasparian's theorem. Most properties known for minimal imperfect graphs hold also for partitionable graphs.

In (P2)–(P11) below, G stands for an (α,ω)-partitionable graph, and $n=\alpha\cdot\omega+1$.

(P2) $\alpha(G)=\alpha$, $\omega(G)=\omega$, $\chi(G)=\omega+1$, $\chi(G-v)=\omega=\omega(G-v)$, *and $\overline{G}$ is an (ω,α)-partitionable graph.*

(P3) *G has exactly n maximum cliques and exactly n maximum stable sets.*

(P4) *Every maximum clique meets all maximum stable sets except exactly one, and every maximum stable set meets all maximum cliques except exactly one.*

(P5) *Every vertex v of G belongs to exactly ω maximum cliques, and the partition of $G - v$ into ω maximum stable sets is unique and consists of the stable sets disjoint from those cliques containing v.*

(P6) *Every vertex v of G belongs to exactly α maximum stable sets, and the partition of $G - v$ into α maximum cliques is unique and consists of the cliques disjoint from those stable sets containing vertex v.*

(P7) *For every vertex v of G and two maximum stable-sets $S_1 \neq S_2$ in the stable sets-partition of $G - v$, the subgraph of G induced by $S_1 \cup S_2 \cup \{v\}$ is 2-connected.*

(P8) *G is $(2\omega - 2)$-connected.*

Properties (P2)–(P6) were proved first by Padberg [846] for minimal imperfect graphs and then by Bland, Huang, and Trotter [110] for partitionable graphs.

(P4) means that the maximum stable sets and cliques can be enumerated as $S_1, \dots, S_n$ and $C_1, \dots, C_n$, respectively, in such a way that $S_i \cap C_j = \emptyset$ if and only if $i = j$. In particular, $\omega(G - S) = \omega(G)$ for any stable set S.

(P5) and (P6), on the other hand, say that maximum stable sets and cliques can be encoded as $S(v, u)$ and $C(v, u)$, respectively, meaning the maximum stable set of $G - v$ that contains u and the maximum clique of $G - v$ that contains u. Note that *the* maximum clique disjoint from $S(x, y)$ is $C(y, x)$.

Property (P7) was proved by Buckingham and Golumbic [164]; it implies a result due to Tucker [1043] saying that the bipartite graph induced by two color classes in a minimal imperfect graph minus a vertex is connected.

(P8) was found by Sebő [962]; it implies that the minimum degree of a minimal imperfect graph is at least $2\omega - 2$, as proved by Olaru [837] and Reed [900]. The paper of Sebő discussed also the structure of cutsets of cardinality $2\omega - 2$.

We need some further notation to describe the next properties of partitionable graphs. In a graph, the vertex x *predominates* the vertex $y \neq x$ if they are adjacent and every maximum clique containing y also contains x, or they are nonadjacent and every maximum stable set containing x also contains y. Two (nonadjacent) vertices form an *even pair* if all chordless paths joining them have even length.

Property (P9) below was shown by Hammer and Maffray [512].

(P9) *G has no pair of vertices x, y such that x predominates y.*

Meyniel [781] (see also [398]) showed that no minimal imperfect graph has an even pair. Later, Bertschi and Reed [105] found that partitionable graphs also have this property.

(P10) (The even-pair lemma) *G has no even pair.*

A *star-cutset* in a graph is a cutset C such that some vertex in C is adjacent to all remaining vertices in C. Star-cutsets can be found in polynomial time [207]. In [207], Chvátal showed the following property of partitionable graphs; it is proved again by Maire [764].

(P11) (The star-cutset lemma) *G has no star-cutsets.*

The star-cutset lemma has been generalized by Olariu in [829]. In the next group of properties (P12)–(P19), G will be an unbreakable graph, as defined below. The term *unbreakable graphs* is due to Chvátal [207].

Definition 14.1.2 *A graph G is* unbreakable *if neither G nor its complement has a star-cutset.*

Chvátal's star-cutset lemma for minimal imperfect graphs and the PGT together imply that the class of unbreakable graphs contains all minimal imperfect graphs. Moreover, (P11) and (P2) imply that all partitionable graphs are unbreakable. However, there are unbreakable graphs that are not partitionable. Note that by definition, the class of unbreakable graphs is closed under taking complements.

(P12) *G has no two vertices* x, y *such that* $N(x) \subseteq \{y\} \cup N(y)$.

(P13) *G has no vertex* v *such that* $G - N[v]$ *is disconnected.*

(P14) *G has no vertex* v *such that* $N(v)$ *induces a disconnected subgraph in* $\overline{G}$.

(P15) *Each induced* P_3 *in G extends into some induced* C_k $(k \geq 4)$.

The first three properties are obvious. Property (P12) means that the Dilworth number of a minimal imperfect graph is equal to the number of its vertices, as first noted by Payan [863]. (P13) and (P14) were first shown for minimal imperfect graphs by Tucker [1043] and Olaru [836] (see also [840]), respectively. (P15) has been proved by Hoàng [555].

Recall that Definitions 1.5.1 and 1.5.2 define a nontrivial module in a graph; in the following, a nontrivial module will be called a *homogeneous set.* The definition of unbreakable graphs yields (P16) directly.

(P16) (The homogeneous set lemma) *G has no homogeneous set.*

For minimal imperfect graphs, this lemma follows from Chvátal's star-cutset lemma, but it also follows from the substitution lemma for perfect graphs (Lemma 2.4.1). To describe further properties of unbreakable graphs we recall the notion of a *wing* in a graph from Definition 1.1.5. Recall that G stands for an unbreakable graph.

(P17) *Every wing in G extends to a* P_4 *in each side.*

(P18) *Every vertex in G is the endpoint of at least two wings.*

The two properties above are due to Olariu [827]; together they imply the following fact proved by Preissmann [876] and Chvátal (unpublished): In an unbreakable graph, all vertices are both endpoints and midpoints of a P_4.

The structure of unbreakable graphs in terms of wings is discussed by Olariu [825, 827, 830]. The last property of minimal imperfect graphs in the context of unbreakable graphs that we shall give here was found by Hayward [528].

(P19) *Every vertex in G belongs to a hole or to an antihole.*

Note that Properties (P12)–(P19) for minimal imperfect graphs are "weaker" than (P1)–(P11), because partitionable graphs are also unbreakable. For a decision whether a graph is indeed a minimal imperfect graph, one would need "deeper" properties. We follow Olariu (see [756]) and call the property (P) *genuine* if all minimal imperfect graphs satisfy (P), but not all partitionable graphs do (hence not all unbreakable graphs). Thus, (P2)–(P19) all are nongenuine properties of minimal imperfect graphs.

In the remaining part of this section, G stands for a minimal imperfect graph.

A *small transversal* in a graph X is a subset $T \subseteq V(X)$, $|T| \leq \alpha(X)+\omega(X)-1$, such that T meets all maximum cliques and all maximum stable sets of X. Note that there are partitionable graphs having a small transversal; see [215]. With this notion, Chvátal pointed out the following property in [202, 203].

(P20) *G has no small transversal.*

An interesting generalization of small transversals, called *small 2-transversals*, was discussed in [403].

Two vertices in a graph are *antitwins* if every other vertex is adjacent to precisely one of them.

(P21) (The antitwin lemma) *G has no antitwins.*

Olariu [821] proved the antitwin lemma and gave a partitionable graph containing antitwins. Later, Maffray [756] showed that for all $\alpha, \omega \geq 5$, there are (α,ω)-partitionable graphs containing antitwins. The antitwin lemma can be generalized as follows. Let Q be a set of vertices in a graph X. A vertex $v \in X - Q$ is *Q-partial* if v is adjacent to some vertex in Q but not to all vertices in Q. Two disjoint subsets Q_1, Q_2 of $V(X)$ form a *homogeneous pair* if $|X - Q_1 - Q_2| \geq 2$, Q_1 or Q_2 has at least two vertices, and no vertex in $X - Q_1 - Q_2$ is Q_1-partial or Q_2-partial. Note that if X has an antitwin, then X contains a homogeneous pair. Chvátal and Shibi have shown in the following property [222].

(P22) (The homogeneous pair lemma) *G has no homogeneous pair.*

Homogeneous pairs in a graph can be found in polynomial time [367]. Another generalization of the antitwin lemma is given by Hoàng [560]. Two vertices $x \neq y$ of a graph G form a *three-pair* if all chordless paths in $G - xy$ have exactly three edges (the edge xy can be missing in G). Note that antitwins in G form a three-pair in G. Thus, the following generalizes the antitwin lemma.

(P23) (The three-pair lemma) *G has no three-pair.*

Recent study of *critical edges* in minimal imperfect graphs [961, 963, 962, 771] yields several genuine properties of Berge minimal imperfect graphs. An edge xy of G is *critical* if $\alpha(G-xy)=\alpha(G)+1$; a pair of nonadjacent vertices x,y is *co-critical* if in $\overline{G}$ the edge xy is critical. One can see that xy is critical if and only if there exists a stable set S not containing x,y of cardinality $\alpha(G)-1$ such that $S\cup\{x\}$ and $S\cup\{y\}$ are maximum stable sets. An analogous statement holds for cocritical pairs. Sebő [963] proved the following property.

(P24) *If G is a minimal imperfect Berge graph, G cannot have a set of vertices $\{v_1,v_2,\ldots,v_k\}$ such that v_1v_k is an edge and $\{v_i,v_{i+1}\}$ is a cocritical pair for $i=1,\ldots,k-1$.*

We conclude this section by giving two conjectures concerning two genuine properties that all minimal imperfect graphs may have. The first one is due to Chvátal [207]. A partition of $V(G)$ into four disjoint, nonempty sets A,B,C,D is a *skew partition* if all edges between A and B are present and no edge between C and D exists.

Conjecture 14.1.1 (The skew-partition conjecture) *No minimal imperfect graph has a skew partition.*

The validity of the above conjecture would imply the star-cutset lemma (see (P11)). The papers of Hoàng [560] and Cornuéjols and Reed [259] contain some partial results in this direction. The result in [259], saying that minimal imperfect graphs different from odd holes cannot have a cutset that induces a complete multipartite subgraph, generalizes a former result of Tucker [1045], saying that minimal imperfect graphs different from odd holes cannot have a stable cutset.

The second conjecture is due to Meyniel and Olariu [782] (see also [560]). Two vertices $x\neq y$ in G form an *odd pair* if all chordless paths in $G-xy$ have an odd number of edges (the edge xy can be missing in G).

Conjecture 14.1.2 (The odd pair conjecture) *No minimal imperfect graph has an odd pair.*

The validity of the odd pair conjecture would imply the three-pair lemma (see (P23)). In [782] it is shown that the odd pair conjecture is equivalent to the following statement:

> A minimal imperfect graph G contains an edge not belonging to a triangle if and only if G is an odd hole.

In view of this, the following fact, proved by De Simone and Galluccio [303], may be of interest.

Proposition 14.1.1 *Every minimal imperfect Berge graph has a spanning subgraph H such that H is minimal imperfect and Berge, and each edge of H belongs to some triangle in H.*

For more conjectures on minimal imperfect graphs (in terms of critical edges) see Sebő [963]. Notice that the truth of the SPGC would imply all these conjectures.

14.2 Some equivalent versions of the SPGC

Each of the conjectures stated in this section is equivalent to the SPGC (Conjecture 2.1.1).

Let C_n^k denote the graph with vertices $v_1, v_2, \ldots, v_n$ and edges $v_i v_j$ for $|i-j| \leq k \pmod{n}$. Note that for $k \geq 2$, C_{2k+1} is an odd hole and C_{2k+1}^{k-1} is an odd antihole. Properties (P3)–(P6) of minimal imperfect graphs tell us that the maximum cliques of a minimal imperfect graph G form a (spanning) subgraph in G that has a structure very close to $C_{\alpha(G)\cdot\omega(G)+1}^{\omega(G)-1}$. In fact, Chvátal [202, 203] showed that the following is an equivalent version of the SPGC.

Conjecture 14.2.1 *Every minimal imperfect graph G with $\alpha(G) = \alpha$ and $\omega(G) = \omega$ has a spanning subgraph isomorphic to $C_{\alpha\cdot\omega+1}^{\omega-1}$.*

Trivially, the SPGC is true if and only if for every minimal imperfect graph G, *each* edge in G or *each* edge in $\overline{G}$ belongs to no triangle. The following reformulation of the SPGC, due to De Simone and Galluccio [303], is less trivial; it is also interesting in connection with the odd pair conjecture (Conjecture 14.1.2; see also remarks thereafter).

Conjecture 14.2.2 *Every minimal imperfect graph or its complement has an edge that belongs to no triangle.*

The next equivalent reformulation of the SPGC is related to the even pair lemma (see Property (P10)). A graph is *minimal even pair-free* if it has no even pair but each of its proper induced subgraphs has an even pair or is a clique. Hougardy [582] observed that odd holes and odd antiholes are exactly those graphs G with the property that G and $\overline{G}$ are minimal even pair free. This allows him to show the equivalence between the SPGC and the following.

Conjecture 14.2.3 *Every minimal imperfect graph is minimal even pair free.*

In studying critical edges in the context of minimal imperfect graphs, Markosjan, Gasparian, and Markosjan [771] and Sebő [963] obtained several equivalent forms of the SPGC. Two very interesting conjectures will be listed here; they are due to Sebő [963].

Conjecture 14.2.4 *Every minimal imperfect graph G has a maximum clique Q such that $G - Q$ is uniquely colorable.*

Conjecture 14.2.5 *Every minimal imperfect graph has a maximum clique in which the critical edges form a connected graph.*

See [963] for more information. Olaru [838, 839] showed the equivalence of the following conjecture to the SPGC.

Conjecture 14.2.6 *Every minimal imperfect graph is minimal strongly imperfect.*

Here a graph is minimal strongly imperfect if it is not strongly perfect (see Definition 2.3.1) but all its proper induced subgraphs are strongly perfect. Note that all conjectures equivalent to the SPGC so far are stated in terms of minimal imperfect graphs. Since the original version of the SPGC does not deal with minimal imperfect graphs, any equivalent form that also does not deal with minimal imperfect graphs may be interesting. Such results have recently been found by Markasjan, Gasparian, and Markosjan [771], Markosjan et al. [772], and Cai and Corneil [173]. The following conjecture was posed in [173].

Conjecture 14.2.7 *Every Berge graph is either perfectly* 2*-transversable or perfectly* 3*-transversable.*

There are many other alternative forms of the SPGC (all in terms of minimal imperfect graphs). They were found by Galeana-Sánchez [412], Giles and Trotter [443], Olariu [827], Olaru and Sachs [840], and Fouquet et al. [403]. Olariu [827] gave seven conjectures equivalent to the SPGC in terms of wings in minimal imperfect graphs (see Properties (P17), (P18)). One of them reads as follows: *The SPGC is true if and only if in each minimal imperfect graph, the spanning subgraph formed by all its wings is minimally* 2*-connected.*

14.3 Large classes of perfect graphs

Many large classes of perfect graphs were introduced by either restricting the notion of perfection (for example, strongly perfect graphs, absorbantly perfect graphs, or superperfect graphs; see the perfection chapter (chapter 2)) or using properties that minimal imperfect graphs cannot have. We shall describe here three typical classes defined of the latter type.

Quasi-parity graphs Meyniel [781] introduced the class of quasi parity graphs when he proved the even pair lemma (see Property (P10)). He called a graph G *quasi parity* if for each induced subgraph H of G, H or $\overline{H}$ has an even pair.

G is *strict quasi parity* if each induced subgraph of G is complete or has an even pair. Clearly, strict quasi-parity graphs are in particular quasi parity. The even pair lemma and Property (P1) together show that quasi-parity graphs are perfect.

Several known classes of perfect graphs are (strict) quasi parity. Examples include Meyniel graphs [781] (hence all parity graphs, all i-triangulated, i.e., Gallai graphs), perfectly orderable graphs [781] (hence all chordal graphs, all comparability graphs), weakly chordal graphs [530] (hence all complements of chordal graphs). All these graphs are perfectly contractile graphs [104, 366], which form a large subclass of strict quasi-parity graphs. Here, following Bertschi [104], a graph G is called *perfectly contractile* if for all induced subgraphs H of G, there is a sequence $H_0 = H, H_1, \ldots, H_k$ such that H_k is a clique, and H_{j+1} is obtained from H_j by contracting an even pair.

Other examples of (strict) quasi-parity graphs are opposition graphs [569], wing-triangulated graphs [585], bull-free Berge graphs [292], and skeletal graphs [545].

In [781] Meyniel asked whether all strongly perfect graphs are (strict) quasi-parity. Even more is conjectured in [366]:

Conjecture 14.3.1 *Every strongly perfect graph is perfectly contractile.*

As mentioned above, two important subclasses of strongly perfect graphs, namely Meyniel graphs (even quasi-Meyniel graphs [547, 366]) and perfectly orderable graphs, are perfectly contractile. Of course there are perfect graphs that are not quasi parity, for instance the line graph of the bipartite graph $K_{3,3} - e$ obtained from $K_{3,3}$ by deleting an edge. No polynomial-time algorithm is known for recognizing (strict) quasi-parity graphs. Note, however, that testing whether a given pair of vertices in a graph is an even pair is co–$\mathbb{NP}$-complete, as shown by Bienstock [107].

Everett et al. [366] discuss quasi-parity graphs and perfectly contractile graphs in detail.

Predomination-perfect graphs A graph G is *predomination perfect*, or in short, *preperfect*, if every induced subgraph of G has a predominant vertex (see Property (P9)). Preperfect graphs were introduced by Hammer and Maffray [512]; by (P9), they are perfect.

The class of preperfect graphs seems to be a very large class of perfect graphs, though very little is known about that class. However, Hammer and Maffray [512] have shown that all parity graphs (hence all bipartite graphs), and all i-triangulated graphs (hence all chordal graphs) are preperfect. They posed the following conjecture.

Conjecture 14.3.2 *All quasi-parity graphs are preperfect.*

Notice that the line graph of $K_{3,3} - e$ is preperfect but not quasi parity. Line graphs of bipartite graphs that are not preperfect and minimal with this property are characterized by Tuza and Wagler [1050].

The class bip* When Chvátal proved the star-cutset lemma (see Property (P11)), he gave a method for generating a large class of perfect graphs from a given one as follows. Let $\mathcal{G}$ be an arbitrary graph class. The new class $\mathcal{G}^*$ is defined recursively by the following two rules:

- If $G \in \mathcal{G}$ then $G \in \mathcal{G}^*$.
- If G or $\overline{G}$ has a star-cutset, and if $G - v \in \mathcal{G}^*$ for all vertices v of G, then $G \in \mathcal{G}^*$.

Property (P11) and the PGT together show that if $\mathcal{G}$ is a class of perfect graphs, then so is $\mathcal{G}^*$. Let **bip** denote the class of all bipartite graphs. **bip*** contains several known classes of perfect graphs: Meyniel graphs (hence all parity graphs and all i-triangulated graphs), perfectly orderable graphs (hence all chordal graphs, all comparability graphs) [207], weakly chordal graphs [526], opposition graphs [822], and alternately orientable graphs [556]. There are line graphs of bipartite graphs not belonging to **bip***. Meyniel asked whether **bip*** is a subclass of quasi-parity graphs. Even more is conjectured in [366]:

Conjecture 14.3.3 *All members in* **bip*** *are perfectly contractile.*

We note that by a result of Hayward [526], **triv*** is exactly the class of weakly chordal graphs; here **triv** denotes the class of all graphs with at most two vertices.

The paper [207] surveys graph operations preserving perfection. Besides the well-known *vertex substitution* and *complementation* due to Lovász (see the PGT and Lemma 2.4.1), other operations are *clique bonding, amalgamation, 2-amalgamation* (see section 11.6), and *split* and *generalized split decompositions* (see the decomposition chapter (chapter 12)). Using such a graph operation, one can obtain a new class of perfect graphs from a given one.

Another way to generate a large class of perfect graphs is proposed by Hujter and Tuza [600] (see also [1049]). In particular, they called a graph G *cograph contraction* if G is obtained from a cograph H by contracting some disjoint stable sets in H, and then joining between all "contracted vertices" by edges. By definition, all cographs are cograph contractions, and it is proved in [600] that cograph contractions are perfect. These graphs are characterized in [712].

We now address a "nonconventional" method of defining classes of perfect graphs. Let Φ be a graph operator that preserves the existence of odd holes and odd antiholes, as well as being closed under induced subgraphs, i.e.,

- If H is an induced subgraph of G, then $\Phi(H)$ is an induced subgraph of $\Phi(G)$.
- If G or $\overline{G}$ has an odd hole, then $\Phi(G)$ or $\overline{\Phi(G)}$ has an odd hole.

Examples are the complement-graph operator, the line graph operator, and the total graph operator. Let Φ be a graph operator with the properties described above. Then a graph G is *Φ-Berge* if $\Phi(G)$ is Berge.

Clearly, Φ-Berge graphs are necessarily Berge. Moreover, depending on the concrete operator Φ, Φ-Berge graphs may have more properties than Berge graphs, which would lead to a proof that the SPGC holds for these Φ-Berge graphs. In the case that Φ is the complement-graph operator, this problem coincides with the SPGC.

Some classes of perfect graphs defined by graph operators are the following:

1. The *Gallai graph* $\Gamma(G)$ of a graph G has all edges of G as its vertices; two edges $e \neq e'$ are adjacent in $\Gamma(G)$ if, in G, $e\Gamma e'$ holds (that is, e and e' are incident but do not form a triangle in G; see Definitions 2.4.5 and 4.2.2). Notice that there is a different notion of Gallai graphs in the sense of i-triangulated graphs. G is said to be *Gallai-perfect* if $\Gamma(G)$ has no (induced) odd holes. Sun [1007] showed that Gallai-perfect graphs are perfect. Gallai-perfect graphs form a large class of perfect graphs; it contains all bipartite graphs, all complements of bipartite graphs, and all line graphs of bipartite graphs.

2. For fixed $1 \leq k \leq 3$, the *k-overlap graph* $\mathcal{O}_k(G)$ has all induced P_4s of G as its vertices; two vertices are adjacent in $\mathcal{O}_k(G)$ if the corresponding P_4s in G have exactly k vertices in common. In [564], Hoàng, Hougardy, and Maffray showed that if $\mathcal{O}_3(G)$ is bipartite, then G is perfect. Related results are presented in [564, 532]. They are motivated by the semistrong perfect graph theorem on the P_4-structure of perfect graphs (see section 2.2). The next graph operator is also motivated by the P_4-structure of perfect graphs.

3. The *wing-graph* $W(G)$ of a graph G has the edges of G as its vertices; two edges are adjacent in $W(G)$ if they are the end-edges (called *wings*) of an induced P_4 in G.

Hoàng [559] proposed the following.

Problem 14.3.1 *Prove that G is perfect if $W(G)$ has no odd holes.*

Graphs considered in the above problem are called *wing perfect.* The problem is open even for the case in which $W(G)$ is bipartite; the graphs of this special case are sometimes called *Hoàng graphs* (see [213, 935]). Partial results toward resolving Problem 14.3.1 are presented in [559, 935, 585]. In [585] it is shown that if $W(G)$ is chordal, then G is strict quasi parity, hence perfect. Recently, Peterson [870] considered the clique graph operator $K(G)$; he proved that G is perfect whenever $K(G)$ is bipartite.

14.4 Graph classes satisfying the SPGC

The SPGC has been proved for several graph classes. The classes we shall address here split into three (not necessarily disjoint) groups: those defined by forbidden induced subgraphs, those defined by graph-valued functions or intersection models, and others.

14.4.1 F-free graphs

A popular way to attack the SPGC consists of showing that the conjecture is true for graphs not containing a given (usually small) graph F as an induced subgraph. Note that by the PGT, the SPGC holds for F-free graphs if and only if it holds for $\overline{F}$-free graphs. Thus, we do not list the corresponding classes of $\overline{F}$-free graphs.

For the case $|F| = 4$, i.e., F has four vertices, the SPGC is proved for

- *P_4-free graphs* or *cographs* by Seinsche [966] and Jung [635],
- *paw-free graphs* by Olariu [823] and Meyniel [780],
- *$K_{1,3}$-free (or claw-free) graphs* by Parthasarathy and Ravindra [860], Hsu [590] (see [599]), and Giles and Trotter [443]. A polynomial-time recognition algorithm for $K_{1,3}$-free perfect graphs is given by Chvátal and Sbihi [223];
- *$(K_4 - e)$-free (or diamond-free) graphs* by Tucker [1047] and Conforti [236] (the proof given by Parthasarathy and Ravindra [861] contains an error as pointed out in [1047]). A polynomial-time recognition algorithm for (K_4-e)-free perfect graphs is given by Fonlupt and Zemirline [399];
- *K_4-free graphs* by Tucker [1043, 1046, 1048]. The complexity of recognition of K_4-free perfect graphs is still open.

The only open case for $|F| = 4$ is $F = C_4$ (equivalently, $F = 2K_2$).

Problem 14.4.1 *Prove the SPGC for C_4-free graphs.*

The papers of De Simone and Galluccio [303] and of Xue [1093] contain partial results for this problem.

For the case $|F| = 5$, the SPGC is proved for

- *bull-free graphs* by Chvátal and Sbihi [222]. Bull-free perfect graphs can be recognized in polynomial time as shown by Reed and Sbihi [902];
- *dart-free graphs* by Sun [1007],
- *chair-free graphs* by Sassano [944].

The SPGC is still open for other forbidden graphs F with five vertices. The most investigated case is that of the P_5.

Problem 14.4.2 *Prove the SPGC for P_5-free graphs.*

Notice that by the PGT, Problem 14.4.1 is a special case of Problem 14.4.2. Partial results for Problem 14.4.2 are given by Olariu [824], Maffray and Preissmann [758], and Barré and Fouquet [68]. In [68] the SPGC is proved for (P_5, F)-free graphs, where F is any connected configuration on five vertices not containing an induced $2K_2$. Further results for graphs described by (more than one) forbidden configurations can be found in [179, 303, 527, 549].

14.4.2 Graph-valued functions and intersection models

Another way to choose a restricted class of graphs in attacking the SPGC makes use of graph operators. For example, one might ask for the validity of the SPGC for line graphs and for total graphs. Depending on the concrete operator Φ, the image $\Phi(G)$ may have more properties than G, which would lead to a proof that the SPGC holds for $\Phi(G)$. Thus, the problem for a given graph operator Φ is as follows: Prove the SPGC for the class of the images under the operator Φ. In the case that Φ is the graph complement operator, this problem coincides with the SPGC. The SPGC is proved for

- *line graphs*, by the results on $K_{1,3}$-free graphs [860, 443];
- *M-graphs*, by De Simone and Mannino [305]. These are a kind of undirected line graphs of (simple) directed graphs and arise in the following way: given $\vec{G} = (V, A)$, the M-graph $M(\vec{G})$ of $\vec{G}$ has A as its vertex set and two distinct arcs $(u, v), (w, x)$ are adjacent in $M(\vec{G})$ if and only if $v \in \{w, x\}$;
- *total graphs*, by Rao and Ravindra [892]. The total graph $T(G)$ of a graph $G = (V, E)$ has $V \cup E$ as its vertex set, and two distinct elements $x, y \in V \cup E$ are adjacent in $T(G)$ if and only if x, y are adjacent vertices in G, or x, y are incident edges in G, or one of x, y is an edge in G and the other is an end vertex of that edge;
- *triangle graphs* or 3-*line graphs*, by Le [708]. Here the triangle graph $L_3(G)$ of a graph G has the triangles of G as vertices; two triangles are adjacent in $L_3(G)$ if

they have an edge in common. Triangle graphs include the line graphs of bipartite graphs; this makes triangle graphs interesting in connection with perfect graphs (see section 14.3).

Two interesting problems in this direction are as follows.

Problem 14.4.3 *Prove the SPGC for the class of all p-powers G^p of graphs G ($p \geq 1$ fixed).*

Problem 14.4.4 *Prove the SPGC for the class of Gallai graphs $\Gamma(G)$ of graphs G.*

Notice that in Problem 14.4.3, one gets the SPGC for $p = 1$, and for all graphs G there exists an integer $p = p(G)$ such that G^p is perfect (even a complete graph). Partial results for chordal graphs are given by Chang and Nemhauser [184]. In the case $p = 2$, the problem is solved for all *subdivisions* $S(G)$ by the result of Rao and Ravindra [896] for total graphs and by the fact that in this case $S(G)^2$ coincides with the total graph of $S(G)$. The subdivision $S(G)$ is the graph resulting from G by subdividing every edge of G. A positive solution for Problem 14.4.4 would imply the result due to Sun saying that Gallai-perfect graphs are perfect (see the previous section). This discussion and a partial result on Problem 14.4.4 can be seen in [709, 711].

Line graphs and triangle graphs are examples of graph classes defined by intersection models. The SPGC holds for further intersection graphs addressed below:

- *intersection graphs of arcs of a circle* or *circular-arc graphs*, by a result of Tucker [1042];
- *intersection graphs of chords of a circle* or *circle graphs*, by a result of Buckingham and Golumbic [164, 165];
- *edge intersection graphs of paths in a tree* or *EPT graphs*, by a result of Golumbic and Jamison [458]. They also proved that the recognition problem for EPT graphs is $\mathbb{NP}$-complete [457]; cf. Theorem 4.4.4;
- *intersection graphs of subtrees in a cactus* or *cactus subtree graphs*, by a result of Gavril [430].

The papers Ravindra [895], Chvátal [211], Akiyama and Chvátal [12], Hoàng, Hougardy, and Maffray [564], Kahn [637], and Le [710, 711] deal with graph-valued functions in connection with perfect graphs. Hertz [543, 544, 545, 546] describes some classes of perfect graphs using quite interesting graph operations. Prisner [883] gives a survey on graph-valued functions with emphasis on line graphs and their generalizations.

14.4.3 Other graph classes

The validity of the SPGC has been shown for the following graph classes:

- *planar graphs*, by Tucker [1040]; [173] contains a new proof. Hsu [592] gave a polynomial-time recognition algorithm for perfect planar graphs;

- *graphs embeddable on the torus* or *toroidal graphs*, by Grinstead [475];
- *graphs with maximum degree at most* 6, by Grinstead [474];
- *normal fraternally orientable graphs* (see Definition 5.6.2 for fraternal orientation), by Galeana-Sánchez [412];
- *slightly triangulated graphs* (graphs without C_k, $k \geq 5$, and such that every induced subgraph has a vertex whose neighborhood is P_4-free, see Definition 7.2.2), by Maire [764];
- *split-neighborhood graphs* (graphs such that every induced subgraph has a vertex whose neighborhood induces a split graph), by Maffray and Preissmann [760];
- *pretty graphs* (graphs such that every induced subgraph has a vertex whose neighborhood contains no P_4 and $2K_2$), by Maffray, Porto, and Preissmann [757];
- *weakly diamond-free graphs* (graphs such that every induced subgraph H has a vertex of degree at most $2 \cdot \omega(H) - 1$ and its neighborhood is $(K_4 - e)$-free), by Ait Haddadene and Gravier [9];
- *degenerate graphs* (graphs such that every induced subgraph H has a vertex of degree at most $\omega(H) + 1$), by Ait Haddadene and Maffray [10];
- 2-*split graphs* (graphs partitionable into two split graphs), by Hoàng and Le [567]. 2-split graphs and perfect 2-split graphs can be recognized in polynomial time; see [142] and [567].

Further classes satisfying the SPGC are discussed in [179, 469].

In view of the P_4-structure of perfect graphs, the following problem is interesting, as it unifies several results in [218, 554, 220, 482].

Problem 14.4.5 *Prove the SPGC for graphs whose vertex set can be partitioned into two subsets, each of which induces a P_4-free graph.*

Graphs considered in the above problem are called *P_4-bipartite* in [568]. It turns out that many well-structured graph classes consist of P_4-bipartite graphs only; examples include parity graphs (hence bipartite graphs, distance-hereditary graphs), split graphs, P_4-reducible graphs, P_4-sparse graphs, P_4-tile graphs, cograph contractions, and complements of all these graphs. Hoàng and Le [566] solved Problem 14.4.5 for some particular cases. They proved that the SPGC is true for (P_4-bipartite) graphs G containing a set T of vertices such that T induces a threshold graph and meets every P_4 of G in a midpoint or every P_4 of G in an endpoint.

We shall close this section with two remarks. Our first remark concerns proof techniques of the results mentioned in this and the previous sections. As we have seen, several graph classes have been proved to be perfect or to satisfy the SPGC, and several properties of minimal imperfect graphs have been found. In view of this variety of facts, one may suspect that a suitable combination of some of these known facts would be enough to prove the SPGC. A negative discussion in this direction is given by Rusu [936, 937]. However, a popular way to prove that a graph class $\mathcal{G}$ is perfect or satisfies the SPGC consists of showing that any graph G of $\mathcal{G}$ either belongs to one of the classes that are known to be perfect or to satisfy the SPGC, or G has some properties that are not fulfilled by any minimal imperfect graph. Almost all results given in this and the previous section have been obtained in this way.

Our second remark concerns classes of graphs complete for the SPGC. A class is *complete for a conjecture* if the truth of the conjecture on this restricted class implies that the conjecture is true in general. Corneil [244] gives several complete graph classes for the SPGC. Among others, some complete graph classes are k-connected graphs for any positive integer k, graphs that are both Eulerian and Hamiltonian, self-complementary graphs, and regular graphs.

14.5 Two semistrong perfect graph conjectures

Any conjecture on perfect graphs that would be implied by the truth of the SPGC, and whose truth in turn would imply the PGT, is called a *semistrong* perfect graph conjecture. An example is Chvátal's conjecture on the P_4-structure of perfect graphs, which was proved by Reed; see Theorem 2.2.1. We shall give here two conjectures having this semistrong property concerning perfect graphs. The first one is due to Cameron, Edmonds, and Lovász [178] and is related to Theorem 2.1.3.

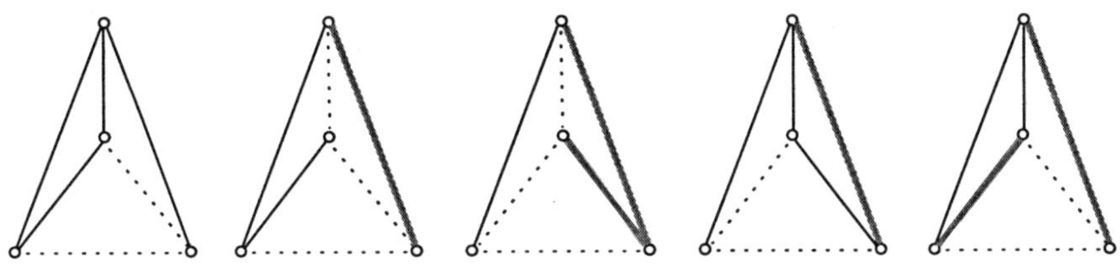

Figure 14.1: Forbidden edge 3-colored K_4s.

Conjecture 14.5.1 *Let the edges of a complete graph be colored with three colors in such a way that no 3-colored K_4 shown in Figure 14.1 occurs. Suppose that each of the graphs formed by two colors is perfect. Then so is the graph formed by the third color.*

As noted by Cameron, Edmonds, and Lovász [178], the SPGC would imply the conjecture above. Clearly, on the other hand, the latter would imply the PGT. The paper of Cameron, Edmonds, and Lovász [178] was motivated by a paper of Gallai [416]; see the remarks in [178]. For another formulation of Conjecture 14.5.1, see [626].

The second conjecture is stated in terms of *Gallai graphs* $\Gamma(G)$; see Definition 2.4.5 and section 3 of this chapter.

Conjecture 14.5.2 *A graph G is perfect if and only if for all induced subgraphs H of G, $\chi(\Gamma(H)) \leq \alpha(H)$ and $\chi(\Gamma(\overline{H})) \leq \omega(H)$.*

This conjecture is due to Le [709, 711]. Clearly, it would imply the PGT, and it is pointed out in [709, 711] that the conjecture would be implied by the SPGC. The necessity described in the conjecture above follows from Proposition 2.4.3 of the perfection chapter (chapter 2).

Notice that by Proposition 2.4.3 and the PGT, $\chi(\Delta(H)) \leq \omega(H)$ and $\chi(\Delta(\overline{H})) \leq \alpha(H)$ hold for all induced subgraphs H of a perfect graph. This property does not characterize perfect graphs, as the odd hole C_5 shows.

14.6 The weakened strong perfect graph conjecture

The SPGC is even open in the following much weaker form.

Conjecture 14.6.1 (The weakened strong perfect graph conjecture) *There exists a function f such that for all Berge graphs G, $\chi(G) \leq f(\omega(G))$.*

Such a function f necessarily satisfies $f(x) \geq x$. The SPGC states that one can choose $f(x) = x$, the best possible choice for f one could hope for. The weakened strong perfect graph conjecture was made in 1987 by Gyárfás [487]. In view of Problems 14.4.1 and 14.4.2, the following results may be of interest, though the bounds lie too far from the presumable bound. The first one is due to Wagon [1074], the second one is due to Gyárfás [487].

- For all $\overline{C_4}$-free (not necessarily Berge) graphs G, $\chi(G) \leq \frac{1}{2}\omega(G)(\omega(G)+1)$.
- For all P_5-free (not necessarily Berge) graphs G, $\chi(G) \leq 4^{\omega(G)-1}$.

The paper of Gyárfás [487] contains many results and problems related to Conjecture 14.6.1; see also [490, 626, 655, 656].

Appendix A

Recognition

Abbreviations for time complexity are listed in the table below.

Abbreviation	Meaning
LIN	linear time $\mathcal{O}(n+m)$, n the number of nodes and m the number of edges of the graph G
$\mathbb{P}$	polynomial time
$\mathbb{NP}$-c	$\mathbb{NP}$-complete
$\mathcal{O}(MM)$	proportional to the time amount of matrix multiplication
α	the best known exponent of n to perform an $n \times n$ matrix multiplication (approximately 2.376)
?	the recognition complexity seems to be unknown
??	the recognition complexity is an open problem
???	the recognition complexity is an important open problem

If there are several references then (at least) the leftmost reference reaches the given time bound.

Class	Time complexity	References/remarks
almost trees (k)	$\mathbb{P}$	for fixed k
alternately orientable	$\mathcal{O}(nm)$	[431, 556]
AT-free	$\mathcal{O}(n^3)$	[253]
Berge	???	(in co-$\mathbb{NP}$)
biconvex	LIN	
bigeodetic	?	
bipartite	LIN	
bipartite permutation	LIN	[995]
bipolarizable	$\mathcal{O}(n^{\alpha+1})$	[571, 996]
bip*	?	
Birkhoff	?	

Class	Time complexity	References/remarks
bithreshold	$\mathcal{O}(n^3)$	[297, 514]
block	LIN	
bounded tolerance	??	
bridged	$\mathcal{O}(n^{\alpha+1})$	[376, 715]
brittle	$\mathcal{O}(\min(m^2, n^3 \log^2 n))$	[946, 653, 996]
bull-free perfect	$\mathcal{O}(n^5)$	[902]
$(C_4, 2K_2)$-free	LIN	[759]
cactus	LIN	
chordal	LIN	[928, 1021]
chordal bipartite	$\mathcal{O}(\min(m \log n, n^2))$	[991, 850, 744]
chordal comparability	LIN	[598, 751]
circle	$\mathcal{O}(n^2)$	[990, 411, 138, 808]
circular arc	$\mathcal{O}(n^2)$	[358, 595, 1044]
circular permutation	LIN	[999, 931]
claw-free	$\mathcal{O}(m^{(\alpha+1)/2)}) \approx \mathcal{O}(m^{1.69})$	[674]
claw-free perfect	$\mathbb{P}$	[223]
clique	?	
clique-Helly	$\mathcal{O}(nm^2)$	[321, 1016]
clique separable	$\mathbb{P}$	[1082, 427]
cographs	LIN	[255]
comparability	$\mathcal{O}(MM)$	[988, 807, 450]
convex	LIN	
co-chordal	LIN	[558]
co-interval	LIN	[994]
co-threshold tolerance	$\mathbb{P}$	
cop-win	$\mathcal{O}(nm)$	
diamond-free	$\mathcal{O}(n^\alpha + m^{3/2})$	[674]
diamond-free perfect	$\mathbb{P}$	[399]
D-graphs	$\mathcal{O}(nm)$	[562]
Dilworth-number bounded		
$k = 1$:	LIN	(cf. threshold)
fixed k:	$\mathcal{O}(n^2)$	[384]
in general:	$\mathcal{O}(n^{2.5})$	[570]
directed path	LIN	[313, 426]
distance-hereditary	LIN	[511, 56]
domination	??	[281]
domino	LIN	[673]
domishold	$\mathcal{O}(\max(m, \log n))$	[767, 85]
dually chordal	LIN	[331, 146]
doubly chordal	LIN	[331, 146, 799]
line graphs	LIN	[934, 721]
EPT	$\mathbb{NP}$-c	[458, 1013]

Class	Time complexity	References/remarks
extended P_4-laden	LIN	[437]
extended P_4-reducible	LIN	[440]
extended P_4-sparse	LIN	[440]
forest-perfect	LIN	[151]
Gallai	$\mathcal{O}(nm)$	[933, 168]
genus-k bounded	$\mathbb{P}$ $\mathbb{NP}$-c	for fixed k [386] for unbounded k [1025]
geodetic	$\mathbb{P}$	
good	$\mathcal{O}(n^3)$	
Halin	LIN	(communicated by Sysło)
Helly	$\mathcal{O}(n^2m)$	[63, 321]
Helly circular-arc	$\mathbb{P}$	[994]
hereditary median	$\mathbb{P}$	[48]
hereditary modular	$\mathbb{P}$	[48]
hereditary pseudomodular	$\mathcal{O}(n^4)$	[57]
HHD-free	$\mathcal{O}(n^3)$	[565, 573]
HHDS-free	??	
hole-free	$\mathbb{P}$	[993]
homogeneously orderable	$\mathcal{O}(n^3)$	[148]
interval	LIN	[127, 688, 597, 499]
interval bigraph	$\mathbb{P}$	[801]
interval-regular	$\mathbb{P}$	[55]
k-outerplanar	$\mathbb{P}$	
k-trees	LIN	(as for chordal)
K_4-free	$\mathcal{O}(m^{(\alpha+1)/2)}) \approx \mathcal{O}(m^{1.69})$	[674]
K_4-free perfect	??	
locally perfect	??	
matrogenic	LIN	[1052]
matroidal	LIN	[1052]
Matula perfect	$\mathbb{P}$	[219]
maxibrittle	$\mathbb{P}$	[878]
median	$\mathcal{O}(\min m\sqrt{n}, n^{1.5}\log n)$	[500]
Meyniel	$\mathcal{O}(m^2)$	[932, 169]
minimal imperfect	???	
modular	$\mathbb{P}$	
murky	$\mathbb{P}$	
N^*-perfect	$\mathbb{P}$	[228]
neighborhood Helly	$\mathcal{O}(n^2m)$	[321]
neighborhood perfect	??	
opposition	??	
outerplanar	LIN	[785]
p-connected	LIN	[37]

Class	Time complexity	References/remarks
p-tree	LIN	[37]
P_4-bipartite	$\mathbb{NP}$-c	[4, 568]
P_4-brittle	$\mathbb{P}$	[566]
P_4-comparability	$\mathcal{O}(n^4)$, $\mathcal{O}(n^2m)$	[572], [894]
P_4-extendible	LIN	[574]
P_4-indifference	$\mathcal{O}(n^2m)$	[894]
P_4-laden	LIN	[437]
P_4-lite	$\mathbb{P}$	
P_4-reducible	LIN	[617]
P_4-simplicial	$\mathcal{O}(n^3\log^2 n)$	[571, 996]
P_4-sparse	LIN	[621]
P_4-tidy	LIN	[439]
parity	LIN	[224, 170]
partial k-trees	$\mathbb{P}$	for fixed k [26]
partitionable	??	
perfect	???	(in co-$\mathbb{NP}$, see [175])
perfect elimination bip	$\mathcal{O}(n^3)$	[447, 456]
perfectly orderable	$\mathbb{NP}$-c	[783, 561]
permutation	LIN	[776, 989, 807]
planar	LIN	[580, 127, 194]
planar perfect	$\mathbb{P}$	[592]
pretty	LIN	[757]
proper circular arc	LIN	[300, 536, 1038]
proper interval	LIN	[247, 300, 293]
proper tolerance	??	
pseudomodular graphs	$\mathbb{P}$	[57]
ptolemaic	LIN	
quasi-brittle	$\mathbb{P}$	[834]
quasi-parity	??	
semi–P_4-sparse	LIN	[401]
series-parallel	LIN	[1064]
slightly triangulated	$\mathbb{P}$	
split	LIN	[518]
split-neighborhood	$\mathbb{P}$	[760]
strict 2-threshold	LIN	[871, 761]
strict quasi-parity	??	
string	$\mathbb{NP}$-hard	[690]
strongly chordal	$\mathcal{O}(\min(n^2, m\log n))$	[991, 850, 744, 373]
strongly orderable	$\mathcal{O}(nm)$	[325]
strongly perfect	??	[84]
strong tree-cographs	$\mathbb{P}$	[1026]
superbrittle	$\mathbb{P}$	[878]

Class	Time complexity	References/remarks
superfragile	$\mathbb{P}$	[878]
superperfect	??	
threshold	LIN	[216, 509]
2-split	$\mathbb{P}$	[142]
2-split perfect	$\mathbb{P}$	[567]
2-threshold	$\mathcal{O}(n^3)$	[749, 893, 1001]
t-interval	$\mathbb{NP}$-c	[1080] (already for $t = 2$)
threshold signed	LIN	[87]
threshold tolerance	$\mathbb{P}$	[795]
tolerance	??	
toroidal	$\mathbb{P}$	(cf. genus-k bounded graphs)
totally unimodular	$\mathcal{O}(n^3)$	[1035]
trapezoid	$\mathcal{O}(n^2)$	[752, 231, 495, 496]
trees	LIN	
tree-cographs	$\mathbb{P}$	[1026]
tree-perfect	LIN	[151]
trivially perfect	LIN	[451]
unbreakable	$\mathbb{P}$	[207]
undirected path	$\mathcal{O}(mn)$	[947, 428]
unigraphs	LIN	[1056]
unit circular arc	$\mathbb{P}$	[1041, 994]
unit tolerance	??	
weak bipolarizable	LIN	[826]
weakly chordal	$\mathcal{O}(n^4)$	[997, 530]
weakly geodetic	$\mathbb{P}$	
Welsh–Powell opposition	$\mathcal{O}(n^4)$	[820]
Welsh–Powell perfect	$\mathbb{P}$	[219]
X–star-chordal	LIN	[499]

Appendix B

Containment Relationships

The inclusions mentioned here represent only a selection of inclusions that seem to be important. We try to give here only inclusions that do not follow from others by transitivity.

Inclusions without reference are obvious from the definition of the classes or from "folklore" theorems.

Many of the inclusions follow from characterizations of the classes by using some simple rules such as the following ones:

- $\mathcal{C}_1 \cap \mathcal{C}_2 \subseteq \mathcal{C}_i$, $i \in \{1, 2\}$;
- $\mathcal{C}_i \subseteq \mathcal{C}_1 \cup \mathcal{C}_2$, $i \in \{1, 2\}$;
- if $\mathcal{C}_1 \subseteq \mathcal{C}_2$, then *co*-$\mathcal{C}_1 \subseteq$ *co*-$\mathcal{C}_2$ (this will be explicitly mentioned only for a few classes, where the class as well as the complement class are important as, e.g., for comparability graphs);
- if $\mathcal{C}_1 \subseteq \mathcal{C}_2$ and $\mathcal{C}_2$ is closed under complementation, then *co*-$\mathcal{C}_1 \subseteq \mathcal{C}_2$ (therefore it will be noticed for classes if they are closed under complementation);
- hereditary $\mathcal{C} \subseteq \mathcal{C}$ (this holds for hereditary in the induced as well as in the isometric sense);
- if $\mathcal{C}_1 \subseteq \mathcal{C}_2$, then hereditary $\mathcal{C}_1 \subseteq$ hereditary $\mathcal{C}_2$;
- if $\mathcal{C}_1 = \mathcal{F}_1$-free and $\mathcal{C}_2 = \mathcal{F}_2$-free and $\mathcal{F}_2 \subseteq \mathcal{F}_1$, then $\mathcal{C}_1 \subseteq \mathcal{C}_2$;
- if $\mathcal{C}_1 = \mathcal{F}_1$-free and $\mathcal{C}_2 = \mathcal{F}_2$-free and for every graph G in $\mathcal{F}_2$ there is a graph in $\mathcal{F}_1$ that is a subgraph in G, then $\mathcal{C}_1 \subseteq \mathcal{C}_2$; good examples are
  - every sun S_k, $k \geq 3$, contains a gem; thus gem-free graphs are sun free),
  - every odd sun S_{2k+1}, $k \geq 2$, contains a net, i.e., $\overline{S_3}$; thus net-free graphs contain no odd sun S_{2k+1}, $k \geq 2$;
- for some parameterized classes, $\mathcal{C}_k \subseteq \mathcal{C}_{k+1}$ holds for all k (examples are Dilworth number, interval number, boxicity, etc.);

- if $\mathcal{C}_1 \cap \mathcal{C}_2 = \mathcal{C}_1 \cap \mathcal{C}_4$ and $\mathcal{C}_2 \subseteq \mathcal{C}_3 \subseteq \mathcal{C}_4$, then $\mathcal{C}_1 \cap \mathcal{C}_2 = \mathcal{C}_1 \cap \mathcal{C}_3$.

$(2K_2, C_4)$-free

= pseudosplit

closed under complementation

P_4-brittle

⊂ perfectly orderable [566]

⊂ P_4-bipartite

P_4-comparability

⊂ perfectly orderable [572]

P_4-extendible

⊂ P_4-tidy [439]

P_4-laden

⊂ brittle [437]

⊂ extended P_4-laden [437]

P_4-lite

= C_5-free ∩ P_4-tidy [439]

⊂ P_4-bipartite

⊂ P_4-laden [437]

⊂ weakly chordal

P_4-reducible

= P_4-extendible ∩ P_4-sparse [439]

= $(C_5, P_5, \overline{P_5}, P, \overline{P}, \text{fork}, \overline{\text{fork}}, S_3, \overline{S_3})$-free [440]

closed under complementation

⊂ extended P_4-reducible [440]

⊂ forest-perfect [151]

⊂ HHDS-free

⊂ permutation (P. Damaschke, personal communication)

P_4-sparse

$=$ $(C_5, P_5, \overline{P_5}, P, \overline{P}, \text{fork}, \overline{\text{fork}})$-free [620]

closed under complementation

$\subset$ P_4-lite [618]

$\subset$ alternately orientable [553]

$\subset$ extended P_4-sparse [440]

$\subset$ superbrittle [620]

$\subset$ Welsh–Powell opposition [835]

S_3-free $\cap$ chordal

$=$ C_4-free $\cap$ induced-hereditary pseudomodular [57]

$=$ induced-hereditary Helly [321]

$\subset$ hereditary pseudomodular [57]

$(S_3, \overline{S_3})$-free $\cap$ chordal

$=$ τ_k-perfect for all $k \geq 2$ [184]

$\subset$ chordal $\cap$ odd-sun-free [184]

$(S_3, \overline{S_3})$-free $\cap$ split

$=$ $(2K_2, C_4, C_5, S_3, \overline{S_3})$-free

closed under complementation

2-threshold

$\subset$ alternately orientable [556]

$\subset$ perfectly orderable [219]

$\subset$ weakly chordal [516]

absolute bipartite retract

$=$ half-disk Helly [52]

$=$ modular $\cap$ open-neighborhood-Helly [52]

$=$ dismantlable $\cap$ open-neighborhood-Helly [53]

absorbantly perfect

$\subset$ perfect [512]

alternately orientable

⊂ bip* [556]

alternately orientable ∩ co-comparability

⊂ trapezoid [382]

biconvex

⊂ convex [155]

bipartable

⊂ perfectly orderable [544]

bipartite

= odd-cycle-free [683]

= perfect ∩ triangle-free

⊂ P_4-brittle

⊂ comparability

⊂ Gallai

⊂ Gallai-perfect [1007]

⊂ parity

⊂ unimodular [452]

bipartite ∩ distance-hereditary

= (6,2)-chordal ∩ bipartite

⊂ chordal bipartite

bipartite ∩ permutation

= AT-free ∩ bipartite

⊂ biconvex [995]

bipartite ∩ X-chordal ∩ X-conformal

⊂ X–star-chordal [75]

bipolarizable

⊂ opposition

⊂ weak bipolarizable [826]

bip*

$\subset$ perfect [207]

bithreshold

$\subset$ 2-split [517]

$\subset$ bipartable [544]

block

$=$ ptolemaic $\cap$ weakly geodetic [647, 587]

$\subset$ geodetic

bounded tolerance

$=$ intersection graphs of parallelograms (squares) [124, 382]

bridged

$=$ hereditary dismantlable [25]

$=$ (C_4, C_5)-free $\cap$ cop-win [25]

$\subset$ hereditary weakly modular [192]

bridged $\cap$ Helly

$=$ bridged $\cap$ clique-Helly [65]

$=$ (C_4, C_5)-free $\cap$ Helly [321]

brittle

closed under complementation

$\subset$ domination [281]

$\subset$ quasi-brittle [834]

cactus

$=$ almost tree (1)

$\subset$ outerplanar

chordal

$\subset$ bridged

$\subset$ extended circle [601]

$\subset$ Gallai

$\subset$ good [570]

$\subset$ HHP-free

$\subset$ slightly triangulated [764]

chordal bipartite

$=$ bipartite $\cap$ hereditary X-chordal $\cap$ X-conformal [332, 146]

$=$ bipartite $\cap$ weakly chordal

$=$ (6,1)-chordal $\cap$ bipartite

$=$ hereditary perfect elimination bipartite [456]

$=$ bipartite $\cap$ co-perfectly orderable [210]

$\subset$ bipartite $\cap$ X-chordal $\cap$ X-conformal

$\subset$ hereditary modular [48]

chordal $\cap$ comparability

$\subset$ comparability graphs of dimension-4 posets [751, 654]

$\subset$ strongly chordal [130]

chordal $\cap$ EPT

$\subset$ undirected path [1013]

chordal $\cap$ odd-sun–free

$=$ chordal $\cap$ neighborhood-perfect [720]

circle

$\subset$ extended circle [601]

circular arc

$\subset$ 2-interval

$\subset$ extended circle [601]

circular permutation

$=$ containment graphs of circular arcs [931]

$\subset$ comparability

clique-Helly

$\subset$ clique [508]

clique separable

$\subset$ alternately orientable [581]

$\subset$ preperfect [581]

$\subset$ strict quasi parity [581]

co-bithreshold

$\subset$ 2-threshold [514]

co-chordal

$=$ absolutely perfect [84]

$\subset$ good [570]

$\subset$ weakly chordal

$\subset$ slightly triangulated [764]

co-comparability

$\subset$ AT-free [461]

cograph

$=$ P_4-free

$=$ comparability graphs of series-parallel posets

closed under complementation

$\subset$ P_4-brittle

$\subset$ P_4-reducible

$\subset$ cograph contraction [600]

$\subset$ distance-hereditary

$\subset$ slightly triangulated [765]

$\subset$ tree-perfect [151]

cograph contraction

$\subset$ weakly chordal

$\subset$ co–P_4-brittle [712]

co-interval

$=$ co-chordal $\cap$ comparability [444]

$\subset$ containment graph of circles [388]

comparability

= containment graphs

⊂ alternately orientable [556]

⊂ P_4-comparability

⊂ superperfect [575], [452, p. 203]

comparability ∩ split

= $(2K_2, C_4, C_5, S_3, \overline{S_3}$,co-rising sun)-free [416, 394]

convex

⊂ interval bigraph [801]

co-proper interval

= comparability graphs of semiorders [906]

⊂ comparability graphs of dimension-3 posets [889]

co-threshold tolerance

⊂ strongly chordal [794]

⊂ tolerance [794]

co-trivially perfect

= $(2K_2, P_4)$-free [451, 1090]

= cograph ∩ co-interval

Dilworth 3

⊂ good [570]

Dilworth 4

⊂ perfect [863]

directed path

⊂ strongly chordal [371]

⊂ undirected path

dismantlable

= cop-win [818]

distance-hereditary

= (5,2)-crossing-chordal [586]

= HHDG-free [56]

⊂ circle

⊂ HHDS-free [148]

⊂ parity [586]

domination

closed under complementation [281]

⊂ weakly chordal [281]

domino

= (claw, gem, W_4)-free [674]

doubly chordal

= clique-chordal ∩ Helly chordal [331, 146]

= chordal ∩ dually chordal [331, 146]

dually chordal

= clique-chordal ∩ clique-Helly [331]

⊂ Helly

⊂ homogeneously orderable [148]

extended P_4-reducible

= $(P_5, \overline{P_5}, P, \overline{P}, \text{fork}, \overline{\text{fork}}, S_3, \overline{S_3})$-free [440]

⊂ P_4-bipartite

⊂ P_4-extendible [439]

⊂ extended P_4-sparse [440]

extended P_4-sparse

= $(P_5, \overline{P_5}, P, \overline{P}, \text{fork}, \overline{\text{fork}})$-free [440]

⊂ P_4-tidy [439]

⊂ semi–P_4-sparse [401]

forest-perfect

closed under complementation

$\subset$ weakly chordal [151]

Gallai

$=$ i-triangulated

$=$ (5,2)-odd-noncrossing-chordal [415]

$\subset$ clique-separable [415]

$\subset$ Meyniel

$\subset$ preperfect [512]

Gallai-perfect

$\subset$ perfect [1007]

geodetic

$\subset$ bigeodetic

$\subset$ weakly geodetic

good

$=$ quasi-triangulated

closed under complementation

$\subset$ brittle [570]

Halin

$\subset$ 2-outerplanar

$\subset$ partial 3-tree [1086, 1087, 350, 117]

Helly

$=$ absolute reflexive retract [537, 817, 886] (see also [57, 53])

$=$ clique-Helly $\cap$ dismantlable [65]

$=$ neighborhood-Helly $\cap$ pseudomodular [57, 63]

Helly chordal

$=$ chordal $\cap$ clique-Helly

Helly circular-arc

$\subset$ circular-arc

hereditary Matula perfect

= HHD-bicycle–free [219]

hereditary median

= $K_{2,3}$-free $\cap$ hereditary modular [48, 47, 804]

hereditary-modular

= bipartite $\cap$ bridged [48]

= hereditary absolute bipartite retract [48]

hereditary N^*-perfect

$\subset$ HHG-free [228]

hereditary V-perfect

$\subset$ murky

hereditary weakly modular

= isometric-HH–free [189]

hereditary Welsh–Powell opposition

= $(C_5, P_5, \overline{P_5}, P)$-free [826]

$\subset$ maxibrittle

$\subset$ opposition [835]

hereditary Welsh–Powell perfect

$\subset$ murky

$\subset$ perfectly orderable [219]

HH-free

= house-free $\cap$ weakly chordal

$\subset$ domination [281]

$\subset$ hereditary weakly modular [192]

HHD-free

= (5,2)-chordal

$\subset$ brittle [820]

$\subset$ Meyniel (see Proposition 3.1.1)

$\subset$ preperfect [581]

HHDS-free

$=$ hereditary homogeneously orderable [148]

$\subset$ hereditary pseudomodular [57]

HHP-free

$\subset$ weak bipolarizable

interval

$=$ AT-free $\cap$ chordal [723]

$=$ boxicity 1

$=$ C_4-free $\cap$ co-comparability [444]

$=$ chordal $\cap$ co-comparability [444]

$\subset$ bounded tolerance [461]

$\subset$ circular-arc

$\subset$ co-threshold tolerance [794]

$\subset$ directed-path

$\subset$ PI

$\subset$ unimodular [452]

interval bigraph

$\subset$ chordal bipartite [801]

interval $\cap$ co-interval

$=$ chordal $\cap$ co-chordal $\cap$ comparability $\cap$ co-comparability

$=$ permutation $\cap$ split

$=$ split $\cap$ threshold signed [86]

$=$ $(2K_2, C_4, C_5, S_3, \overline{S_3}$,rising sun,co-rising sun)-free

closed under complementation

$\subset$ comparability $\cap$ split

line

= (claw, $W_5, \overline{A}, K_5 - e, \overline{P_2 \cup P_3}, \overline{C_4 \cup 2K_1}, \overline{P_2 \cup P_3}, \overline{R}, \overline{\text{twin} - \text{house}}, \overline{\text{twin} - C_5}$)-free (see Theorem 7.1.8)

⊂ EPT

line graphs of bipartite graphs

= (claw, diamond, odd-hole)-free (see Corollary 7.1.4)

⊂ domino [674]

⊂ Gallai-perfect [1007]

locally perfect

⊂ perfect [875]

matrogenic

closed under complementation

⊂ unigraph [1052, 768]

matroidal

= C_5-free ∩ matrogenic [866]

closed under complementation

⊂ chordal ∪ cochordal [866]

Matula perfect

⊂ perfectly orderable [219]

maxibrittle

= ($C_5, P_5, \overline{P_5}$,fish,co-fish)-free [878]

closed under complementation

median

⊂ modular [47]

⊂ pseudomedian [60]

⊂ quasi-median [62]

Meyniel

= (5,2)-odd–chordal [780]

= very strongly perfect [553, 555]

⊂ bip* [207]

⊂ locally perfect [543]

⊂ perfectly contractile [547]

⊂ strongly perfect [896]

Meyniel ∩ co-Meyniel

= $(C_5, P_5, \overline{P_5})$-free [553]

= HHD-free ∩ co-HHD–free

closed under complementation

⊂ HHD-free

⊂ murky

⊂ hereditarily Welsh–Powell perfect [219]

minimally imperfect

closed under complementation

⊂ partitionable [736]

modular

⊂ bipartite [48]

⊂ pseudomodular [57]

⊂ weakly modular [51, 61, 189]

multitolerance

⊂ perfect [855, 856]

murky

= $(C_5, P_6, \overline{P_6})$-free [527]

closed under complementation

⊂ perfect

N*-perfect

⊂ perfectly orderable [228]

neighborhood-Helly

$\subset$ clique-Helly

neighborhood-perfect

$\subset$ Berge [720]

opposition

$\subset$ bip* [820, 822]

$\subset$ strict quasi parity [569]

outerplanar

$\subset$ circle

$\subset$ series-parallel [341, 1075]

parity

$=$ (5,2)–odd-crossing-chordal

$\subset$ P_4-bipartite [170]

$\subset$ Meyniel

$\subset$ preperfect [512]

partitionable

$\subset$ unbreakable [207]

perfect

$=$ $\mathcal{C}(G)$-perfect

$=$ perfectly 1-transversable

closed under complementation [735]

$\subseteq$ Berge

$\subseteq$ kernel solvable [133]

$\subset$ normal [685]

perfectly contractile

$\subset$ strict quasi parity [104]

perfectly orderable

$\subset$ bip* [207]

$\subset$ perfectly contractile [366]

$\subset$ strongly perfect [205, 340]

permutation

$=$ circle graph with equator [452]

$=$ comparability $\cap$ co-comparability [345]

$=$ comparability of dimension-2 poset [345]

$=$ containment graph of intervals [345]

closed under complementation

$\subset$ bounded tolerance [460]

$\subset$ circle

$\subset$ containment graph of circles [1031]

$\subset$ PI

PI

$\subset$ PI*

PI*

$\subset$ trapezoid

planar

$=$ genus 0

$\subset$ $\mathcal{P}_3$-perfect [173]

$\subset$ 3-interval [959]

power-chordal

$=$ chordal $\cap$ clique-chordal

preperfect

closed under complementation

$\subset$ perfect [512]

proper circular-arc

$\subset$ circle [454, p. 192]

proper interval

= unit interval [906]

= $(\text{claw}, S_3, \overline{S_3})$-free ∩ chordal [1078]

= claw-free ∩ interval [1078]

= astral-triple–free [606]

⊂ P_4-bipartite [568]

⊂ unit circular-arc

proper-tolerance

⊂ bounded-tolerance [460, 461]

pseudomedian

⊂ pseudomodular [60]

pseudomodular

= 3-Helly [57]

ptolemaic

= chordal ∩ distance-hereditary [588]

= chordal ∩ gem-free [588]

⊂ directed-path [276]

quasi-brittle

⊂ perfectly orderable [834]

quasi-parity

closed under complementation

⊂ perfect

semi–P_4-sparse

= $(P_5, \overline{P_5}, \overline{\text{fork}})$-free [401]

series-parallel

= K_4-minor–free [341]

= partial 2-tree [1075]

⊂ extended circle [601]

$\subset$ planar [341]

short-chorded

$\subset$ Berge [745]

slightly triangulated

$\subset$ perfect [765]

split

$=$ chordal $\cap$ co-chordal [394]

$=$ $(2K_2, C_4, C_5)$-free [394]

closed under complementation

$\subset$ P_4-brittle

$\subset$ polar [1053, 1054, 1055]

strict 2-threshold

$\subset$ 2-threshold

$\subset$ comparability [516]

strict quasi parity

$\subset$ quasi-parity

strongly chordal

$=$ chordal $\cap$ sun-free [372]

$=$ hereditary dually chordal [146]

$\subset$ generalized strongly chordal [277]

strongly orderable

$=$ generalized strongly chordal

$\subset$ weakly chordal [277]

strongly perfect

$\subset$ absorbantly perfect

strong tree-cograph

$\subset$ Birkhoff [1026]

$\subset$ tree-cograph [1026]

superbrittle

$= (C_5, P_5, \overline{P_5}, A, \overline{A}, \text{parapluie}, \text{parachute})$-free [878]

$\subset$ Meyniel $\cap$ co-Meyniel [553]

superfragile

$= (C_4, P_4, \text{dart})$-free [878]

superperfect

$\subset$ perfect [575], [454, p. 203]

threshold

$= (2K_2, C_4, P_4)$-free [216]

$=$ Dilworth 1 [217, 397]

$=$ trivially perfect $\cap$ co-trivially perfect

$=$ comparability graphs of threshold orders [216]

$=$ cograph $\cap$ split

$=$ cograph $\cap$ interval $\cap$ co-interval

closed under complementation

$\subset$ 2-threshold

$\subset$ matroidal [866]

$\subset$ co-threshold tolerance [794]

threshold-signed

$=$ Dilworth 2 [86]

closed under complementation [86]

$\subset$ boxicity-2 [86]

$\subset$ interval $\cup$ cointerval

$\subset$ maxibrittle

$\subset$ permutation [86]

tolerance

$\subset$ alternately orientable [461]

$\subset$ coperfectly orderable [461]

⊂ multitolerance [855, 856]

⊂ weakly chordal [461]

toroidal

= genus 1

trapezoid

= co-comparability graphs of posets of interval dimension 2

= bounded multitolerance [854, 856]

⊂ extended circle [601]

⊂ weakly chordal [246]

tree

⊂ 2-interval [1032, 472]

⊂ bipartite ∩ distance-hereditary

⊂ block

⊂ boxicity-2

⊂ cactus

⊂ perfect ∩ planar

⊂ strong tree-cograph [1026]

⊂ tree-perfect

tree-cograph

⊂ forest-perfect

tree-perfect

closed under complementation

⊂ forest-perfect

trivially perfect

= (C_4, P_4)-free [451, 1090]

= chordal ∩ cograph

= cograph ∩ interval

= comparability graphs of arborescence orders [1090, 1091]

= intersection of nested intervals

unbreakable

⊂ p-connected [37]

undirected path

⊂ chordal

unimodular

⊂ perfect [91]

unit circular-arc

⊂ proper circular-arc [1041]

unit tolerance

⊂ proper tolerance [124]

***V*-perfect**

⊂ Welsh–Powell perfect [228]

weak bipolarizable

= HHDA-free [826]

weakly chordal

closed under complementation

⊂ bip* [526]

⊂ perfectly contractile [366]

Welsh–Powell perfect

⊂ perfectly orderable [219]

wing-triangulated

⊂ strict quasi parity [585]

X–star-chordal

⊂ bipartite ∩ X-chordal [75]

Bibliography

[1] J. ABELLO, O. EGECIOGLU, Visibility graphs of staircase polygons with uniform step length, *Internat. J. Comput. Geom. Appl.*, 3 (1993), 27–37.

[2] J. ABELLO, O. EGECIOGLU, K. KUMAR, Visibility graphs of staircase polygons and the weak Bruhat order I: from visibility graphs to maximal chains, *Discrete Comput. Geom.*, 14 (1995), 331–358.

[3] J. ABELLO, H. LIN, S. PISUPATI, On visibility graphs of simple polygons, *Congres. Numer.*, 90 (1992), 119–128.

[4] D. ACHLIOPTAS, The complexity of G-free colourability, *Discrete Math.*, 165/166 (1997) 21–30.

[5] R. AHARONI, R. HOLZMAN, Fractional kernels in digraphs, *J. Combin. Theory B*, 73 (1998), 1–6.

[6] A.V. AHO, J.E. HOPCROFT, J.D. ULLMAN, *The Design and Analysis of Computer Algorithms*, Addison–Wesley, Reading, MA (1974).

[7] M. AIGNER, E. TRIESCH, Reconstructing a graph from its neighbourhood list, *Combin. Probab. Comput.*, 2 (1993), 103–113.

[8] M. AIGNER, E. TRIESCH, Realizability and uniqueness in graphs, *Discrete Math.*, 136 (1994), 3–20.

[9] H. AIT HADDADENE, S. GRAVIER, On weakly diamond-free Berge graphs, *Discrete Math.*, 159 (1996), 237–240.

[10] H. AIT HADDADENE, F. MAFFRAY, Coloring perfect degenerate graphs, *Discrete Math.*, 163 (1997), 211–215

[11] M. AJTAI, J. KOMLÓS, E. SZEMERÉDI, Sorting in $c \log n$ parallel steps, *Combinatorica*, 3 (1983), 1–19.

[12] J. AKIYAMA, V. CHVÁTAL, Packing graphs perfectly, *Discrete Math.*, 85 (1990), 247–255.

[13] M.O. ALBERTSON, K.L. COLLINS, Duality and perfection for edges in cliques, *J. Combin. Theory B*, 36 (1984), 298–309.

[14] V. E. ALEXEJEW, Polynomial-time algorithm for the maximum stable set problem on chair-free graphs (in Russian), manuscript, University of Nishni Novgorod, 1998.

[15] N. ALON, Eigenvalues and expanders, *Combinatorica*, 6 (1986), 83–96.

[16] N. ALON, N. KAHALE, A spectral technique for coloring random 3-colorable graphs, *SIAM J. Comput.*, 26 (1997), 1733–1748.

[17] N. ALON, N. KAHALE, Approximating the independence number via the ϑ-function, *Math. Programming*, 80 (1998), 253–264.

[18] N. ALON, P. SEYMOUR, R. THOMAS, A separator theorem for graphs with an excluded minor and its applications, *J. Amer. Math. Soc.*, 3 (1990), 801–808.

[19] R. ANAND, H. BALAKRISHNAN, C. PANDU RANGAN, Treewidth of distance-hereditary graphs, manuscript (1994).

[20] T. ANDREAE, On superperfect noncomparability graphs, *J. Graph Theory*, 9 (1985), 523–532.

[21] T. ANDREAE, On the unit interval number of a graph, *Discrete Appl. Math.*, 22 (1988/89), 1–7.

[22] T. ANDREAE, Some results on visibility graphs, *Discrete Appl. Math.*, 40 (1992), 5–17.

[23] T. ANDREAE, U. HENNIG, A. PARRA, On a problem concerning tolerance graphs, *Discrete Appl. Math.*, 46 (1993), 73–78.

[24] R.P. ANSTEE, M. FARBER, Characterizations of totally balanced matrices, *J. Algorithms*, 5 (1984), 215–230.

[25] R.P. ANSTEE, M. FARBER, On bridged graphs and cop-win graphs, *J. Combin. Theory B*, 44 (1988), 22–28.

[26] S. ARNBORG, D.G. CORNEIL, A. PROSKUROWSKI, Complexity of finding embeddings in a k-tree, *SIAM J. Alg. Discrete Methods*, 8 (1987), 277–284.

[27] S. ARNBORG, B. COURCELLE, A. PROSKUROWSKI, D. SEESE, An algebraic theory of graph reduction, *J. Assoc. Comput. Mach.*, 40 (1993), 1134–1164.

[28] S. ARNBORG, J. LAGERGREN, D. SEESE, Easy problems for tree-decomposable graphs, *J. Algorithms*, 12 (1991), 308–340.

[29] S. ARNBORG, A. PROSKUROWSKI, Linear time algorithms for $\mathbb{NP}$-hard problems restricted to partial k-trees, *Discrete Appl. Math.*, 23 (1989), 11–24.

[30] S. ARNBORG, A. PROSKUROWSKI, Characterization and recognition of partial 3-trees, *SIAM J. Alg. Discrete Methods*, 7 (1986), 305–314.

[31] S. ARNBORG, A. PROSKUROWSKI, D.G. CORNEIL, Forbidden minors characterization of partial 3-trees, *Discrete Math.*, 80 (1990), 1–19.

[32] T. ASANO, Difficulty of the maximum independent set problem on intersection graphs of geometric objects, in Proc. Sixth International Confence on the Theory and Applications of Graphs, Western Michigan University, Y. Alavi, G. Chartrand, O.R. Oellerman, A.J. Schwenk, eds., John Wiley, New York (1988).

[33] M.D. ATKINSON, On computing the number of linear extensions of a tree, *Order*, 7 (1990), 23–25.

[34] F. AURENHAMMER, J. HAGAUER, W. IMRICH, Cartesian graph factorization at logarithmic cost per edge, *Comput. Complexity*, 2 (1992), 331–349.

[35] G. AUSIELLO, A. D'ATRI, M. MOSCARINI, Chordality properties on graphs and minimal conceptual connections in semantic data models, *J. Comput. System Sci.*, 33 (1986), 179–202.

[36] L. AUSLANDER, S. PARTER, On embedding graphs in the sphere, *J. Math. Mech.*, 10 (1961), 517–523.

[37] L. BABEL, *On the P_4-structure of graphs*, Habilitation thesis, TU München, Munich (1997).

[38] L. BABEL, A. BRANDSTÄDT, V.B. LE, Recognizing the P_4-structure of bipartite graphs, to appear in *Discrete Appl. Math.*

[39] L. BABEL, S. OLARIU, On the isomorphism of graphs with few P_4s, in 21st Intern. Workshop on Graph-Theoretic Concepts in Comp. Sci. WG '95, M. Nagl, ed., *Lecture Notes in Comput. Sci.*, 1017 (1995), 24–36.

[40] L. BABEL, S. OLARIU, A new characterization of P_4-connected graphs, in 22nd Intern. Workshop on Graph-Theoretic Concepts in Comp. Sci. WG'96, *Lecture Notes in Comput. Sci.*, 1197 (1996), 17–30.

[41] G. BACSÓ, E. BOROS, V. GURVICH, F. MAFFRAY, M. PREISSMANN, On minimally imperfect graphs with circular symmetry, *J. Graph Theory*, 29 (1998), 209–222.

[42] G. BACSÓ, Z. TUZA, A characterization of graphs without long induced paths, *J. Graph Theory*, 14 (1990), 455–464.

[43] G. BACSÓ, Z. TUZA, Dominating cliques in P_5-free graphs, *Period. Math. Hung.*, 21 (1990), 303–308.

[44] B.S. BAKER, Approximation algorithms for $\mathbb{NP}$-complete problems on planar graphs, *J. ACM*, 41 (1994), 153–180.

[45] K.A. BAKER, P.C. FISHBURN, F.S. ROBERTS, Partial orders of dimension 2, *Networks*, 2 (1971), 11–28.

[46] R. Balakrishnan, P. Paulraja, Powers of chordal graphs, *J. Austral. Math. Soc., Ser. A*, 35 (1983), 211–217.

[47] H.-J. Bandelt, Characterizing median graphs, manuscript, 1987.

[48] H.-J. Bandelt, Hereditary modular graphs, *Combinatorica*, 8 (1988), 149–157.

[49] H.-J. Bandelt, Neighbourhood–Helly powers, manuscript, (1992).

[50] H.-J. Bandelt, Graphs with edge–preserving majority functions, *Discrete Math.*, 103 (1992), 1-5.

[51] H.-J. Bandelt, V.D. Chepoi, A Helly theorem in weakly modular space, *Discrete Math.*, 160 (1996), 25–39.

[52] H.-J. Bandelt, A. Dählmann, H. Schütte, Absolute retracts of bipartite graphs, *Discrete Appl. Math.*, 16 (1987), 191–215.

[53] H.-J. Bandelt, M. Farber, P. Hell, Absolute reflexive retracts and absolute bipartite retracts, *Discrete Appl. Math.*, 44 (1993), 9–20.

[54] H.-J. Bandelt, A. Henkmann, F. Nicolai, Powers of distance-hereditary graphs, *Discrete Math.*, 145 (1995), 37–60.

[55] H.-J. Bandelt, H.M. Mulder, Interval-regular graphs of diameter two, *Discrete Math.*, 50 (1984), 117–134.

[56] H.-J. Bandelt, H.M. Mulder, Distance-hereditary graphs, *J. Combin. Theory B*, 41 (1986), 182–208.

[57] H.-J. Bandelt, H.M. Mulder, Pseudo-modular graphs, *Discrete Math.*, 62 (1986), 245–260.

[58] H.-J. Bandelt, H.M. Mulder, Three interval conditions for graphs, *Ars Combin.*, 29 (1990), 213–223.

[59] H.-J. Bandelt, H.M. Mulder, Metric characterization of parity graphs, *Discrete Math.*, 91 (1991), 221–230.

[60] H.-J. Bandelt, H.M. Mulder, Pseudo-median graphs: decomposition via amalgamation and Cartesian multiplication, *Discrete Math.*, 94 (1991), 161–180.

[61] H.-J. Bandelt, H.M. Mulder, Cartesian factorization of interval-regular graphs having no long isometric odd cycles, in *Graph Theory, Combinatorics, and Applications*, Vol. 1, Y. Alavi, G. Chartrand, R. Oellermann, A.J. Schwenk, eds., John Wiley, New York (1991), 55–75.

[62] H.-J. Bandelt, H.M. Mulder, E. Wilkeit, Quasi-median graphs and algebras, *J. Graph Theory*, 18 (1994), 681–703.

[63] H.-J. Bandelt, E. Pesch, Dismantling absolute retracts of reflexive graphs, *European J. Combin.*, 10 (1989), 211–220.

[64] H.-J. Bandelt, E. Pesch, Efficient characterizations of n-chromatic absolute retracts, *J. Combin. Theory B*, 53 (1991), 5–31.

[65] H.-J. Bandelt, E. Prisner, Clique graphs and Helly graphs, *J. Combin. Theory B*, 51 (1991), 34–45.

[66] H.-J. Bandelt, M. van de Vel, Superextensions and the depth of median graphs, *J. Combin. Theory A*, 57 (1991), 187–202.

[67] J. Bang-Jensen, P. Hell, On chordal proper circular arc graphs, *Discrete Math.*, 128 (1994), 395–398.

[68] V. Barré, J.-L. Fouquet, On minimal imperfect graphs without induced P_5, manuscript, Université du Maine, Le Mans, 1996.

[69] J.-P. Barthélemy, J. Constantin, Median graphs, parallelism and posets, *Discrete Math.*, 111 (1993), 49–63.

[70] D. Bauer, H.J. Broersma, H.J. Veldman, Not every 2-tough graph is Hamiltonian, Memorandum No. 1400, Universiteit Twente, Enschede, the Netherlands (1997).

[71] D. BAUER, S.L. HAKIMI, E. SCHMEICHEL, Recognizing tough graphs is NP-hard, *Discrete Appl. Math.*, 28 (1990), 191–195.

[72] S. BAUMANN, A linear algorithm for the homogeneous decomposition of graphs, Tech. Report M-9615, Institut für Mathematik, TU München, Munich (1996)

[73] C. BEERI, R. FAGIN, D. MAIER, A. MENDELZON, J.A. ULLMAN, M. YANNAKAKIS, Properties of acyclic database schemas, in Proc. 13th Ann. ACM Sympos. on Theory of Comp. (1981), 355–362.

[74] C. BEERI, R. FAGIN, D. MAIER, M. YANNAKAKIS, On the desirability of acyclic database schemes, *J. Assoc. Comput. Mach.*, 30 (1983), 479–513.

[75] H. BEHRENDT, A. BRANDSTÄDT, Domination and the use of maximum neighbourhoods, Schriftenreihe des Fachbereichs Mathematik der Universität Duisburg, Duisburg, Germany, SM-DU-204 (1992).

[76] L.W. BEINEKE, Derived graphs of digraphs, in *Beiträge zur Graphentheorie*, H. Sachs, H.-J. Voss, H.-J. Walter, eds. Teubner, Leipzig (1968), 17–33.

[77] L.W. BEINEKE, Characterization of derived graphs, *J. Combin. Theory*, 9 (1970), 129–135.

[78] L.W. BEINEKE, R.E. PIPPERT, The enumeration of labelled 2-trees, *Notices Amer. Math. Soc.*, 15 (1968), 384.

[79] L.W. BEINEKE, R.E. PIPPERT, The number of labelled k-dimensional trees. *J. Combin. Theory*, 6 (1969), 200–205.

[80] L.W. BEINEKE, R.E. PIPPERT, Properties and characterizations of k-trees, *Mathematika*, 18 (1971), 141–151.

[81] S. BELLANTONI, I. BEN–ARROYO HARTMAN, T. PRZYTYCKA, S. WHITESIDES, Grid intersection graphs and boxicity, *Discrete Math.*, 114 (1993), 41–49.

[82] M. BELLARE, O. GOLDREICH, S. GOLDWASSER, Randomness in interactive proofs, *Comput. Complexity*, 3 (1993), 319–354.

[83] I. BEN-ARROYO HARTMAN, I. NEWMAN, R. ZIV, On grid intersection graphs, *Discrete Math.*, 87 (1991), 41–52.

[84] C. BENZAKEN, Y. CRAMA, P. DUCHET, P.L. HAMMER, F. MAFFRAY, More characterizations of triangulated graphs, *J. Graph Theory*, 14 (1990), 413–422.

[85] C. BENZAKEN, P.L. HAMMER, Linear separation of domination sets in graphs, *Ann. Discrete Math.*, 3 (1978), 1–10.

[86] C. BENZAKEN, P.L. HAMMER, D. DE WERRA, Threshold characterization of graphs with Dilworth number two, *J. Graph Theory*, 9 (1985), 245–267.

[87] C. BENZAKEN, P.L. HAMMER, D. DE WERRA, Split graphs of Dilworth number 2, *Discrete Math.*, 55 (1985), 123–128.

[88] C. BERGE, Les problèmes de colorations en théorie des graphes, *Publ. Inst. Stat. Univ. Paris*, 9 (1960), 123–160.

[89] C. BERGE, Färbung von Graphen, deren sämtliche bzw. deren ungerade Kreise starr sind, *Wiss. Zeitschr. Martin-Luther-Univ. Halle-Wittenberg*, 10 (1961), 114–115.

[90] C. BERGE, *Graphs and Hypergraphs*, North–Holland, Amsterdam (1985)

[91] C. BERGE, Perfect graphs, *Studies in Graph Theory Part I*, D.R. Fulkerson, ed., Vol. 11 of MAA Studies in Mathematics, The Mathematical Association of America (1973), 1–22.

[92] C. BERGE, The q-perfect graphs. Part I: the case $q = 2$, *Colloquia Math. Soc. János Bolyai*, 60 (1991), 67–75.

[93] C. BERGE, The q-perfect graphs. Part II, *Le Matematiche*, 47 (1992), 205–211.

[94] C. BERGE, The q-perfect graphs, RUTCOR Research Report, Rutgers University, New Brunswick, NJ, RRR 23-92 (1992).

[95] C. BERGE, Motivations and history of some of my conjectures, *Discrete Math.*, 165/166 (1997), 61–70.

[96] C. BERGE, V. CHVÁTAL (eds.), *Topics on Perfect Graphs, Ann. Discrete Math.*, 21, North–Holland, Amsterdam (1984).

[97] C. BERGE, P. DUCHET, Probleme Seminaire du Lundi, Tech. Report M.S.H. 54 Bd. Raspail 75006 Paris (1983).

[98] C. BERGE, P. DUCHET, Strongly perfect graphs, *Ann. Discrete Math.*, 21 (1984), 57–61.

[99] C. BERGE, P. DUCHET, Perfect graphs and kernels, *Bull. Inst. Math. Acad. Sinica,* 16 (1988), 263–274.

[100] C. BERGE, M. LAS VERGNAS, Sur un théorème du type König pour hypergraphes, *Ann. New York Acad. Sci.,* 175 (1970), 32–40.

[101] M.W. BERN, E.L. LAWLER, A.L. WONG, Linear time computation of optimal subgraphs of decomposable graphs, *J. Algorithms,* 8 (1987), 216–235.

[102] P.A. BERNSTEIN, N. GOODMAN, Power of natural semijoins, *SIAM J. Comput.*, 10 (1981), 751–771.

[103] P. BERTOLAZZI, G. DIBATTISTA, C. MANNINO, R. TAMASSIA, Optimal upward planarity testing of single-source digraphs, *SIAM J. Comput.*, 27 (1998), 132–169.

[104] M. BERTSCHI, Perfectly contractile graphs, *J. Combin. Theory B,* 50 (1990), 222–230.

[105] M. BERTSCHI, B.A. REED, A note on even pairs, *Discrete Math.,* 65 (1987), 317–318; Erratum, *Discrete Math.,* 71 (1988) p.187.

[106] E. BIBELNIEKS, P.M. DEARING, Neighbourhood subtree tolerance graphs, *Discrete Appl. Math.,* 43 (1993), 13–26.

[107] D. BIENSTOCK, On the complexity of testing for odd induced holes and odd induced paths, *Discrete Math.,* 90 (1991), 85–92; Corrigendum, *Discrete Math.,* 102 (1992), 109.

[108] R.E. BIXBY, A composition for perfect graphs, *Ann. Discrete Math.*, 21 (1984), 221–224.

[109] J.R.S. BLAIR, B. PEYTON, An introduction to chordal graphs and clique trees, in *Graph Theory and Sparse Matrix Computation*, A. George et al., eds., Springer, New York (1993), 1–29.

[110] R.G. BLAND, H.-C. HUANG, L.E. TROTTER, JR., Graphical properties related to minimal imperfection, *Discrete Math.,* 27 (1979), 11–22.

[111] Z. BLÁZSIK, M. HUJTER, A. PLUHÁR, ZS. TUZA, Graphs with no induced C_4 and $2K_2$, *Discrete Math.,* 115 (1993), 51–55.

[112] M. BLIDIA, P. DUCHET, F. MAFFRAY, On kernels in perfect graphs, *Combinatorica,* 13 (1993), 231–233.

[113] L.M. BLUMENTHAL, *Theory and Applications of Distance Geometry*, Oxford University Press, London (1953).

[114] H.L. BODLAENDER, Classes of graphs with bounded treewidth, Technical Report RUU-CS-86-22, Dept. of Computer Sci., Utrecht University, Utrecht, the Netherlands (1986)

[115] H.L. BODLAENDER, Dynamic programming on graphs with bounded treewidth, in Proc. 15th International Colloqu. on Automata, Languages and Programming, *Lecture Notes in Comput. Sci.*, 317 (1988), 105–119.

[116] H.L. BODLAENDER, Some classes of graphs with bounded treewidth, *Bull. EATCS,* 36 (1988), 116–126.

[117] H.L. BODLAENDER, Planar graphs with bounded treewidth, Technical Report RUU-CS-88-14, Dept. of Computer Sci., Utrecht University, Utrecht, the Netherlands (1988)

[118] H.L. BODLAENDER, Achromatic number is $\mathbb{NP}$–complete for cographs and interval graphs, *Inform. Process. Lett.,* 31 (1989), 135–138.

[119] H.L. BODLAENDER, A tourist guide through treewidth, *Acta Cybernet.*, 11 (1993), 1–23.

[120] H.L. BODLAENDER, A linear time algorithm for finding tree-decompositions of small treewidth, *SIAM J. Comput.*, 25 (1996), 1305–1317.

[121] H.L. Bodlaender, T. Kloks, D. Kratsch, Treewidth and pathwidth of permutation graphs, *Lecture Notes in Comput. Sci.*, 700 (1993), 114–125; *SIAM J. Discrete Math.*, 8 (1995), 606–616.

[122] H.L. Bodlaender, R.H. Möhring, The pathwidth and treewidth of cographs, *SIAM J. Discrete Math.*, 6 (1993), 181–188.

[123] H.L. Bodlaender, D.M. Thilikos, Treewidth for graphs with small chordality, *Discrete Appl. Math.*, 79 (1997), 45–61.

[124] K.P. Bogart, P.C. Fishburn, G. Isaak, L. Langley, Proper and unit tolerance graphs, *Discrete Appl. Math.*, 60 (1995), 99–117.

[125] B. Bollobás, *Random Graphs*, Academic Press, New York (1985).

[126] J.A. Bondy, U.S.R. Murty, *Graph Theory with Applications*, MacMillan, London (1976)

[127] K.S. Booth, G.S. Lueker, Testing for the consecutive ones property, interval graphs, and graph planarity using PQ-tree algorithms, *J. Comput. System Sci.*, 13 (1976), 335–379.

[128] R.B. Borie, R.G. Parker, C.A. Tovey, Deterministic decomposition of recursive graph classes, *SIAM J. Discrete Math.*, 4 (1991), 481–501.

[129] R.B. Borie, R.G. Parker, C.A. Tovey, Automatic generation of linear-time algorithms from predicate calculus descriptions of problems on recursively constructed graph classes, *Algorithmica*, 7 (1992), 555–581.

[130] R.B. Borie, J.P. Spinrad, Construction of a simple elimination sceme for a chordal comparability graph in linear time, *Discrete Appl. Math.*, 91 (1999), 287–292.

[131] C.F. Bornstein, J.L. Szwarcfiter, On clique convergent graphs, *Graphs Combin.*, 11 (1995), 213–220.

[132] E. Boros, O. Cepek, On perfect $(0, \pm 1)$ matrices, *Discrete Math.*, 165/166 (1997), 81–100.

[133] E. Boros, A.V. Gurvich, Perfect graphs are kernel solvable, *Discrete Math.*, 159 (1996), 35–55.

[134] E. Boros, A.V. Gurvich, Stable effectivity functions and perfect graphs, RUTCOR Research Report, Rutgers University, New Brunswick, NJ, RRR 23–95 (1995)

[135] E. Boros, A.V. Gurvich, A. Vasin, Stable families of coalitions and normal hypergraphs, *Math. Social Sci.*, 34 (1997), 107–123.

[136] A. Bouchet, Characterizing and recognizing circle graphs, in Proc. 6th Yugoslav Seminar on Graph Theory, Dubrovnik (1985), 57–69.

[137] A. Bouchet, A polynomial algorithm for recognizing circle graphs (in French), *C.R. Acad. Sci. Paris, Sér. I Math.*, 300 (1985), 569–572.

[138] A. Bouchet, Reducing prime graphs and recognizing circle graphs, *Combinatorica*, 7 (1987), 243–254.

[139] A. Bouchet, Circle graph obstructions, *J. Combin. Theory B*, 60 (1994), 107–144.

[140] V. Bouchitté, I. Todinca, Minimal triangulations for graphs with "few" minimal seperators, Research Report 98-24, École Normale Supérieure de Lyon, Lyon, France (1998).

[141] A. Brandstädt, Classes of bipartite graphs related to chordal graphs, *Discrete Appl. Math.*, 32 (1991), 51–60.

[142] A. Brandstädt, Partitions of graphs into one or two independent sets and cliques, *Discrete Math.*, 152 (1996), 47–54; Corrigendum, *Discrete Math.*, 186 (1998), 295.

[143] A. Brandstädt, V.D. Chepoi, F.F. Dragan, The algorithmic use of hypertree structure and maximum neighbourhood orderings, *Discrete Appl. Math.*, 82 (1998), 43–77.

[144] A. Brandstädt, V.D. Chepoi, F.F. Dragan, Perfect elimination orderings of chordal powers of graphs, *Discrete Math.*, 158 (1996), 273–278.

[145] A. Brandstädt, V.D. Chepoi, F.F. Dragan, Clique r-domination and clique r-packing problems on dually chordal graphs, *SIAM J. Discrete Math.*, 10 (1997), 109–127.

[146] A. BRANDSTÄDT, F.F. DRAGAN, V.D. CHEPOI, V.I. VOLOSHIN, Dually chordal graphs, *SIAM J. Discrete Math.*, 11 (1998), 437–455.

[147] A. BRANDSTÄDT, F.F. DRAGAN, V.B. LE, T. SZYMCZAK, On stable cutsets in graphs, manuscript (1998).

[148] A. BRANDSTÄDT, F.F. DRAGAN, F. NICOLAI, Homogeneously orderable graphs, *Theoret. Comput. Sci.*, 172 (1997), 209–232.

[149] A. BRANDSTÄDT, F.F. DRAGAN, F. NICOLAI, LexBFS-orderings and powers of chordal graphs, *Discrete Math.*, 171 (1997) 27–42.

[150] A. BRANDSTÄDT, V.B. LE, Recognizing the P_4-structure of block graphs, Preprint CS–01–98, Universität Rostock, 1997. To appear in *Discrete Appl. Math.*

[151] A. BRANDSTÄDT, V.B. LE, Tree- and forest-perfect graphs, Preprint CS–03–98, Universität Rostock, 1997. To appear in *Discrete Appl. Math.*

[152] A. BRANDSTÄDT, V.B. LE, S. OLARIU, Linear-time recognition of the P_4-structure of trees, RUTCOR Research Report, Rutgers University, New Brunswick, NJ, RRR 19–96 (1996)

[153] A. BRANDSTÄDT, V.B. LE, S. OLARIU, Efficiently recognizing the P_4-structure of trees and of bipartite graphs without short cycles, manuscript, Universität Rostock, 1997. To appear in *Graphs Combin.*

[154] A. BRANDSTÄDT, V.B. LE, T. SZYMCZAK, Duchet-type theorems for powers of HHD-free graphs, *Discrete Math.*, 177 (1997), 9–16.

[155] A. BRANDSTÄDT, J. SPINRAD, L. STEWART, Bipartite permutation graphs are bipartite tolerance graphs, *Congres. Numer.*, 58 (1987), 165–174.

[156] H. BREU, D.G. KIRKPATRICK, Unit disk graph recognition is $\mathbb{NP}$–hard, *Comput. Geom.*, 9 (1998), 3–24.

[157] H. BREU, D.G. KIRKPATRICK, On the complexity of recognizing intersection and touching graphs of disks, *Lecture Notes in Comput. Sci.*, 1027, 88–98.

[158] G. BRIGHTWELL, On the complexity of diagram testing, *Order*, 10 (1993), 297–303.

[159] H. BROERSMA, E. DAHLHAUS, T. KLOKS, A linear time algorithm for minimum fill-in and treewidth for distance-hereditary graphs, *Lecture Notes in Comput. Sci.*, 1335 (1997), 109–117.

[160] H. BROERSMA, T. KLOKS, D. KRATSCH, H. MÜLLER, Independent sets in asteroidal triple-free graphs, *Lecture Notes in Comput. Sci.* 1256 (1997), 760–770.

[161] J. BROUSEK, Z. RYJÁVCEK, O. FAVARON, Forbidden subgraphs, Hamiltonicity, and closure in claw-free graphs, *Discrete Math.*, 196 (1999), 29–50.

[162] A.E. BROUWER, P. DUCHET, A. SCHRIJVER, Graphs whose neighbourhoods have no special cycles, *Discrete Math.*, 47 (1983), 177–182.

[163] J.I. BROWN, D.G. CORNEIL, A.R. MAHJOUB, A note on K_i-perfect graphs, *J. Graph Theory*, 14 (1990), 333–340.

[164] M.A. BUCKINGHAM, M.C. GOLUMBIC, Partitionable graphs, circle graphs and the Berge strong perfect graph conjecture, *Discrete Math.*, 44 (1983), 45–54.

[165] M.A. BUCKINGHAM, M.C. GOLUMBIC, Recent results on the strong perfect graph conjecture, *Ann. Discrete Math.*, 20 (1984), 75–82.

[166] P. BUNEMAN, A note on the metric properties of trees, *J. Combin. Theory B*, 1 (1974), 48–50.

[167] P. BUNEMAN, A characterization of rigid circuit graphs, *Discrete Math.*, 9 (1974), 205–212.

[168] M. BURLET, *Etude algorithmique de certaines classes de graphes parfaits*, Ph.D. thesis (These troisieme cycle), Grenoble, France (1981).

[169] M. BURLET, J. FONLUPT, Polynomial algorithm to recognize a Meyniel graph, *Ann. Discrete Math.*, 21 (1984), 225–252.

[170] M. BURLET, J.P. UHRY, Parity graphs, *Ann. Discrete Math.*, 21 (1984), 253–277.

[171] H. Busemann, *The Geometry of Geodesics*, Academic Press, New York (1955)

[172] H. Busemann, B.B. Phadke, Peakless and monotone functions on G-spaces, *Tsukuba J. Math.*, 7 (1983), 105–135.

[173] L. Cai, D.G. Corneil, A generalization of perfect graphs–i-perfect graphs, *J. Graph Theory*, 23 (1996), 87–103.

[174] L. Cai, D.G. Corneil, A. Proskurowski, A generalization of line graphs: (X,Y)-intersection graphs, *J. Graph Theory*, 21 (1996), 267–287.

[175] K.B. Cameron, *Polyhedral and algorithmic ramifications of antichains*, Ph.D. thesis, University of Waterloo, Canada (1982).

[176] K.B. Cameron, A min-max relation for the partial q-colourings of a graph. Part II: Box Perfection, *Discrete Math.*, 74 (1989), 15–27.

[177] K. Cameron, J. Edmonds, Lambda decompositions, *J. Graph Theory*, 26 (1997), 9–16.

[178] K.B. Cameron, J. Edmonds, L. Lovász, A note on perfect graphs, *Period. Math. Hung.*, 17 (1986), 173–175.

[179] O.M. Carducci, The strong perfect graph conjecture holds for diamonded odd cycle-free graphs, *Discrete Math.*, 110 (1992), 17–34.

[180] G.J. Chang, *k-domination and graph covering problems*, Ph.D. thesis, School of OR and IE, Cornell University, Ithaca, NY (1982).

[181] G.J. Chang, Labeling algorithms for domination problems in sun-free chordal graphs, *Discrete Appl. Math.*, 22 (1988), 21–34.

[182] G.J. Chang, M. Farber, Z. Tuza, Algorithmic aspects of neighborhood numbers, *SIAM J. Discrete Math.*, 6 (1993), 24–29.

[183] G.J. Chang, G.L. Nemhauser, The k-domination and k-stability problem on sun-free chordal graphs, *SIAM J. Alg. Discrete Methods*, 5 (1984), 332–345.

[184] G.J. Chang, G.L. Nemhauser, Covering, packing and generalized perfection, *SIAM J. Alg. Discrete Methods*, 6 (1985), 109–132.

[185] G. Chartrand, H.J. Gavlas, M. Schultz, Convergent sequences of iterated H-line graphs, *Discrete Math.*, 147 (1995), 73–86.

[186] M. Cheah, D.G. Corneil, On the structure of trapezoid graphs, *Discrete Appl. Math.*, 66 (1996), 109–133.

[187] G. Chen, M.S. Jacobson, A. Kézdy, J. Lehel, Tough enough chordal graphs are Hamiltonian, *Networks*, 31 (1998), 29–38.

[188] V.D. Chepoi, *d-convex sets in graphs*, Disseration thesis (in Russian), Moldova State University, Chişinău (1986).

[189] V.D. Chepoi, Classifying graphs by metric triangles (in Russian), *Metody Diskretnogo Analiza*, 49 (1989), 75–93.

[190] V.D. Chepoi, Peakless functions on graphs, *Discrete Appl. Math.*, 73 (1997), 175–189.

[191] V.D. Chepoi, Bridged graphs are cop-win graphs: an algorithmic proof, *J. Combin. Theory B*, 69 (1997), 97–100.

[192] V.D. Chepoi, On distance-preserving and domination elimination orderings, *SIAM J. Discrete Math.*, 11 (1998), 414–436.

[193] Zh.A. Chernyak, A.A. Chernyak, About recognizing (α,β) classes of polar graphs, *Discrete Math.*, 62 (1986), 133–138.

[194] N. Chiba, T. Nishizeki, S. Abe, T. Ozawa, A linear algorithm for embedding planar graphs using PQ-trees, *J. Comput. System Sci.*, 30 (1985), 54–76.

[195] A. Chmeiss, P. Jegou, A generalization of chordal graphs and the maximum clique problem, *Inform. Process. Lett.*, 62 (1997), 61–66.

[196] S.-H. CHOI, S.Y. SHIN, K.-Y. CHWA, Characterizing and recognizing the visibility graph of a funnel shaped polygon, *Algorithmica*, 14 (1995), 27–51.

[197] C.A. CHRISTEN, S.M. SELKOW, Some perfect colouring properties of graphs, *J. Combin. Theory B*, 27 (1979), 49–59.

[198] F.R.K. CHUNG, Separator theorems and their applications, in *Paths, Flows and VLSI-Layout*, B. Korte, L. Lovász, H. J. Prömel, A. Schrijver, eds., Springer, Berlin (1990), 17–34.

[199] F.R.K. CHUNG, *Spectral Graph Theory, CBMS Regional Conference Series in Mathematics*, 92 (1997).

[200] V. CHVÁTAL, Tough graphs and Hamiltonian circuits, *Discrete Math.*, 5 (1973), 215–228.

[201] V. CHVÁTAL, On certain polytopes associated with graphs, *J. Combin. Theory B*, 18 (1975), 138–154.

[202] V. CHVÁTAL, On the strong perfect graph conjecture, *J. Combin. Theory B*, 20 (1976), 139–141.

[203] V. CHVÁTAL, An equivalent version of the strong perfect graph conjecture, in *Topics on Perfect Graphs* (C. Berge, V. Chvátal, eds.), *Ann. Discrete Math.*, 21 (1984), 193–196.

[204] V. CHVÁTAL, A semi-strong perfect graph conjecture, in *Topics on Perfect Graphs* (C. Berge, V. Chvátal, eds.), *Ann. Discrete Math.*, 21 (1984), 279–280.

[205] V. CHVÁTAL, Perfectly ordered graphs, in *Topics on Perfect Graphs* (C. Berge, V. Chvátal, eds.), *Ann. Discrete Math.*, 21 (1984), 63–65.

[206] V. CHVÁTAL, Recognizing decomposable graphs, *J. Graph Theory*, 8 (1984), 51–53.

[207] V. CHVÁTAL, Star-cutsets and perfect graphs, *J. Combin. Theory B*, 39 (1985), 189–199.

[208] V. CHVÁTAL, On the P_4-structure of perfect graphs III. Partner Decompositions, *J. Combin. Theory B*, 43 (1987), 349–353.

[209] V. CHVÁTAL, Perfect graphs, *Surveys in Combinatorics*, C. Whitehead, ed., *LMS Lect. Notes Series* 123, Cambridge University Press, Cambridge (1987).

[210] V. CHVÁTAL, A class of perfectly orderable graphs, Research Report 89573-OR, University of Bonn (1989).

[211] V. CHVÁTAL, Which line graphs are perfectly orderable?, *J. Graph Theory*, 5 (1990), 555–558.

[212] V. CHVÁTAL, Which claw-free graphs are perfectly orderable?, *Discrete Appl. Math.*, 44 (1993), 39–63.

[213] V. CHVÁTAL, Problems concerning perfect graphs, Collection distributed at the DIMACS workshop on Perfect Graphs, Princeton University (1993), http://dimacs.rutgers.edu/pub/perfect/problems.tex.

[214] V. CHVÁTAL, In praise of Claude Berge, *Discrete Math.*, 165/166 (1997), 3–9.

[215] V. CHVÁTAL, R.L. GRAHAM, A.F. PEROLD, J.H. WHITESIDES, Combinatorial designs related to the strong perfect graph conjecture, *Discrete Math.*, 26 (1979), 83–92.

[216] V. CHVÁTAL, P.L. HAMMER, Set-packing and threshold graphs, Research Report, Comp. Sci. Dept. University of Waterloo, Canada CORR 73-21 (1973).

[217] V. CHVÁTAL, P.L. HAMMER, Aggregation of inequalities in integer programming, *Ann. Discrete Math.*, 1 (1977), 145–162.

[218] V. CHVÁTAL, C.T. HOÀNG, On the P_4-structure of perfect graphs I. Even decompositions, *J. Combin. Theory B*, 39 (1985), 209–219.

[219] V. CHVÁTAL, C.T. HOÀNG, N.V.R. MAHADEV, D. DE WERRA, Four classes of perfectly orderable graphs, *J. Graph Theory*, 11 (1987), 481–495.

[220] V. CHVÁTAL, W.J. LENHART, N. SBIHI, Two-colourings that decompose perfect graphs, *J. Combin. Theory B*, 49 (1990), 1–9.

[221] V. CHVÁTAL, I. RUSU, A note on graphs with no long holes, presented at DIMACS Workshop on Perfect Graphs (1993).

[222] V. Chvátal, N. Sbihi, Bull-free Berge graphs are perfect, *Graphs Combin.*, 3 (1987), 127–139.

[223] V. Chvátal, N. Sbihi, Recognizing claw-free perfect graphs, *J. Combin. Theory B*, 44 (1988), 154–176.

[224] S. Cicerone, G. Di Stefano, On the extension of bipartite to parity graphs, in ODSA'97 workshop, Rostock (1997). To appear in *Discrete Appl. Math.*

[225] S. Cicerone, G. Di Stefano, Graph classes between parity and distance-hereditary graphs, ODSA'97 workshop, Rostock (1997). To appear in *Discrete Appl. Math.*

[226] S. Cicerone, G. Di Stefano, Graphs with bounded induced distance, *Lecture Notes in Comput. Sci.*, 1517 (1998), 177–191.

[227] B.N. Clark, C.J. Colbourn, D.S. Johnson, Unit disk graphs, *Discrete Math.*, 86 (1990), 165–177.

[228] M. Cochand, D. de Werra, Generalized neighbourhoods and a class of perfectly orderable graphs, *Discrete Appl. Math.*, 15 (1986), 213–220.

[229] O. Cogis, Une caractérisation des graphes orientés de dimension Ferrers égale à 2, Université Pierre et Marie Curie (Paris VI), C.N.R.S., Paris (1979).

[230] O. Cogis, Ferrers digraphs and threshold graphs, *Discrete Math.*, 38 (1982), 33–46.

[231] O. Cogis, On the Ferrers dimension of a digraph, *Discrete Math.*, 38 (1982), 47–52.

[232] C.J. Colbourn, On testing isomorphism of permutation graphs, *Networks,* 11 (1981), 13–21.

[233] P. Colley, *Visibility graphs of uni-monotone polygons*, Masters thesis, Dept. of Computer Science, University of Waterloo, Canada, (1991).

[234] P. Colley, in Recognizing visibility graphs of uni-monotone polygons, Proc. 4th Canadian Conference on Computational Geometry (1992), 29–34.

[235] P. Colley, A. Lubiw, J. Spinrad, Visibility graphs of towers, *Computational Geometry Theory* 7 (1997), 161–172.

[236] M. Conforti, $K_4 - e$ free graphs and star cutsets, *Lecture Notes in Math.*, 1403 (1989), 236–253.

[237] M. Conforti, D.G. Corneil, A.R. Mahjoub, K_i-covers II. K_i-perfect graphs, *J. Graph Theory,* 11 (1987), 569–584.

[238] M. Conforti, G. Cornuéjols, A. Kapoor, K. Vuskovic, Balanced $0, \pm 1$ matrices, parts 1–2, preprints, Carnegie Mellon University, Pittsburgh, PA (1993).

[239] M. Conforti, G. Cornuéjols, R. Rao, Decomposition of balanced $(0, 1)$ matrices, parts I–VII, preprints, Carnegie Mellon University, Pittsburgh, PA (1991).

[240] M. Conforti, R. Rao, Structural properties of restricted and strongly unimodular matrices, *Math. Programming,* 38 (1987), 17–27.

[241] M. Conforti, R. Rao, Testing balancedness and perfection of linear matrices, *Math. Programming,* Ser. A, 61 (1993), 1–18.

[242] D. Coppersmith, U. Vishkin, Solving $\mathbb{NP}$-hard problems in almost trees: vertex cover, *Discrete Appl. Math.*, 10 (1985), 27–45.

[243] T.H. Cormen, C.E. Leiserson, R.L. Rivest, *Introduction to Algorithms*, MIT Press, Cambridge, MA (1990).

[244] D.G. Corneil, Families of graphs complete for the strong perfect graph conjecture, *J. Graph Theory,* 10 (1986), 33–40.

[245] D.G. Corneil, J. Fonlupt, Stable set bonding in perfect graphs and parity graphs, *J. Combin. Theory B*, 59 (1993), 1–14.

[246] D.G. Corneil, P.A. Kamula, Extensions of permutation and interval graphs, *Congres. Numer.*, 58 (1987), 267–275.

[247] D.G. Corneil, H. Kim, S. Natarajan, S. Olariu, A.P. Sprague, Simple linear time recognition of unit interval graphs, *Inform. Process. Lett.*, 55 (1995), 99–104.

[248] D.G. Corneil, D.G. Kirkpatrick, Families of recursively defined perfect graphs, *Congres. Numer.*, 39 (1983), 237–246.

[249] D.G. Corneil, H. Lerchs, L. Stewart-Burlingham, Complement reducible graphs, *Discrete Appl. Math.*, 3 (1981), 163–174.

[250] D.G. Corneil, S. Olariu, L. Stewart, Asteroidal triple-free graphs, *SIAM J. Discrete Math.*, 10 (1997), 399–430.

[251] D.G. Corneil, S. Olariu, L. Stewart, Computing a dominating pair in an asteroidal triple-free graph in linear time, *Lecture Notes in Comput. Sci.*, 955 (1995), 358–368.

[252] D.G. Corneil, S. Olariu, L. Stewart, A linear time algorithm to compute a dominating path in AT-free graphs, *Inform. Process. Lett.*, 54 (1995), 253–257.

[253] D.G. Corneil, S. Olariu, L. Stewart, Linear time algorithms for dominating pairs in asteroidal triple-free graphs, *Lecture Notes in Comput. Sci.*, 944 (1995), 292–303.

[254] D.G. Corneil, S. Olariu, L. Stewart, The ultimate interval graph recognition algorithm?, in ACM–SIAM Symposium on Discrete Algorithms '98 (1998), 175–180.

[255] D.G. Corneil, Y. Perl, L. Stewart, Cographs: recognition, application and algorithms, *Congres. Numer.*, 43 (1984), 249–258.

[256] D.G. Corneil, Y. Perl, L.K. Stewart, A linear recognition algorithm for cographs, *SIAM J. Comput.*, 14 (1985), 926–934.

[257] G. Cornuéjols, W.H. Cunningham, Compositions for perfect graphs, *Discrete Math.*, 55 (1985), 245–254.

[258] G. Cornuéjols, D. Naddef, W.R. Pulleyblank, Halin graphs and the traveling salesman problem, *Math. Programming*, 26 (1983), 287–294.

[259] G. Cornuéjols, B. Reed, Complete multi-partite cutsets in minimal imperfect graphs, *J. Combin. Theory B*, 59 (1993), 191–198.

[260] C. Coullard, A. Lubiw, Distance visibility graphs, *Internat. J. Comput. Geom. Appl.*, 2 (1992), 349–362.

[261] B. Courcelle, Graphs and monadic second order logic: some open problems, *Bull. EATCS*, 49 (1993), 110–124.

[262] A. Cournier, M. Habib, A new linear algorithm for modular decomposition, *Lecture Notes in Comput. Sci.*, 787 (1994) 68-84.

[263] M.B. Cozzens, The $\mathbb{NP}$-completeness of the boxicity of a graph, manuscript (1982).

[264] M.B. Cozzens, L.L. Kelleher, Dominating cliques in graphs, *Discrete Math.*, 86 (1990), 101–116.

[265] M.B. Cozzens, F.S. Roberts, Computing the boxicity of a graph by covering its complement by cointerval graphs, *Discrete Appl. Math.*, 6 (1983), 217–228.

[266] M.B. Cozzens, F.S. Roberts, On dimensional properties of graphs, *Graphs Combin.*, 5 (1989), 29–46.

[267] Y. Crama, P.L. Hammer, T. Ibaraki, Strong unimodularity for graphs and hypergraphs, *Discrete Appl. Math.*, 15 (1986), 221–239.

[268] Y. Crama, P.L. Hammer, T. Ibaraki, Packing, covering and partitioning problems with strongly unimodular constraint matrices, *Math. Oper. Res.*, 15 (1990), 258–267.

[269] W.H. Cunningham, *A combinatorial decomposition theory*, Ph.D. thesis, University of Waterloo, Canada (1973).

[270] W.H. Cunningham, Decomposition of directed graphs, *SIAM J. Alg. Discrete Methods*, 3 (1982), 214–228.

[271] W.H. Cunningham, J. Edmonds, A combinatorial decomposition theory, *Canad. J. Math.*, 32 (1980), 734–765.

[272] D.M. Cvetković, M. Doob, H. Sachs, *Spectra of Graphs, Theory and Applications*, Academic Press, Berlin (1980).

[273] D.M. CVETKOVIĆ, M. DOOB, I. GUTMAN, A. TORGAŠEV, *Recent Results in the Theory of Graph Spectra*, *Ann. Discrete Math.*, 36, North–Holland, Amsterdam (1988).

[274] I. CZISZÁR, J. KÖRNER, L. LOVÁSZ, K. MARTON, G. SIMONYI, Entropy splitting for antiblocking pairs and perfect graphs, *Combinatorica*, 10 (1990), 27–40.

[275] I. DAGAN, M.C. GOLUMBIC, R.Y. PINTER, Trapezoid graphs and their coloring, *Discrete Appl. Math.*, 21 (1988), 35–46.

[276] E. DAHLHAUS, *Chordale Graphen im besonderen Hinblick auf parallele Algorithmen*, Habilitation thesis, Universität Bonn (1991).

[277] E. DAHLHAUS, Generalized strongly chordal graphs, manuscript (1993).

[278] E. DAHLHAUS, Efficient parallel and linear time sequential split decomposition, *Lecture Notes in Comput. Sci.*, 880 (1994), 171–180.

[279] E. DAHLHAUS, P. DUCHET, On strongly chordal graphs, *Ars Combin.*, 24B (1987), 23–30.

[280] E. DAHLHAUS, J. GUSTEDT, R.M. MCCONNELL, Efficient and practical modular decomposition, Tech. Report TU Berlin FB Mathematik, 524/1996 (1996), in 8th ACM–SIAM Symposium on Discrete Algorithms (1997), 26–35.

[281] E. DAHLHAUS, P.L. HAMMER, F. MAFFRAY, S. OLARIU, On domination elimination orderings and domination graphs, *Lecture Notes in Comput. Sci.*, 903 (1994), 81–92.

[282] E. DAHLHAUS, M. KARPINSKI, Matching and multidimensional matching in chordal and strongly chordal graphs, *Discrete Math. Appl.*, 84 (1998), 79–91.

[283] R.C. DALANG, L.E. TROTTER, JR., D. DE WERRA, On randomized stopping points and perfect graphs, *J. Combin. Theory B*, 45 (1988), 320–344.

[284] P. DAMASCHKE, Forbidden ordered subgraphs, in *Festschrift zu Ehren von Gerhard Ringel*, R. Bodendiek, R. Henn, eds., *Topics in Combinatorics and Graph Theory*, Physica Verlag, Heidelberg (1990), 219–229.

[285] P. DAMASCHKE, Hamiltonian-hereditary graphs, manuscript (1990).

[286] P. DAMASCHKE, Distances in cocomparability graphs and their powers, *Discrete Appl. Math.*, 35 (1992), 67–72.

[287] P. DAS, Unidigraphic and unigraphic degree sequences through uniquely realizable integer-pair sequences, *Discrete Math.*, 45 (1983), 45–59.

[288] A. D'ATRI, M. MOSCARINI, Distance-hereditary graphs, Steiner trees, and connected domination, *SIAM J. Comput.*, 17 (1988), 521–538.

[289] A. D'ATRI, M. MOSCARINI, On hypergraph acyclicity and graph chordality, *Inform. Process. Lett.*, 29 (1988), 271–274.

[290] A. D'ATRI, M. MOSCARINI, A. SASSANO, The Steiner tree problem and homogeneous sets, *Lecture Notes in Comput. Sci.*, 324 (1988), 249–261.

[291] D.P. DAY, O.R. OELLERMANN, H.C. SWART, Steiner distance-hereditary graphs, *SIAM J. Discrete Math.*, 7 (1994), 437–442.

[292] C. DE FIGUEIREDO, F. MAFFRAY, O. PORTO, On the structure of bull-free perfect graphs, *Graphs Combin.*, 13 (1997), 31–55.

[293] C. DE FIGUEIREDO, J. MEIDANIS, C.P. DE MELLO, A linear time algorithm for proper interval graph recognition, *Inform. Process. Lett.*, 56 (1995), 179–184.

[294] B. DE FLUITER, *Algorithms for graphs of small treewidth*, Ph.D. thesis, University of Utrecht, Utrecht, the Netherlands (1997).

[295] H. DE FRAYSSEIX, A characterization of circle graphs, *European J. Combin.*, 5 (1984), 223–238.

[296] H. DE FRAYSSEIX, P. OSSONA DE MENDEZ, Planarity and edge poset dimension, *European J. Combin.*, 17 (1996), 731–740.

[297] S. DE AGOSTINO, R. PETRESCHI, A. STERBINI, An $\mathcal{O}(n^3)$ recognition algorithm for bithreshold graphs, *Algorithmica*, 17 (1997), 416–425.

[298] P. DEGANO, R. DE NICOLA, U. MONTANARI, Partial ordering descriptions and observations of nondeterministic processes, *Lecture Notes in Comput. Sci.*, 354 (1989), 438–466.

[299] D. DEGIORGI, A new linear algorithm to detect a line graph and output its root graph, Tech. Report 148, Inst. für Theoretische Informatik, ETH Zürich, Zürich, Switzerland (1990).

[300] X. DENG, P. HELL, J. HUANG, Linear time representation algorithms for proper circular arc graphs and proper interval graphs, *SIAM J. Comput.*, 25 (1996), 390–403.

[301] J.S. DEOGUN, D. KRATSCH, Diametral path graphs, *Lecture Notes in Comput. Sci.*, 1017 (1995), 344–357.

[302] U. DERIGS, O. GOECKE, R. SCHRADER, Bisimplicial edges, Gaussian elimination and matchings in bipartite graphs, Intern. Workshop on Graph–Theoretic Concepts in Comp. Sci. Trauner Verlag, Berlin, W. Germany, (1984), 79–87.

[303] C. DE SIMONE, A. GALLUCCIO, New classes of Berge perfect graphs, *Discrete Math.*, 131 (1994), 67–79.

[304] C. DE SIMONE, J. KÖRNER, On the road to perfection: normal graphs, manuscript, (1996).

[305] C. DE SIMONE, C. MANNINO, Easy instances of the plant location problem, manuscript, (1996).

[306] C. DE SIMONE, A. SASSANO, Stability number of bull- and chair-free graphs, *Discrete Appl. Math.*, 41 (1993), 121–129.

[307] W. DEUBER, X. ZHU, Circular colorings of weighted graphs, *J. Graph Theory*, 23 (1996), 365–376.

[308] D. DE WERRA, On line perfect graphs, *Math. Programming*, 15 (1978), 236–238.

[309] D. DE WERRA, On some characterizations of totally unimodular matrices, *Math. Programming*, 20 (1981), 14–21.

[310] D. DE WERRA, A. HERTZ, On perfectness of sums of graphs, *Discrete Math.*, 195 (1999), 93–101.

[311] R. DIESTEL, Simplicial decomposition of graphs—some uniqueness results, *J. Combin. Theory B*, 42 (1987), 133–145.

[312] R. DIESTEL, *Graph Decompositions. A Study in Infinite Graph Theory*, Clarendon Press, Oxford (1990).

[313] P.F. DIETZ, *Intersection graph algorithms*, Ph.D. thesis, Comp. Sci. Dept., Cornell University Ithaka, NY (1984).

[314] R.P. DILWORTH, A decomposition theorem for partially ordered sets, *Ann. Math. Ser. 251* (1950), 161–166.

[315] G. DING, Recognizing the P_4-structure of a tree, *Graphs Combin.*, 10 (1994), 323–328.

[316] G. DIRAC, On rigid circuit graphs, *Abh. Math. Sem. Univ. Hamburg*, 25 (1961), 71–76.

[317] G. DIRAC, A property of 4-chromatic graphs and some remarks on critical graphs, *J. London Math. Soc.*, 27 (1952), 71–76.

[318] H.N. DJIDJEV, A separator theorem, *CR Acad. Bulgar. Sci.*, 34 (1981), 643–645.

[319] D. DOBKIN, H. EDELSBRUNNER, Searching for empty convex polygons, *Algorithmica*, 5 (1990), 561–571.

[320] J.-P. DOIGNON, On realizable biorders and the biorder dimension of a relation, *J. Math. Psychology*, 28 (1984), 73–109.

[321] F.F. DRAGAN, *Centers of graphs and the Helly property* (in Russian), Dissertation thesis, Moldova State University, Chişinău, Moldova (1989).

[322] F.F. DRAGAN, Conditions of coincidence of the local and global minima of the eccentricity function on graphs and the Helly property (in Russian), *Res. Appl. Math. Inform.* (1990), 49–56.

[323] F.F. DRAGAN, HT-graphs: centers, connected r-domination and Steiner trees, *Comput. Sci. J. Moldova*, 1 (1993), 64–83.

[324] F.F. DRAGAN, On greedy matching ordering and greedy matched graphs, *Lecture Notes in Comput. Sci.*, 1335 (1997), 184–198.

[325] F.F. DRAGAN, Strongly orderable graphs, manuscript, University of Rostock (1997). To appear in *Discrete Appl. Math.*

[326] F.F. DRAGAN, A. BRANDSTÄDT, r-Dominating cliques in graphs with hypertree structure, *Discrete Math.*, 162 (1996), 93–108.

[327] F.F. DRAGAN, F. NICOLAI, LexBFS-orderings of distance-hereditary graphs, Schriftenreihe des Fachbereichs Mathematik der Universität Duisburg, Duisburg, Germany, SM–DU–303 (1995).

[328] F.F. DRAGAN, F. NICOLAI, LexBFS-orderings and powers of HHD-free graphs, Schriftenreihe des Fachbereichs Mathematik der Universität Duisburg, Duisburg, Germany, SM–DU–322 (1996).

[329] F.F. DRAGAN, F. NICOLAI, A. BRANDSTÄDT, Convexity and HHD-free graphs, *SIAM J. Discrete Math.*, 12 (1999), 119–135.

[330] F.F. DRAGAN, F. NICOLAI, A. BRANDSTÄDT, Powers of HHD-free graphs, *Intern. J. Computer Math.*, 69 (1998), 217–242.

[331] F.F. DRAGAN, C.F. PRISACARU, V.D. CHEPOI, Location problems in graphs and the Helly property (in Russian) (1987) (appeared partially in *Diskretnaja Matematika*, 4 (1992), 67–73).

[332] F.F. DRAGAN, V.I. VOLOSHIN, Incidence graphs of biacyclic hypergraphs, *Discrete Appl. Math.*, 68 (1996), 259–266.

[333] P. DUCHET, Propriété de Helly et problèmes de représentation, Colloqu. Internat. CNRS 260, Problemes Combinatoires et Theorie du Graphs, Orsay, France (1976), 117–118.

[334] P. DUCHET, *Representations, noyaux en théorie des graphes et hypergraphes*, Doct. Diss. Université Paris VI, Paris, France (1979).

[335] P. DUCHET, Classical perfect graphs, *Ann. Discrete Math.*, 21 (1984), 67–96.

[336] P. DUCHET, Parity graphs are kernel-m-solvable, *J. Combin. Theory B*, 43 (1987), 121–126.

[337] P. DUCHET, Convex sets in graphs II: minimal path convexity, *J. Combin. Theory B*, 44 (1988), 307–316.

[338] P. DUCHET, Hypergraphs, in *Handbook of Combinatorics*, Vol. 1, R.L. Graham et al., eds., Elsevier, Amsterdam (1995), 381–432.

[339] P. DUCHET, H. MEYNIEL, Ensembles convèxes dans les graphes I, théorèmes de Helly et de Radon pour graphes et surfaces, *European J. Combin.*, 4 (1983), 127–132.

[340] P. DUCHET, S. OLARIU, Graphes parfaitement ordonnables généralises, *Discrete Math.*, 90 (1991), 99–101.

[341] R.J. DUFFIN, Topology of series-parallel networks, *J. Math. Analys. Appl.*, 10 (1965), 303–318.

[342] D. DUFFUS, M. GINN, V. RÖDL, On the computational complexity of ordered subgraph recognition, *Random Structures Algorithms*, 7 (1995), 223–268.

[343] D. DUFFUS, R.J. GOULD, M.S. JACOBSON, Forbidden subgraphs and the Hamiltonian theme, in *The Theory and Applications of Graphs*, Wiley, New York, (1981), 297–316.

[344] R. DUKE, Types of cycles in hypergraphs, *Ann. Discrete Math.*, 27 (1985), 399–418.

[345] B. DUSHNIK, E.W. MILLER, Partially ordered sets, *Amer. J. Math.*, 63 (1941), 600–610.

[346] C. EBENEGGER, P.L. HAMMER, D. DE WERRA, Pseudo-Boolean functions and stability of graphs, *Ann. Discrete Math.*, 19 (1984), 83–98.

[347] G. EHRLICH, S. EVEN, R.E. TARJAN, Intersection graphs of curves in the plane, *J. Combin. Theory B*, 21 (1976), 8–20.

[348] H. EL-GINDY, *Hierarchical decomposition of polygons with applications*, Ph.D. Thesis, McGill University, Montreal (1985).

[349] C.C. ELGOT, J.B. WRIGHT, Series-parallel graphs and lattices, *Duke Math. J.*, 26 (1959), 325.

[350] E.S. ELMALLAH, C.J. COLBOURN, Partial k-tree algorithms, *Congres. Numer.*, 64 (1988), 105–119.

[351] E.S. ELMALLAH, L.K. STEWART, Independence and domination in polygon graphs, *Discrete Math.*, 44 (1993), 65–77.

[352] E.S. ELMALLAH, L.K. STEWART, Polygon graph recognition, *J. Algorithms,* 26 (1998), 101–140.

[353] D. EPPSTEIN, Parallel recognition of series parallel graphs, *Inform. and Comput.*, 98 (1992), 41–55.

[354] P. ERDŐS, T. GALLAI, Graphen mit Punkten vorgeschriebenen Grades, *Math. Lapok,* 11 (1960), 264–272.

[355] P. ERDŐS, A. GOODMAN, L. PÓSA, The representation of a graph by set intersection, *Canad. J. Math.*, 18 (1966), 106–112.

[356] F. ESCALANTE, Über iterierte Clique-Graphen, *Abh. Math. Sem. Univ. Hamburg,* 39 (1973), 59–68.

[357] E.M. ESCHEN, R. HAYWARD, J.P. SPINRAD, R. SRITHARAN, Weakly triangulated comparability graphs, manuscript (1997).

[358] E.M. ESCHEN, J.P. SPINRAD, An $\mathcal{O}(n^2)$ algorithm for circular-arc graph recognition, in Proc. 4th ACM–SIAM Symposium on Discrete Algorithms (1993), 128–137.

[359] E.M. ESCHEN, R. SRITHARAN, A characterization of some graph classes with no long holes, *J. Combin. Theory B,* 65 (1995), 156–162.

[360] S. EVEN, *Graph algorithms*, Computer Science Press, Potomac, MD (1979).

[361] S. EVEN, A. ITAI, Queues, stacks and graphs, *Theory of Machines and Computations*, Z. Kohavi, A. Paz, eds., Academic Press, New York (1971), 71–86.

[362] S. EVEN, A. PNUELI, A. LEMPEL, Permutation graphs and transitive graphs, *J. Assoc. Comput. Mach.,* 19 (1972), 400–410.

[363] H. EVERETT, *Visibility graph recognition*, Ph.D. thesis, University of Toronto, 1990.

[364] H. EVERETT, D. CORNEIL, Recognizing visibility graphs of spiral polygons, *J. Algorithms,* 11 (1990), 1–26.

[365] H. EVERETT, D. CORNEIL, Negative results on characterizing visibility graphs, *Computational Geom.*, 5 (1995), 51–63.

[366] H. EVERETT, C.M.H. DE FIGUEIREDO, C. LINHARES-SALES, F. MAFFRAY, O. PORTO, B.A. REED, Path parity and perfection, *Discrete Math.,* 165/166 (1997), 233–252.

[367] H. EVERETT, S. KLEIN, B. REED, An algorithm for finding homogeneous pairs, *Discrete Appl. Math.,* 72 (1997), 209–218.

[368] H. EVERETT, A. LUBIW, J. O'ROURKE, Recovery of convex hulls from external visibility graphs, manuscript, Smith College (1991).

[369] R. FAGIN, Degrees of acyclicity for hypergraphs and relational database schemes, *J. Assoc. Comput. Mach.,* 30 (1983), 515–550.

[370] R. FAGIN, Acyclic database schemes of various degrees: a painless introduction, *Lecture Notes in Comput. Sci.,* 159 (1983), 65–89.

[371] M. FARBER, Applications of linear programming duality to problems involving independence and domination, Tech. Report 81-31, Dept. of Comp. Sci., Simon Fraser University, Canada (1982); Ph.D. thesis, Rutgers University.

[372] M. FARBER, Characterizations of strongly chordal graphs, *Discrete Math.*, 43 (1983), 173–189.

[373] M. FARBER, Domination, independent domination and duality in strongly chordal graphs, *Discrete Appl. Math.,* 7 (1984), 115–130.

[374] M. FARBER, Bridged graphs and geodesic convexity, *Discrete Math.,* 66 (1987), 249–257.

[375] M. FARBER, R.E. JAMISON, Convexity in graphs and hypergraphs, *SIAM J. Alg. Discrete Methods,* 7 (1986), 433–444.

[376] M. FARBER, R.E. JAMISON, On local convexity in graphs, *Discrete Math.,* 66 (1987), 231–247.

[377] R. FAUDREE, E. FLANDRIN, Z. RYJÁČEK, Claw-free graphs—a survey, *Discrete Math.,* 164 (1997), 87–147.

[378] T. FEDER, P. HELL, J. HUANG, List homomorphisms and circular arc graphs, manuscript (1996).

[379] J. FEIGENBAUM, Directed Cartesian-product graphs have unique factorizations that can be computed in polynomial time, *Discrete Appl. Math.*, 15 (1986), 105–110.

[380] J. FEIGENBAUM, J. HERSHBERGER, A.A. SCHÄFFER, A polynomial time algorithm for finding the prime factors of Cartesian product graphs, *Discrete Appl. Math.*, 12 (1985), 123–138.

[381] J. FEIGENBAUM, A.A. SCHÄFFER, Recognizing composite graphs is equivalent to testing graph isomorphism, *SIAM J. Comput.*, 15 (1986), 619–627.

[382] S. FELSNER, Tolerance graphs and orders, *J. Graph Theory,* 28 (1998), 129–140.

[383] S. FELSNER, P.C. FISHBURN, W.T. TROTTER, Finite three dimensional partial orders which are not sphere orders, manuscript (1997).

[384] S. FELSNER, V. RAGHAVAN, J. SPINRAD, Recognition algorithms for orders of small width and graphs of small Dilworth number, manuscript (1998).

[385] C.M. FIDUCCIA, E.R. SCHEINERMAN, A. TRENK, J.S. ZITO, Dot product representations of graphs, *Discrete Math.*, 181 (1998), 113–138.

[386] I.S. FILOTTI, G.I. MILLER, J. REIF, On determining the genus of a graph in $\mathcal{O}(|V|^{\mathcal{O}(g)})$ steps, in 11th Ann. ACM Sympos. on Theory of Comp. New York (1979), 27–37.

[387] P.C. FISHBURN, *Interval Orders and Interval Graphs*, John Wiley, New York (1985).

[388] P.C. FISHBURN, Interval orders and circle orders, *Order,* 5 (1989), 225–234.

[389] P.C. FISHBURN, W.T. TROTTER, Angle orders, *Order,* 1 (1985), 333–343.

[390] C. FLAMENT, Hypergraphes arborés, *Discrete Math.*, 21 (1978), 223–226.

[391] C. FLOTOW, *Potenzen von Graphen*, Dissertation thesis, Universität Hamburg (1995).

[392] C. FLOTOW, On powers of m-trapezoid graphs, *Discrete Appl. Math.*, 63 (1995), 187–192.

[393] C. FLOTOW, On powers of circular arc graphs and proper circular arc graphs, *Discrete Appl. Math.*, 69 (1996), 199–207.

[394] S. FÖLDES, P.L. HAMMER, Split graphs, *Congres. Numer.*, 19 (1977), 311–315.

[395] S. FÖLDES, P.L. HAMMER, Split graphs having Dilworth number two, *Canad. J. Math.*, 29 (1977), 666–672.

[396] S. FÖLDES, P.L. HAMMER, On a class of matroid producing graphs, in *Colloquia Math. Soc. János Bolyai,* A. Hajnal, V.T. Sós, eds., (1978), 331–352.

[397] S. FÖLDES, P.L. HAMMER, The Dilworth number of a graph, *Ann. Discrete Math.*, 2 (1978), 211–219.

[398] J. FONLUPT, J.P. UHRY, Transformations which preserve perfection and H-perfection of graphs, *Ann. Discrete Math.*, 16 (1982), 83–85.

[399] J. FONLUPT, A. ZEMIRLINE, A polynomial recognition algorithm of $(K_4 - e)$-free perfect graphs, *Rev. Maghrébine Math.*, 2 (1993), 1–26.

[400] L.R. FORD, D.R. FULKERSON, *Flows in Networks*, Princeton University Press, Princeton, NJ (1962).

[401] J.L. FOUQUET, V. GIAKOUMAKIS, On semi-P_4-sparse graphs, *Discrete Math.*, 165/166 (1997), 277–300.

[402] J.L. FOUQUET, V. GIAKOUMAKIS, F. MAIRE, H. THUILLIER, On graphs without P_5 and $\overline{P_5}$, *Discrete Math.*, 146 (1995), 33–44.

[403] J. L. FOUQUET, F. MAIRE, I. RUSU, H. THUILLIER, On transversals in minimal imperfect graphs, *Discrete Math.*, 165/166 (1997), 301–312.

[404] J.C. FOURNIER, Une caractérization des graphes des cordes, *C.R. Acad. Sci. Paris,* 286A (1978), 811–813.

[405] J.C. FOURNIER, M. LAS VERGNAS, Une classe d'hypergraphes bichromatiques, *Discrete Math.*, 2 (1972), 407–410.

[406] D.R. FULKERSON, Blocking and anti-blocking pairs of polyhedra, *Math. Programming,* 1 (1971), 168–194.

[407] D.R. FULKERSON, Anti-blocking polyhedra, *J. Combin. Theory B*, 12 (1972), 50–71.

[408] D.R. FULKERSON, On the perfect graph theorem, in *Mathematical Programming*, T.C. Hu, S.M. Robinson, eds., Academic Press, New York (1973), 69–76.

[409] D.R. FULKERSON, O.A. GROSS, Incidence matrices and interval graphs, *Pacific J. Math.*, 15 (1965), 835–855.

[410] D.R. FULKERSON, A.J. HOFFMAN, R. OPPENHEIM, On balanced matrices, *Math. Programming,* 1 (1974), 120–132.

[411] C.P. GABOR, W.L. HSU, K.J. SUPOWIT, Recognizing circle graphs in polynomial time, *J. Assoc. Comput. Mach.*, 36 (1989), 435–474.

[412] H. GALEANA-SÁNCHEZ, Normal fraternally orientable graphs satisfy the strong perfect graph conjecture, *Discrete Math.*, 122 (1993), 167–177.

[413] J. GALIL, R. KANNAN, E. SZEMERÉDI, On nontrivial separators for k-page graphs and simulations by nondeterministic one-tape Turing machines, *J. Comput. System Sci.*, 38 (1989), 134–149.

[414] P. GALINIER, M. HABIB, C. PAUL, Graphs and their clique graphs, in Proc. 21st Intern. Workshop on Graph–Theoretic Concepts in Comp. Sci. 1995, *Lecture Notes in Comput. Sci.,* 1017 (1995), 358–371.

[415] T. GALLAI, Graphen mit triangulierbaren ungeraden Vielecken, *Magyar Tud. Akad. Mat. Kutado Int. Közl.* 7 (1962), 3–36.

[416] T. GALLAI, Transitiv orientierbare Graphen, *Acta Math. Acad. Sci. Hung.*, 18 (1967), 25–66.

[417] R. GARBE, *Algorithmic Aspects of Interval Orders*, Ph.D. thesis, University of Twente, Enschede, the Netherlands (1994).

[418] M.R. GAREY, R.L. GRAHAM, D.S. JOHNSON, D.E. KNUTH, Complexity results for bandwidth minimization, *SIAM J. Appl. Math.*, 34 (1978), 477–495.

[419] M.R. GAREY, D.S. JOHNSON, *Computers and Intractability—A Guide to the Theory of* $\mathbb{NP}$*-completeness*, W. H. Freeman, San Francisco (1979).

[420] M.R. GAREY, D.S. JOHNSON, L. STOCKMEYER, Some simplified $\mathbb{NP}$-complete graph problems, *Theoret. Comput. Sci.*, 1 (1976), 237–267.

[421] A. GARG, R. TAMASSIA, Upward planarity testing, *Order,* 12 (1995), 109–133.

[422] G.S. GASPARIAN, Minimal imperfect graphs: a simple approach, *Combinatorica*, 16 (1996), 209–212.

[423] F. GAVRIL, Algorithms for a maximum clique and a maximum independent set of a circle graph, *Networks,* 3 (1973), 261–273.

[424] F. GAVRIL, The intersection graphs of subtrees in trees are exactly the chordal graphs, *J. Combin. Theory B*, 16 (1974), 47–56.

[425] F. GAVRIL, Algorithms on circular-arc graphs, *Networks,* 4 (1974), 357–369.

[426] F. GAVRIL, A recognition algorithm for the intersection graphs of directed paths in directed trees, *Discrete Math.,* 13 (1975), 237–249.

[427] F. GAVRIL, Algorithms on clique separable graphs, *Discrete Math.*, 19 (1977), 159–165.

[428] F. GAVRIL, A recognition algorithm for the intersection graphs of paths in trees, *Discrete Math.*, 23 (1978), 211–227.

[429] F. GAVRIL, Intersection graphs of proper subtrees of unicyclic graphs, *J. Graph Theory,* 18 (1994), 615–627.

[430] F. GAVRIL, Intersection graphs of Helly families of subtrees, *Discrete Appl. Math.*, 66 (1996), 45–56.

[431] F. GAVRIL, V. TOLEDANO LAREDO, D. DE WERRA, Chordless paths, odd holes and kernels in graphs without m-obstructions, *J. Algorithms,* 17 (1994), 207–221.

[432] F. GAVRIL, J. URRUTIA, Intersection graphs of concatenable subtrees of graphs, *Discrete Appl. Math.,* 52 (1994), 195–209.

[433] A.M.H. GERARDS, A short proof of Tutte's characterization of totally unimodular matrices, *Linear Algebra Appl.,* 114/115 (1989), 207–212.

[434] S. GHOSH, On recognizing and characterizing visibility graphs of simple polygons, *Lecture Notes in Comput. Sci.,* 318 (1988), 96–104 .

[435] A. GHOUILA-HOURI, Caractérization des graphes non orientés dont on peut orienter les arrêtes de manière à obtenir le graphe d'une relation d'ordre, *C.R. Acad. Sci. Paris*, 254 (1962), 1370–1371.

[436] V. GIAKOUMAKIS, On extended P_4-sparse graphs, in 4th Twente workshop on graphs and combinatorial optimization, Enschede, the Netherlands (1995).

[437] V. GIAKOUMAKIS, P_4-laden graphs: a new class of brittle graphs, *Inform. Process. Lett.,* 60 (1996), 29–36.

[438] V. GIAKOUMAKIS, Décomposition modulaire et problèmes d'optimisation, Document de syntèse d'activité scientifique présenté en vue d'obtenir une habilitation à diriger des recherches en Informatique, Université d'Amiens (1996).

[439] V. GIAKOUMAKIS, F. ROUSSEL, H. THUILLIER, On P_4-tidy graphs, *Discrete Math. Theoret. Comp. Sci.,* 1 (1997), 17–41.

[440] V. GIAKOUMAKIS, J.-M. VANHERPE, On extended P_4-reducible and extended P_4-sparse graphs, *Theoret. Comput. Sci.,* 180 (1997), 269–286.

[441] J.R. GILBERT, J.P. HUTCHINSON, R.E. TARJAN, A separator theorem for graphs of bounded genus, *J. Algorithms,* 5 (1984), 391–407.

[442] J.R. GILBERT, D.J. ROSE, A. EDENBRANDT, A separator theorem for chordal graphs, *SIAM J. Alg. Discrete Methods,* 5 (1984), 306–313.

[443] R. GILES, L.E. TROTTER JR., On stable set polyhedra for $K_{1,3}$-free graphs, *J. Combin. Theory B,* 31 (1981), 313–326.

[444] P.C. GILMORE, A.J. HOFFMAN, A characterization of comparability graphs and of interval graphs, *Canad. J. Math.,* 16 (1964), 539–548.

[445] J. GIMBEL, End vertices in interval graphs, *Discrete Appl. Math.,* 21 (1988), 257–259.

[446] M. GOEMANS, D. WILLIAMSON, Improved approximation algorithms for maximum cut and satisfiability problems using semidefinite programing, *J. Assoc. Comput. Mach.,* 42 (1995), 1115–1145.

[447] L. GOH, D. ROTEM, Recognition of perfect elimination bipartite graphs, *Inform. Process. Lett.,* 15 (1982), 179–182.

[448] P.W. GOLDBERG, M.C. GOLUMBIC, H. KAPLAN, R. SHAMIR, Four strikes against physical mapping of DNA, *J. Comput. Biol.,* 2 (1995), 139–152.

[449] A.J. GOLDSTEIN, An efficient and constructive algorithm for testing whether a graph can be embedded in the plane, Graphs and Combinatorics Conf., Office of Naval Research Logistics Proj., Dept. of Math., Princeton University, Princeton, NJ (1963).

[450] M.C. GOLUMBIC, The complexity of comparability graph recognition and coloring, *Computing,* 18 (1977), 199–208.

[451] M.C. GOLUMBIC, Trivially perfect graphs, *Discrete Math.,* 24 (1978), 105–107.

[452] M.C. GOLUMBIC, *Algorithmic Graph Theory and Perfect Graphs*, Academic Press, New York (1980).

[453] M.C. GOLUMBIC, Containment graphs and intersection graphs, Tech. Report IBM Israel, TR 88.135 (1984).

[454] M.C. GOLUMBIC, Interval graphs and related topics, *Discrete Math.,* 55 (1985), 113–243.

[455] M.C. GOLUMBIC, Algorithmic aspects of intersection graphs and representation hypergraphs, *Graphs Combin.*, 4 (1988), 307–321.

[456] M.C. GOLUMBIC, C.F. GOSS, Perfect elimination and chordal bipartite graphs, *J. Graph Theory*, 2 (1978), 155–163.

[457] M.C. GOLUMBIC, R.E. JAMISON, Edge and vertex intersections of paths in a tree, *Discrete Math.*, 55 (1985), 151–159.

[458] M.C. GOLUMBIC, R.E. JAMISON, The edge intersection graphs of paths in a tree, *J. Combin. Theory B*, 38 (1985), 8–22.

[459] M.C. GOLUMBIC, H. KAPLAN, R. SHAMIR, Graph sandwich problems, *J. Algorithms*, 19 (1995), 449–473.

[460] M.C. GOLUMBIC, C.L. MONMA, A generalization of interval graphs with tolerances, *Congres. Numer.*, 35 (1982), 321–331.

[461] M.C. GOLUMBIC , C.L. MONMA, W.T. TROTTER, Tolerance graphs, *Discrete Appl. Math.*, 9 (1984), 157–170.

[462] M.C. GOLUMBIC, D. ROTEM, J. URRUTIA, Comparability graphs and intersection graphs, *Discrete Math.*, 43 (1983), 37–46.

[463] M.C. GOLUMBIC, E.R. SCHEINERMAN, Containment graphs, posets and related classes of graphs, in Combinatorial Mathematics, G.S. Bloom et al., eds., *Ann. NY Acad. Sci.*, 555 (1985), 192–204.

[464] M.C. GOLUMBIC, R. SHAMIR, Complexity and algorithms for reasoning about time: A graph theoretic approach, *J. Assoc. Comput. Mach.*, 40 (1993), 1108–1133.

[465] J.E GOODMAN, R. POLLACK, B. STURMFELS, Coordinate representation of order types requires exponential storage, in 21st Ann. ACM Sympos. on Theory of Comp. (1989), 405–410.

[466] N. GOODMAN, O. SHMUELI, Syntactic characterization of tree database schemes, *J. Assoc. Comput. Mach.*, 30 (1983), 767–786.

[467] I.M. GORGOS, Characterization of quasitriangulated graphs, Tech. Report , University of Kishinev (1982).

[468] M.H. GRAHAM, On the universal relation, Tech. Report , University of Toronto (1979).

[469] S. GRAVIER, A sequential algorithm for coloring some perfect graphs, Preprint (1997).

[470] C. GREENE, Some partitions associated with a partially ordered set *J. Combin. Theory A*, 20 (1976), 69–79.

[471] C. GREENE, D.J. KLEITMAN, The structure of Sperner k-families, *J. Combin. Theory A*, 20 (1976), 41–68.

[472] J.R. GRIGGS, D.B. WEST, Extremal values of the interval number of a graph, *SIAM J. Alg. Discrete Methods*, 1 (1979), 1–7.

[473] P. GRILLET, Maximal chains and antichains, *Fund. Math.*, 65 (1969), 157–167.

[474] C.M. GRINSTEAD, *The strong perfect graph conjecture for a class of graphs*, Ph.D. thesis, UCLA, (1978).

[475] C.M. GRINSTEAD, The perfect graph conjecture for toroidal graphs, in Topics on Perfect Graphs, C. Berge, V. Chvátal, eds., *Ann. Discrete Math.*, 21 (1984), 97–102.

[476] M. GRÖTSCHEL, L. LOVÁSZ, A. SCHRIJVER, The ellipsoid method and its consequences in combinatorial optimization, *Combinatorica*, 1 (1981), 169–197; Corrigendum: *Combinatorica*, 4 (1984), 291–295.

[477] M. GRÖTSCHEL, L. LOVÁSZ, A. SCHRIJVER, *Geometric Algorithms and Combinatorial Optimization*, Springer, Berlin, (1988).

[478] Y. GUREVICH, L. STOCKMEYER, U. VISHKIN, Solving $\mathbb{NP}$-hard problems on graphs that are almost trees and an application to facility location problems, *J. Assoc. Comput. Mach.*, 31 (1984), 459–473.

[479] V.A. GURVICH, On repetition-free boolean functions (in Russian), *Uspekhi Mat. Nauk*, 32 (1977), 183–184.

[480] V.A. GURVICH, On the normal forms of positional games, *Soviet Math. Dokl.*, 25 (1982), 572–574.

[481] V.A. GURVICH, A generalized LOVÁSZ inequality for perfect graphs, *Russ. Math. Surv.* 48 (1993), 184–185.

[482] V.A. GURVICH, Biseparated graphs are perfect, *Russ. Acad. Sci. Dokl. Math.*, 48 (1994), 134–141.

[483] V.A. GURVICH, S. HOUGARDY, Partitionable graphs, manuscript (1996).

[484] J. GUSTEDT, On the pathwidth of chordal graphs, *Discrete Appl. Math.*, 45 (1993), 233–248.

[485] M. GUTIERREZ, L. OUBIÑA, Metric characterizations of proper interval graphs and tree-clique graphs, *J. Graph Theory,* 21 (1996), 199–205.

[486] W. GUTJAHR, E. WELZL, G. WOEGINGER, Polynomial graph colorings, *Discrete Appl. Math.*, 35 (1992), 29–45.

[487] A. GYÁRFÁS, Problems from the world surrounding perfect graphs, *Zastosow. Mat.*, 19 (1987), 413–441.

[488] A. GYÁRFÁS, D. KRATSCH, J. LEHEL, F. MAFFRAY, Minimal non-neighbourhood-perfect graphs, *J. Graph Theory,* 21 (1996), 55–66.

[489] A. GYÁRFÁS, J. LEHEL, A Helly type problem in trees, in *Combinatorial Theory and Applications*, P. Erdős et al., eds., *Colloquia Math. Soc. János Bolyai,* 4, North–Holland, Amsterdam (1970), 571–584.

[490] A. GYÁRFÁS, J. LEHEL, Effective on-line coloring of P_5-free graphs, *Combinatorica,* 11 (1991), 181–184.

[491] M. HABIB, *Substitution de structures combinatoires, théorie et algorithmes*, Ph.D. thesis, Université Pierre et Marie Curie, Paris VI (1981).

[492] M. HABIB, R. JEGOU, N-free posets as generalizations of series-parallel posets, *Discrete Appl. Math.*, 12 (1985), 279–291.

[493] M. HABIB, D. KELLY, R.H. MÖHRING, Interval dimension is a comparability invariant, *Discrete Math.*, 88 (1991), 221–229.

[494] M. HABIB, M.C. MAURER, On the X-join decomposition for undirected graphs, *Discrete Appl. Math.*, 3 (1979), 198–207.

[495] M. HABIB, R.H. MÖHRING, A fast algorithm for recognizing trapezoid graphs and partial orders of interval dimension two, preprint (1988).

[496] M. HABIB, R.H. MÖHRING, Recognition of partial orders with interval dimension two via transitive orientation with side constraints, Tech. Report No. 244/90, TU Berlin (1990).

[497] M. HABIB, R.H. MÖHRING, Treewidth of cocomparability graphs and a new order-theoretic parameter, *Order,* 11 (1994), 47–60.

[498] M. HABIB, M. MORVAN, J.-X. RAMPON, On the calculation of transitive reduction-closure of orders, *Discrete Math.*, 111 (1993), 289–303.

[499] M. HABIB, C. PAUL, L. VIENNOT, LexBFS, a partition refining technique. Application to transitive orientation, interval graph recognition and consecutive 1's testing, manuscript (1996).

[500] J. HAGAUER, W. IMRICH, S. KLAVŽAR, Recognizing median graphs in subquadratic time, *Theoret. Comput. Sci.*, 215 (1999), 123–136.

[501] A. HAJNAL, J. SURÁNYI, Über die Auflösung von Graphen in vollständige Teilgraphen, *Ann. Univ. Sci. Budapest, Eötvös Sect. Math.* 1 (1958), 113–121.

[502] G. HAJÓS, Über eine Art von Graphen, *Intern. Math. Nachr.* 11 (1957), Problem 65.

[503] R. HALIN, Simpliziale Zerfällungen beliebiger (endlicher oder unendlicher) Graphen, *Math. Annal.*, 156 (1964), 216–225.

[504] R. HALIN, Studies on minimally n-connected graphs, in *Combinatorial mathematics and its applications*, D.J.A. Welsh, ed., Academic Press, New York (1971), 129–136.

[505] R. HALIN, *Graphentheorie*, Akademie-Verlag, Berlin (1989).

[506] R. HALIN, Some remarks on interval graphs, *Combinatorica,* 2 (1982), 297–304.

[507] R. HALIN, Simplicial decompositions and triangulated graphs, in *Graph Theory and Combinatorics*, B. Bollobás, ed., Academic Press, London (1984).

[508] R.C. HAMELINK, A partial characterization of clique graphs, *J. Combin. Theory,* 5 (1968), 192–197.

[509] P.L. HAMMER, T. IBARAKI, B. SIMEONE, Degree sequences of threshold graphs, *Congres. Numer.*, 21 (1978), 329–355.

[510] P.L. HAMMER, A.K. KELMANS, On universal threshold graphs, *Combin. Probab. Comput.* 3 (1994), 327–344.

[511] P.L. HAMMER, F. MAFFRAY, Completely separable graphs, *Discrete Appl. Math.*, 27 (1990), 85–99.

[512] P.L. HAMMER, F. MAFFRAY, Preperfect graphs, *Combinatorica,* 13 (1993), 199–208.

[513] P.L. HAMMER, F. MAFFRAY, M. PREISSMANN, A characterization of chordal bipartite graphs, RUTCOR Research Report, Rutgers University, New Brunswick, NJ, RRR (1989), 16–89.

[514] P.L. HAMMER, N.V.R. MAHADEV, Bithreshold graphs, *SIAM J. Discrete Math.*, 6 (1985), 497–506.

[515] P.L. HAMMER, N.V.R. MAHADEV, D. DE WERRA, The struction of a graph: application to CN-free graphs, *Combinatorica,* 5 (1985), 141–147.

[516] P.L. HAMMER, N.V.R. MAHADEV, U.N. PELED, Some properties of 2-threshold graphs, *Networks,* 19 (1989), 17–23.

[517] P.L. HAMMER, N.V.R. MAHADEV, U.N. PELED, Bipartite bithreshold graphs, *Discrete Math.*, 119 (1993), 79–96.

[518] P.L. HAMMER, B. SIMEONE, The splittance of a graph, *Combinatorica,* 1 (1981), 275–284.

[519] F. HARARY, On the group of the composition of two graphs, *Duke Math. J.*, 26 (1959), 29–34.

[520] F. HARARY, *Graph Theory*, Addison–Wesley, Reading, MA (1969).

[521] F. HARARY, J.A. KABELL, F.R. MCMORRIS, Bipartite intersection graphs, *Comment. Math. Univ. Carolin.*, 23 (1982), 739–745.

[522] F. HARARY, T.A. MCKEE, The square of a chordal graph, *Discrete Math.*, 128 (1994), 165–172.

[523] I.B. HARTMAN, I. NEWMAN, R. ZIV, On grid intersection graphs, *Discrete Math.*, 87 (1991), 41–52.

[524] T.W. HAYNES, S.T. HEDETNIEMI, P.J. SLATER, eds., *Fundamentals of Domination in Graphs*, Marcel Dekker, New York, Basel (1998), Vol. 208.

[525] T.W. HAYNES, S.T. HEDETNIEMI, P.J. SLATER, eds., *Domination in Graphs: Advanced Topics*, Marcel Dekker, New York, Basel (1998), Vol. 209.

[526] R.B. HAYWARD, Weakly triangulated graphs, *J. Combin. Theory B,* 39 (1985), 200–208.

[527] R.B. HAYWARD, Murky graphs, *J. Combin. Theory B,* 49 (1990), 200–235.

[528] R.B. HAYWARD, Discs in unbreakable graphs, *Graphs Combin.*, 11 (1995), 249–254.

[529] R.B. HAYWARD, Meyniel weakly triangulated graphs—I: co-perfect orderability, *Discrete Appl. Math.*, 73 (1997), 199–210.

[530] R.B. HAYWARD, C. HOÀNG, F. MAFFRAY, Optimizing weakly triangulated graphs, *Graphs Combin.*, 5 (1989), 339–349; Erratum, *Graphs Combin.*, 6 (1990), 33–35.

[531] R.B. HAYWARD, S. HOUGARDY, B. REED, Recognizing P_4-structures, manuscript (1996).

[532] R.B. HAYWARD, W.J. LENHART, On the P_4-structure of perfect graphs IV. Partner graphs, *J. Combin. Theory B,* 48 (1990), 135–139.

[533] X. HE, Efficient parallel algorithms for series-parallel graphs, *J. Algorithms,* 12 (1991), 409–430.

[534] S.T. HEDETNIEMI, R. LASKAR, eds., *Topics on Domination*, *Ann. Discrete Math.*, 48, North–Holland, Amsterdam (1991).

[535] P. HELL, *Rétractions de graphes*, Ph.D. thesis, Université de Montréal (1972).

[536] P. HELL, J. BANG-JENSEN, J. HUANG, Local tournaments and proper circular arc graphs, *Lecture Notes in Comput. Sci.*, 450 (1990), 101–108.

[537] P. HELL, I. RIVAL, Absolute retracts and varieties of reflexive graphs, *Canad. J. Math.*, 39 (1987), 544–567.

[538] P. HELL, F.S. ROBERTS, Analogues of the Shannon capacity of a graph, *Ann. Discrete Math.*, 12 (1982), 155–168.

[539] R.L. HEMMINGER, L.W. BEINEKE, Line graphs and line digraphs, in *Selected Topics in Graph Theory* I, L.W. Beineke, R.T. Wilson, eds., Academic Press, London (1978), 271–305.

[540] P.B. HENDERSON, Y. ZALCSTEIN, A graph-theoretic characterization of the PV_{chunk} class of synchronizing primatives, *SIAM J. Comput.*, 6 (1977), 88–108.

[541] U. HENNIG, *Über Toleranzgraphen*, Diploma thesis, FB Mathematik, FU Berlin (1988).

[542] M.A. HENNING, Irredundance perfect graphs, *Discrete Math.*, 142 (1995), 107–120.

[543] A. HERTZ, Slim graphs, *Graphs Combin.*, 5 (1989), 149–157.

[544] A. HERTZ, Bipartable graphs, *J. Combin. Theory B*, 45 (1988), 1–12.

[545] A. HERTZ, Skeletal graphs—a new class of perfect graphs, *Discrete Math.*, 78 (1989), 291–296.

[546] A. HERTZ, Slender graphs, *J. Combin. Theory B*, 47 (1989), 231–236.

[547] A. HERTZ, A fast algorithm for colouring Meyniel graphs, *J. Combin. Theory B*, 50 (1990), 231–240.

[548] A. HERTZ, Bipolarizable graphs, *Discrete Math.*, 81 (1990), 25–32.

[549] A. HERTZ, Most unbreakable murky graphs are bull-free, *Graphs Combin.*, 9 (1993), 173–175.

[550] A. HERTZ, Polynomially solvable cases for the maximum stable set problem, *Discrete Appl. Math.*, 60 (1995), 195–210.

[551] A. HERTZ, D. DE WERRA, Perfectly orderable graphs are quasiparity: a short proof, *Discrete Math.*, 68 (1988), 111–113.

[552] T. HIRAGUCHI, On the dimension of partially ordered sets, *Sci. Rep. Kanazawa Univ.* 1 (1951), 77–94.

[553] C.T. HOÀNG, *Perfect graphs*, Ph.D. thesis, School of Computer Science, McGill University, Montreal (1985).

[554] C.T. HOÀNG, On the P_4-structure of perfect graphs II. Odd decompositions, *J. Combin. Theory B*, 39 (1985), 220–232.

[555] C.T. HOÀNG, On a conjecture of Meyniel, *J. Combin. Theory B*, 42 (1987), 302–312.

[556] C.T. HOÀNG, Alternating orientation and alternating colouration of perfect graphs, *J. Combin. Theory B*, 42 (1987), 264–273.

[557] C.T. HOÀNG, On the sibling structure of perfect graphs, *J. Combin. Theory B*, 49 (1990), 282–286.

[558] C.T. HOÀNG, Recognition and optimization algorithms for co-triangulated graphs, Tech. Report No. 90637-OR, Forschungsinstitut für Diskrete Mathematik, Bonn, 1990.

[559] C.T. HOÀNG, On the two-edge-colourings of perfect graphs, *J. Graph Theory*, 19 (1995), 271–279.

[560] C.T. HOÀNG, Some properties of minimal imperfect graphs, *Discrete Math.*, 160 (1996), 165–175.

[561] C.T. HOÀNG, On the complexity of recognizing a class of perfectly orderable graphs, *Discrete Appl. Math.*, 66 (1996), 219–226.

[562] C.T. HOÀNG, A note on perfectly orderable graphs, *Discrete Appl. Math.*, 65 (1996), 379–386.

[563] C.T. HOÀNG, On the disk-structure of perfect graphs I. The paw-structure, Tech. Report 97-01-04 Lakehead University, Thunder Bay, Canada (1997).

[564] C.T. HOÀNG, S. HOUGARDY, F. MAFFRAY, On the P_4-structure of perfect graphs V. Overlap graphs, *J. Combin. Theory B*, 67 (1996), 212–237.

[565] C.T. Hoàng, N. Khouzam, On brittle graphs, *J. Graph Theory,* 12 (1988), 391–404.

[566] C.T. Hoàng, V.B. Le, On P_4-transversals of perfect graphs, manuscript (1997).

[567] C.T. Hoàng, V.B. Le, Recognizing perfect 2-split graphs, to appear in *SIAM J. Discrete Math.*

[568] C.T. Hoàng, V.B. Le, P_4-colorings and P_4-bipartite graphs, manuscript (1998).

[569] C.T. Hoàng, F. Maffray, Opposition graphs are strict quasi-parity graphs, *Graphs Combin.*, 5 (1989), 83–85.

[570] C.T. Hoàng, N.V.R. Mahadev, A note on perfect orders, *Discrete Math.*, 74 (1989), 77–84.

[571] C.T. Hoàng, B.A. Reed, Some classes of perfectly orderable graphs, *J. Graph Theory,* 13 (1989), 445–463.

[572] C.T. Hoàng, B.A. Reed, P_4-comparability graphs, *Discrete Math.*, 74 (1989), 173–200.

[573] C.T. Hoàng, P. Sritharan, Finding houses and holes in graphs, manuscript (1999).

[574] W. Hochstättler, H. Schindler, Recognizing P_4-extendible graphs in linear time, Tech. Report 95.188, Universität Köln, (1995).

[575] A.J. Hoffman, A generalization of max flow-min cut, *Math. Programming,* 6 (1974), 352–359.

[576] A.J. Hoffman, On greedy algorithms that succeed, in *Surveys in Combinatorics, London Mathematical Society Lecture Notes Series* 103, Cambridge University Press (1985), 97–112.

[577] A.J. Hoffman, A.W.J. Kolen, M. Sakarovitch, Totally-balanced and greedy matrices, *SIAM J. Alg. Discrete Methods,* 6 (1985), 721–730.

[578] A.J. Hoffman, On eigenvalues and colorings of graphs, in *Graph Theory and Its Applications*, B. Harries, ed., Academic Press, New York (1970), 79–91.

[579] J.E. Hopcroft, R.E. Tarjan, Planarity testing in $\mathcal{O}(|V|\log|V|)$ steps, *Information Process.*, 72, Vol. 1, Foundations and Systems, North–Holland, Amsterdam (1972), 85–90.

[580] J.E. Hopcroft, R.E. Tarjan, Efficient planarity testing, *J. Assoc. Comput. Mach.*, 21 (1974), 549–568.

[581] S. Hougardy, Inclusion diagram of perfect graph classes, manuscript, HU Berlin (1997).

[582] S. Hougardy, Even pairs and the strong perfect graph conjecture, *Discrete Math.*, 154 (1996), 277–278.

[583] S. Hougardy, *On the P_4-structure of perfect graphs*, Dissertation thesis, Fachbereich Informatik, HU Berlin (1995).

[584] S. Hougardy, Perfect graphs with unique P_4-structure, *Discrete Math.*, 166 (1997), 411–420.

[585] S. Hougardy, V.B. Le, A. Wagler, Wing-triangulated graphs are perfect, *J. Graph Theory,* 24 (1997), 25–31.

[586] E. Howorka, A characterization of distance-hereditary graphs, *Quart. J. Math. Oxford,* Ser. 2, 28 (1977), 417–420.

[587] E. Howorka, On metric properties of certain clique graphs, *J. Combin. Theory B,* 27 (1979), 67–74.

[588] E. Howorka, A characterization of ptolemaic graphs, *J. Graph Theory,* 5 (1981), 323–331.

[589] E. Howorka, Betweenness in graphs, *Abstracts Amer. Math. Soc.* 2, 783-06-5 (1981).

[590] W.-L. Hsu, How to color claw-free perfect graphs, *Ann. Discrete Math.*, 11 (1981), 189–197.

[591] W.-L. Hsu, Decomposition of perfect graphs, *J. Combin. Theory B,* 43 (1987), 70–94.

[592] W.-L. Hsu, Recognizing planar perfect graphs, *J. Assoc. Comput. Mach.*, 34 (1987), 255–288.

[593] W.-L. Hsu, $\mathcal{O}(m \cdot n)$ algorithms for the recognition and isomorphism problems on circular-arc graphs, *SIAM J. Comput.*, 24 (1995), 411–439.

[594] W.-L. Hsu, A simple test for interval graphs, in Intern. Workshop on Graph–Theoretic Concepts in Comp. Sci. 1993, *Lecture Notes in Comput. Sci.*, 657, E.W. Mayr, ed. (1993), 11–16.

[595] W.-L. HSU, $\mathcal{O}(M \cdot N)$ algorithms for the recognition and isomorphism problems on circular-arc graphs, *SIAM J. Comput.*, 24 (1995), 411–439.

[596] W.-L. HSU, On-line recognition of interval graphs in $\mathcal{O}(m+n \log n)$ time, *Lecture Notes in Comput. Sci.*, 1120 (1995), 27–38.

[597] W.-L. HSU, T.H. MA, Substitution decomposition on chordal graphs and applications, in ISA'91 Algorithms, W.-L. Hsu, R.T.C. Lee, eds., *Lecture Notes in Comput. Sci.*, 557 (1991), 52–60.

[598] W.-L. HSU, T.H. MA, Fast and simple algorithms for recognizing chordal comparability graphs and interval graphs, *SIAM J. Comput.*, 28 (1999), 1004–1020.

[599] W.-L. HSU, G.L. NEMHAUSER, Algorithms for minimum covering by cliques and maximum clique in claw-free perfect graphs, *Discrete Math.*, 37 (1981), 181–191.

[600] M. HUJTER, Z. TUZA, Precoloring extensions III: Classes of perfect graphs, *Combin. Probab. Comput.*, 5 (1996), 35–56.

[601] C. HUNDACK, H. STAMM-WILBRANDT, Extended circle graphs I, Tech. Report IAI-TR-95-5, Institut für Informatik III, Rheinische Friedrich-Wilhelm-Universität Bonn (1995).

[602] T. IBARAKI, U.N. PELED, Sufficient conditions for graphs to have threshold number 2, *Ann. Discrete Math.*, 11 (1981), 241–268.

[603] K. IIJAMA, Y. SHIBATA, A bipartite representation of a triangulated graph and its chordality, lCS 79-1, Dept. of Comp. Sci., Genma University (1979).

[604] H. IMAI, Finding connected components of an intersection graph of squares in the euclidean plane, *Inform. Process. Lett.*, 15 (1982), 125–128.

[605] H. IMAI, T. ASANO, Finding the connected components and a maximum clique of an intersection graph of rectangles in the plane, *J. Algorithms*, 4 (1983), 310–323.

[606] Z. JACKOWSKI, A new characterization of proper interval graphs, *Discrete Math.*, 105 (1992), 103–109.

[607] M.S. JACOBSON, J. LEHEL, L.M. LESNIAK, ϕ-threshold and ϕ-tolerance chain graphs, *Discrete Appl. Math.*, 44 (1993), 191–203.

[608] M.S. JACOBSON, F.R. MCMORRIS, H.M. MULDER, Tolerance intersection graphs, manuscript (1989).

[609] M.S. JACOBSON, F.R. MCMORRIS, H.M. MULDER, An introduction to tolerance intersection graphs, in Y. Alavi, G. Chartrand, O.R. Oellerman, A.J. Schwenk, eds., *Graph Theory, Combinatorics and Applications*, Vol. 2, John Wiley, New York (1991), 705–723.

[610] M.S. JACOBSON, F.R. MCMORRIS, E. SCHEINERMAN, General results on tolerance intersection graphs, *J. Graph Theory*, 15 (1991), 573–577.

[611] R.E. JAMISON, A development of axiomatic convexity, Tech. Report 481, Clemson University, Clemson, SC (1970).

[612] R.E. JAMISON, *A general theory of convexity*, Ph.D. thesis, University of Washington, Seattle, (1974).

[613] R.E. JAMISON, Convexity and block graphs, *Congres. Numer.*, 33 (1981), 129–142.

[614] R.E. JAMISON, A perspective on abstract convexity: Classifying alignments by varieties, in *Convexity and Related Combinatorial Geometry*, D.C. Kay, M. Breen, eds., Marcel Dekker, New York, Basel (1982), 113–150.

[615] B. JAMISON, S. OLARIU, P_4-reducible graphs—a class of uniquely tree representable graphs, *Studies in Appl. Math.*, 81 (1989), 79–87.

[616] B. JAMISON, S. OLARIU, On the semi-perfect elimination, *Advances in Appl. Math.*, 9 (1988), 364–376.

[617] B. JAMISON, S. OLARIU, A linear-time recognition algorithm for P_4-reducible graphs, *Theoret. Comput. Sci.*, 145 (1995), 329–344.

[618] B. JAMISON, S. OLARIU, A new class of brittle graphs, *Studies in Appl. Math.*, 81 (1989), 89–92.

[619] B. JAMISON, S. OLARIU, On a unique tree representation for P_4-extendible graphs, *Discrete Appl. Math.*, 34 (1991), 151–164.

[620] B. JAMISON, S. OLARIU, A unique tree representation for P_4-sparse graphs, *Discrete Appl. Math.*, 35 (1992), 115–129.

[621] B. JAMISON, S. OLARIU, Recognizing P_4-sparse graphs in linear time, *SIAM J. Comput.*, 21 (1992), 381–406.

[622] B. JAMISON, S. OLARIU, *p*-components and the homogeneous decomposition of graphs, *SIAM J. Discrete Math.*, 8 (1995), 448–463.

[623] K. JANSEN, A characterization for parity graphs and a coloring problem with costs, DIMACS Technical Report 98–7, (1998).

[624] E. JARRETT, On iterated triangular line graphs, in Proc. 7th Conference on Graph Theory, Combin. Algor. and Appl., Kalamazoo 1992 (1995), 589–599.

[625] E.M. JAWHARI, D. MISANE, M. POUZET, Retracts: graphs and ordered sets from the metric point of view, *Combinatorics and ordered sets, Contemp. Math., A.M.S.* (1986), 175–226.

[626] T.R. JENSEN, B. TOFT, *Graph Coloring Problems*, John Wiley, New York (1995).

[627] D.S. JOHNSON, The NP-completeness column: an ongoing guide, *J. Algorithms*, 2 (1981), 393–405.

[628] D.S. JOHNSON, The NP-completeness column: an ongoing guide, *J. Algorithms*, 3 (1982), 88–99, 182–195, 288–300, 381–395.

[629] D.S. JOHNSON, The NP-completeness column: an ongoing guide, *J. Algorithms*, 4 (1983), 87–100, 189–203, 286–300, 397–411.

[630] D.S. JOHNSON, The NP-completeness column: an ongoing guide, *J. Algorithms*, 5 (1984), 147–160, 284–299, 433–447, 595–609.

[631] D.S. JOHNSON, The NP-completeness column: an ongoing guide, *J. Algorithms*, 6 (1985), 145–159, 291–305, 434–451.

[632] D.S. JOHNSON, The NP-completeness column: an ongoing guide, *J. Algorithms*, 7 (1986), 289–305, 584–601.

[633] D.S. JOHNSON, The NP-completeness column: an ongoing guide, *J. Algorithms*, 8 (1987), 285–303, 438–448.

[634] D.S. JOHNSON, The NP-completeness column: an ongoing guide, *J. Algorithms*, 9 (1988), 426–444.

[635] H.A. JUNG, On a class of posets and the corresponding comparability graphs, *J. Combin. Theory B*, 24 (1978), 125–133.

[636] N. KAHALE, Eigenvalues and expansion of regular graphs, *J. Assoc. Comput. Mach.*, 42 (1995), 1091–1106.

[637] J. KAHN, A family of perfect graphs associated with directed path graphs, *J. Combin. Theory B*, 37 (1984), 279–282.

[638] V.B. KALININ, Intersection graphs of curves and segments (in Russian), *Kibernetika*, 3 (1982), 122–123.

[639] V.B. KALININ, On intersection graphs (in Russian), in *Algorithmic Constructions and Their Efficiency, Yaroslav. Gos. Univ. Yaroslavl* 140 (1983), 72–76.

[640] V.B. KALININ, On a question of Berge (in Russian), *Mat. Sametki*, 34 (1983), 131–133.

[641] V.B. KALININ, A sufficient condition for the representability of a graph as the intersection graph of curves in the plane (in Russian), in *Modelling and optimization of computing systems and processes, Yaroslav. Gos. Univ. Yaroslavl* (1988), 37–40.

[642] S. KANNAN, M. NAOR, S. RUDICH, Implicit representations of graphs, *SIAM J. Discrete Math.*, 5 (1992), 596–603.

[643] H. KAPLAN, R. SHAMIR, Pathwidth, bandwidth and completion problems to proper interval graphs with small cliques, *SIAM J. Comput.*, 25 (1996), 540–561.

[644] I.A. KARAPETJAN, Coloring of arc graphs (in Russian), *Akad. Nauk Armjan. SSR Dokl.* 70 (1980), 306–311.

[645] D. KARGER, R. MOTWANI, M. SUDAN, Approximate graph coloring by semidefinite programming, *J. Assoc. Comput. Mach.*, 45 (1998), 246–265.

[646] R.M. KARP, Mapping the genome: some combinatorial problems arising in molecular biology, in 25th Ann. ACM Sympos. on Theory of Comp. (1993), 278–285 .

[647] D.C. KAY, G. CHARTRAND, A characterization of certain ptolemaic graphs, *Canad. J. Math.*, 17 (1965), 342–346.

[648] P.E. KEARNEY, D.G. CORNEIL, Tree powers, *J. Algorithms*, 29 (1998), 111–131.

[649] D. KELLY, On the dimension of partially ordered sets, *Discrete Math.*, 35 (1981), 135–156.

[650] D. KELLY, W.T. TROTTER, JR., Dimension theory for ordered sets, *Ordered sets*, I. Rival, ed., Reidel, Dordrecht (1982), 171–211.

[651] A. KELMANS, The number of trees in graphs I, *Automat. Remote Control,* 26 (1965), 2118–2129.

[652] A. KELMANS, The number of trees in graphs II, *Automat. Remote Control,* 27 (1966), 233–241.

[653] N. KHOUZAM, Masters thesis, McGill University, Montreal (1986).

[654] H.A. KIERSTEAD, J. QIN, W.T. TROTTER, The dimension of cycle-free orders, *Order,* 9 (1992), 103–110.

[655] H.A. KIERSTEAD, V. RÖDL, Applications of hypergraph coloring to coloring graphs not inducing certain trees, *Discrete Math.*, 150 (1996), 187–193.

[656] H.A. KIERSTEAD, W.T. TROTTER, Colorful induced subgraphs, *Discrete Math.*, 101 (1992), 165–169.

[657] T. KIKUNO, N. YOSHIDA, Y. KAKUDA, A linear time algorithm for the domination number of a series-parallel graph, *Discrete Appl. Math.*, 5 (1983), 299–311.

[658] L.M. KIROUSIS, C.H. PAPADIMITRIOU, Interval graphs and searching, *Discrete Math.*, 55 (1985), 181–184.

[659] L.M. KIROUSIS, C.H. PAPADIMITRIOU, Searching and pebbling, *Theoret. Comput. Sci.*, 47 (1986), 205–218.

[660] S. KLAVŽAR, Absolute retracts of split graphs, *Discrete Math.*, 134 (1994), 75–84.

[661] S. KLAVŽAR, H.M. MULDER, Median graphs: characterizations, location theory and related structures, Tech. Report 9641/A, Erasmus University of Rotterdam.

[662] M.M. KLAWE, D.G. CORNEIL, A. PROSKUROWSKI, Isomorphism testing in hookup classes, *SIAM J. Alg. Discrete Methods,* 3 (1982), 260–274.

[663] D.J. KLEITMAN, S.Y. LI, A note on unigraphic sequences, *Studies Appl. Math.* 54 (1975), 283–287.

[664] B. KLINZ, R. RUDOLF, G.J. WOEGINGER, Permutation matrices to avoid forbidden submatrices, *Discrete Appl. Math.*, 60 (1995), 223–248.

[665] B. KLINZ, R. RUDOLF, G.J. WOEGINGER, On the recognition of permuted bottleneck Monge matrices, *Discrete Appl. Math.*, 63 (1995), 43–74.

[666] T. KLOKS, Treewidth of circle graphs, *Lecture Notes in Comput. Sci.*, 762 (1993), 108–117.

[667] T. KLOKS, $K_{1,3}$-free and W_4-free graphs, *Inform. Process. Lett.*, 60 (1996), 221–223 .

[668] T. KLOKS, Treewidth—Computations and approximations, *Lecture Notes in Comput. Sci.*, 842 (1994).

[669] T. KLOKS, H. BODLAENDER, Testing superperfection of k-trees, *Lecture Notes in Comput. Sci.*, 621 (1992), 292–303.

[670] T. KLOKS, D. KRATSCH, Treewidth of chordal bipartite graphs, *J. Algorithms*, 19 (1995), 266–281.

[671] T. KLOKS, D. KRATSCH, Listing all minimal separators of a graph, *SIAM J. Comput.*, 27 (1998), 605–613.

[672] T. KLOKS, D. KRATSCH, Computing a perfect edge without vertex elimination ordering of a chordal bipartite graph, *Inform. Process. Lett.*, 55 (1995), 11–16.

[673] T. KLOKS, D. KRATSCH, H. MÜLLER, Dominoes, *Lecture Notes in Comput. Sci.*, 903 (1995), 106–120.

[674] T. KLOKS, D. KRATSCH, H. MÜLLER, Finding and counting small induced subgraphs efficiently, *Lecture Notes in Comput. Sci.*, 1017 (1995), 14–23.

[675] T. KLOKS, D. KRATSCH, J. SPINRAD, Treewidth and pathwidth of cocomparability graphs of bounded dimension, *Computing Science Notes* 93/46, Eindhoven University of Technology (1993).

[676] T. KLOKS, D. KRATSCH, J. SPINRAD, On treewidth and minimum fill-in of asteroidal triple-free graphs, *Theoret. Comput. Sci.*, 175 (1997), 309–335.

[677] D.E. KNUTH, The sandwich theorem, *Electronic J. Comb.*, 1 (1994). (http://www.combinatorics.org/)

[678] M. KOEBE, Spider graphs—a new class of intersection graphs, unpublished manuscript

[679] M. KOEBE, Colouring of spider graphs, R. Bodendiek, R. Henn, eds., *Topics in Combinatorics and Graph Theory*, Physica Verlag, Heidelberg (1990), 435–441.

[680] M. KOEBE, On a new class of intersection graphs, in *Fourth Czechoslovakian Symposium on Combinatorics, Graphs and Complexity*, J. Nešetřil, M. Fiedler, eds., Elsevier, Amsterdam (1992), 141–143.

[681] P. KOEBE, Kontaktprobleme der konformen Abbildung, *Berichte über die Verhandlungen der Sächsischen Akademie der Wissenschaften*, Leipzig, Math.–Physikal. Klasse, 88 (1936), 141–164.

[682] D. KÖNIG, Über Graphen und ihre Anwendungen auf Determinantentheorie und Mengenlehre, *Math. Annal.*, 77 (1916), 453–465.

[683] D. KÖNIG, Theorie der endlichen und unendlichen Graphen, *Akademische Verlagsgesellschaft Leipzig* (1936).

[684] J. KÖRNER, G. SIMONYI, Z. TUZA, Perfect couples of graphs, *Combinatorica*, 12 (1992), 179–192.

[685] J. KÖRNER, An extension of the class of perfect graphs, *Studia Math. Hungar.*, 8 (1973), 405–409.

[686] M. KOREN, Pairs of sequences with a unique realization by bipartite graphs, *J. Combin. Theory B*, 21 (1976), 224–234.

[687] M. KOREN, Sequences with a unique realization by simple graphs, *J. Combin. Theory B*, 21 (1976), 235–244.

[688] N. KORTE, R.H. MÖHRING, An incremental linear-time algorithm for recognizing interval graphs, *SIAM J. Comput.*, 18 (1989), 68–81.

[689] J. KRATOCHVÍL, String graphs I. The number of critical nonstring graphs is infinite, *J. Combin. Theory B*, 52 (1991), 53–66.

[690] J. KRATOCHVÍL, String graphs II. Recognizing string graphs is $\mathbb{NP}$-hard, *J. Combin. Theory B*, 52 (1991), 67–78.

[691] J. KRATOCHVÍL, A special planar satisfiability problem and a consequence of its $\mathbb{NP}$-completeness, *Discrete Appl. Math.*, 52 (1994), 233–252.

[692] J. KRATOCHVÍL, Intersection graphs of noncrossing arc-connected sets in the plane, *Lecture Notes in Comput. Sci.*, 1190 (1996), 257–270.

[693] J. KRATOCHVÍL, M. GOLJAN, P. KUČERA, *String Graphs*, Academia, Prague (1986).

[694] J. KRATOCHVÍL, J. MATOUŠEK, Intersection graphs of segments, *J. Combin. Theory B*, 62 (1994), 289–315.

[695] J. KRATOCHVÍL, T. PRZYTYCKA, Grid intersection and box intersection graphs on surfaces, *Lecture Notes in Comput. Sci.*, 1027 (1995), 365–372.

[696] J. KRATOCHVÍL, ZS. TUZA, Intersection dimensions of graph classes, *Graphs Combin.*, 10 (1994), 159–168.

[697] D. KRATSCH, *The structure of graphs and the design of efficient algorithms*, Habilitation thesis, Friedrich-Schiller-Universität, Jena (1995).

[698] D. KRATSCH, Dominating pair graphs—In the world surrounding asteroidal triple-free graphs, ODSA'97 workshop, Rostock, 1997.

[699] D. KRATSCH, *Domination in Graphs*, Volume 2: Advances, Chapter 8: Algorithms, T. Haynes, S. Hedetniemi, P. Slater, eds., Marcel Dekker, New York, Basel (1998).

[700] T. KRATZKE, B. REZNICK, D. WEST, Eigensharp graphs: decomposition into complete bipartite subgraphs, *Trans. Amer. Math. Soc.*, 308 (1988), 637–653.

[701] J. KRAUSZ, Démonstration nouvelle d'une théorème de Whitney sur les réseaux, *Math. Fiz. Lapok*, 50 (1943), 75–85.

[702] C. KURATOWSKI, Sur le probleme des corbes gauches en topologie, *Fund. Math.*, 15 (1930), 271–283.

[703] C.W.H. LAM, S. SWIERCZ, L. THIEL, E. REGENER, A computer search for (α, ω)-graphs, *Congres. Numer.*, 27 (1979), 285–289.

[704] R. LASKAR, D. SHIER, On powers and centers of chordal graphs, *Discrete Appl. Math.*, 6 (1983), 139–147.

[705] E.L. LAWLER, Graphical algorithms and their complexity, *Math. Centre Tracts*, 81 (1976), 3–32.

[706] E.L. LAWLER, J.K. LENSTRA, A.H.G. RINNOOY KAN, D.B. SHMOYS, *The Traveling Salesman Problem*, John Wiley, New York (1990).

[707] V.B. LE, *Transitive Kantenforcierung und perfekte Graphen*, Diploma Thesis, TU Berlin (1988).

[708] V.B. LE, Perfect k-line graphs and k-total graphs, *J. Graph Theory*, 17 (1993), 65–73.

[709] V.B. LE, *Gallai graphs and their iteration behavior*, Dissertation Thesis, TU Berlin (1994).

[710] V.B. LE, Cycle-perfect graphs are perfect, *J. Graph Theory*, 23 (1996), 351–353.

[711] V.B. LE, Gallai graphs and anti-Gallai graphs, *Discrete Math.*, 159 (1996), 179–189.

[712] V.B. LE, A good characterization of cograph contractions, to appear in *J. Graph Theory*.

[713] V. B. LE, Bipartite-perfect graphs, manuscript (1999).

[714] V.B. LE, E. PRISNER, Facet graphs of pure simplical complexes and related concepts, Preprint Heft 19, Math. Seminar, Univ. Hamburg (1992).

[715] V. B. LE, J. SPINRAD, Consequences of an algorithm for bridged graphs, manuscript (1999).

[716] B. LECLERC, B. MONJARDET, Orders C.A.C, *Fund. Math.*, (1973), 11–22.

[717] J. LEHEL, Helly hypergraphs and abstract interval structures, *Ars Combin.*, 16A (1983), 239–253.

[718] J. LEHEL, A characterization of totally balanced hypergraphs, *Discrete Math.*, 57 (1985), 59–65.

[719] J. LEHEL, Neighbourhood-perfect line graphs, *Graphs Combin.*, 10 (1994), 353–361.

[720] J. LEHEL, Z. TUZA, Neighbourhood perfect graphs, *Discrete Math.*, 61 (1986), 93–101.

[721] P.G.H. LEHOT, An optimal algorithm to detect a line graph and output its root graph, *J. Assoc. Comput. Mach.*, 21 (1974), 569–575.

[722] C.E. LEISERSON, Area-efficient graph layouts (for VLSI), Tech. Report CMU-CS-80-138, Carnegie-Mellon University, Pittsburgh, PA (1978).

[723] C. LEKKERKERKER, D. BOLAND, Representation of finite graphs by a set of intervals on the real line, *Fund. Math.*, 51 (1962), 45–64.

[724] A. LEMPEL, S. EVEN, I. CEDERBAUM, An algorithm for planarity testing of graphs, *Theory of Graphs: Int. Symp. Rome*, P. Rosenstiehl, ed., Gordon and Breach, New York (1966), 215–232.

[725] H. LERCHS, On cliques and kernels, Tech. Report Dept. of Comput. Sci., University of Toronto (1971).

[726] H. LERCHS, On the clique-kernel structure of graphs, Tech. Report Dept. of Comput. Sci., University of Toronto (1972).

[727] S.Y.R. LI, Graphic sequences with unique realization, *J. Combin. Theory B*, 19 (1975), 42–68.

[728] R. LIN, S. OLARIU, An NC recognition algorithm for cographs, *J. Parallel Dist. Comput.*, 13 (1991), 76–90.

[729] Y.-L. LIN, S.S. SKIENA, Algorithms for square roots of graphs, *SIAM J. Discrete Math.*, 8 (1995), 99–118.

[730] Y.-L. LIN, S.S. SKIENA, Complexity aspects of visibility graphs, *Internat. J. Comput. Geom.*, 5 (1995), 289–312.

[731] R.J. LIPTON, R.E. TARJAN, A separator theorem for planar graphs, *SIAM J. Appl. Math.*, 36 (1979), 177–189.

[732] J. LIU, H. ZHOU, Dominating subgraphs in graphs with some forbidden structures, *Discrete Math.*, 135 (1994), 163–168.

[733] M. LOEBL, S. POLJAK, A hierarchy of totally unimodular matrices, *Discrete Math.*, 76 (1989), 241–246.

[734] P.J. LOOGES, S. OLARIU, Optimal greedy algorithms for indifference graphs, *Comput. Math. Appl.*, 25 (1993), 15–25.

[735] L. LOVÁSZ, Normal hypergraphs and the perfect graph conjecture, *Discrete Math.*, 2 (1972), 253–267.

[736] L. LOVÁSZ, A characterization of perfect graphs, *J. Combin. Theory B*, 13 (1972), 95–98.

[737] L. LOVÁSZ, *Combinatorial Problems and Exercises*, North–Holland, Amsterdam (1979).

[738] L. LOVÁSZ, On the Shannon capacity of a graph, *IEEE Trans. Information Theory IT*, IT-25 (1979), 1–7.

[739] L. LOVÁSZ, Perfect graphs, in *Selected Topics in Graph Theory* 2, L.W. Beineke, R.J. Wilson, eds., Academic Press, New York (1983), 55–87.

[740] L. LOVÁSZ, An algorithmic theory of numbers, graphs, and convexity, *CMBS Regional Conf. Series in Appl. Math.* SIAM, Philadelphia (1986).

[741] L. LOVÁSZ, Stable sets and polynomials, *Discrete Math.*, 124 (1994), 137–153.

[742] L. LOVÁSZ, M.D. PLUMMER, *Matching Theory*, Ann. Discrete Math. 29, North–Holland, Amsterdam (1986).

[743] A. LUBIW, *Γ-free matrices*, Masters thesis, Dept. of Combin. and Optim., University of Waterloo, Canada (1982).

[744] A. LUBIW, Doubly lexical orderings of matrices, *SIAM J. Comput.*, 16 (1987), 854–879.

[745] A. LUBIW, Short-chorded and perfect graphs, *J. Combin. Theory B*, 51 (1991), 24–33.

[746] A. LUBOTZKY, Discrete groups, expanding graphs, and invariant measures, *Progress in Mathematics* 125, Birkhäuser Verlag, Basel (1995).

[747] A. LUBOTZKY, R. PHILLIPS, P. SARNAK, Ramanujan graphs, *Combinatorica*, 8 (1988), 261–277.

[748] R.D. LUCE, Semiorders and a theory of utility discrimination, *Economica*, 24 (1956), 178–191.

[749] T.H. MA, On the threshold dimension 2 graphs, Tech. Report 11529, Inst. of Information Science, Academia Sinica, Nankang, Taipei, Republic of China (1993).

[750] T.H. MA, J.P. SPINRAD, An $\mathcal{O}(n^2)$ algorithm for undirected split decomposition, *J. Algorithms*, 16 (1994), 145–160.

[751] T.H. MA, J.P. SPINRAD, Cycle-free partial orders and chordal comparability graphs, *Order*, 8 (1991), 175–183.

[752] T.H. MA, J.P. SPINRAD, On the 2-chain subgraph cover and related problems, *J. Algorithms*, 17 (1994), 251–268.

[753] P.A. MACMAHON, The combination of resistances, *The Electrician*, 28 (1892), 601–602.

[754] F. MAFFRAY, On kernels in i-triangulated graphs, *Discrete Math.*, 61 (1986), 247–251.

[755] F. MAFFRAY, Kernels in perfect line graphs, *J. Combin. Theory B,* 55 (1992), 1–8.

[756] F. MAFFRAY, Antitwins in partitionable graphs, *Discrete Math.*, 112 (1993), 275–278.

[757] F. MAFFRAY, O. PORTO, M. PREISSMANN, A generalization of simplicial elimination orderings, *J. Graph Theory,* 23 (1996), 203–208.

[758] F. MAFFRAY, M. PREISSMANN, Perfect graphs with no P_5 and no K_5, *Graphs Combin.*, 10 (1994), 179–184.

[759] F. MAFFRAY, M. PREISSMANN, Linear recognition of pseudo-split graphs, *Discrete Appl. Math.*, 52 (1994), 307–312.

[760] F. MAFFRAY, M. PREISSMANN, Split-neighbourhood graphs and the strong perfect graph conjecture, *J. Combin. Theory B,* 63 (1995), 294–309.

[761] N.V.R. MAHADEV, U.N. PELED, Strict 2-threshold graphs, *Discrete Appl. Math.,* 21 (1988), 113–131.

[762] N.V.R. MAHADEV, U.N. PELED, *Threshold Graphs and Related Topics*, *Ann. Discrete Math.,* 56, North–Holland, Amsterdam (1995).

[763] N.V.R. MAHADEV, U.N. PELED, F. SUN, Equistable graphs, *J. Graph Theory,* 18 (1994), 281–299.

[764] F. MAIRE, Slightly triangulated graphs are perfect, *Graphs Combin.*, 10 (1994), 263–268.

[765] F. MAIRE, Polyominoes and perfect graphs, *Inform. Process. Lett.*, 50 (1994), 57–61.

[766] M.V. MARATHE, H. BREU, H.B. HUNT III, S.S. RAVI, D.J. ROSENKRANTZ, Simple heuristics for unit disk graphs, *Networks,* 25 (1995), 59–68.

[767] P. MARCHIORO, A. MORGANA, Structure and recognition of domishold graphs, *Discrete Math.,* 50 (1984), 239–251.

[768] P. MARCHIORO, A. MORGANA, R. PETRESCHI, B. SIMEONE, Degree sequences of matrogenic graphs, *Discrete Math.,* 51 (1984), 46–61.

[769] E. MARCZEWSKI, Sur deux propriétés des classes d'ensembles, *Fund. Math.*, 33 (1945), 303–307.

[770] G.A. MARGULIS, Explicit group-theoretical constructions of combinatorial schemes and their applications to the design of expanders and concentrators, *Prob. Peredaci Inf.*, 24 (1988), 51–60.

[771] S.E. MARKOSJAN, G.S. GASPARIAN, A.S. MARKOSJAN, On a conjecture of Berge, *J. Combin. Theory B,* 56 (1992), 97–107.

[772] S. MARKOSJAN, G. GASPARIAN, I. KARPETJAN, A. MARKOSJAN, On essential components and critical sets of a graph, *Discrete Math.*, 178 (1998), 137–153.

[773] S.E. MARKOSJAN, G.S. GASPARIAN, B. REED, β-perfect graphs, *J. Combin. Theory B,* 67 (1996), 1–11.

[774] D.W. MATULA, k-components, clusters, and slicings in graphs, *SIAM J. Appl. Math.,* 22 (1972), 459–480.

[775] R.M. MCCONNELL, J. SPINRAD, Linear-time modular decomposition and efficient transitive orientation of comparability graphs, in 5th ACM–SIAM Symposium on Discrete Algorithms (1994), 536–545.

[776] R.M. MCCONNELL, J. SPINRAD, Linear-time transitive orientation, in 8th ACM–SIAM Symposium on Discrete Algorithms (1997), 19–25.

[777] T.A. MCKEE, How chordal graphs work, *Bulletin of the ICA*, 9 (1993), 27–39.

[778] T.A. MCKEE, F.R. MCMORRIS, *Topics in Intersection Graph Theory*, SIAM Monographs on Discrete Math. and Appl. 2, SIAM, Philadelphia, 1999.

[779] F.R. MCMORRIS, D.R. SHIER, Representing chordal graphs on $K_{1,n}$, *Comment. Math. Univ. Carolin.*, 24 (1983), 489–494.

[780] H. MEYNIEL, The graphs whose odd cycles have at least two chords, in Topics on Perfect Graphs, C. Berge, V. Chvátal, eds., *Ann. Discrete Math.,* 21 (1984), 115–119.

[781] H. MEYNIEL, A new property of critical imperfect graphs and some consequences, *European J. Combin.*, 8 (1987), 313–316.

[782] H. MEYNIEL, S. OLARIU, A new conjecture about minimal imperfect graphs, *J. Combin. Theory B*, 47 (1989), 244–247.

[783] M. MIDDENDORF, F. PFEIFFER, On the complexity of recognizing perfectly orderable graphs, *Discrete Math.*, 80 (1990), 327–333.

[784] G.J. MINTY, On maximal independent sets of vertices in claw-free graphs, *J. Combin. Theory B*, 28 (1980), 284–304.

[785] S.L. MITCHELL, Linear algorithms to recognize outerplanar and maximal outerplanar graphs, *Inform. Process. Lett.*, 9 (1979), 229–232.

[786] R.H. MÖHRING, Algorithmic aspects of comparability graphs and interval graphs, in *Graphs and Order*, I. Rival, ed., Reidel, Dordrecht (1985), 41–101.

[787] R.H. MÖHRING, Algorithmic aspects of the substitution decomposition in optimization over relations, set systems and boolean functions, *Ann. Oper. Res.*, 14 (1985), 195–225.

[788] R.H. MÖHRING, Computationally tractable classes of ordered sets, in *Algorithms and Order*, I. Rival, ed., Kluwer Acaddemic, Dordrecht (1989), 105–193.

[789] R.H. MÖHRING, Graph problems related to gate matrix layout and PLA folding, in *Computational Graph Theory, Computing*, G. Tinhofer, E. Mayr, H. Noltemeier, M. Syslo, eds., Suppl. 7 (1990), 17–51.

[790] R.H. MÖHRING, Triangulating graphs without asteroidal triples, *Discrete Appl. Math.*, 64 (1996), 281–287.

[791] R.H. MÖHRING, F.J. RADERMACHER, Substitution decomposition for discrete structures and connections with combinatorial optimization, *Ann. Discrete Math.*, 19 (1984), 257–356.

[792] B. MOHAR, S. POLJAK, Eigenvalues in combinatorial optimization, in *IMA Volumes in Mathematics and its Applications*, Vol. 50, R. Brualdi, S. Friedland, V. Klee, eds., Springer, Berlin (1993), 108–149.

[793] B. MONIEN, The bandwidth minimization problem for caterpillars with hair length 3 is $\mathbb{NP}$-complete, *SIAM J. Alg. Discrete Methods,* 7 (1986), 505–512.

[794] C.L. MONMA, B. REED, W.T. TROTTER, JR., A generalization of threshold graphs with tolerances, *Congres. Numer.*, 55 (1986), 187–197.

[795] C.L. MONMA, B. REED, W.T. TROTTER, JR., Threshold tolerance graphs, *J. Graph Theory,* 12 (1988), 343–362.

[796] C.L. MONMA, V.K. WEI, Intersection graphs of paths in a tree, *J. Combin. Theory B*, 41 (1986), 141–181.

[797] J.W. MOON, The number of labelled k-trees, *J. Combin. Theory,* 6 (1969), 196–199.

[798] M. MOSCARINI, Alpha-graphs, Steiner trees and connected domination, manuscript (1987).

[799] M. MOSCARINI, Doubly chordal graphs, Steiner trees and connected domination, *Networks,* 23 (1993), 59–69.

[800] R. MOTWANI, M. SUDAN, Computing roots of graphs is hard, *Discrete Appl. Math.*, 54 (1994), 81–88.

[801] H. MÜLLER, Recognizing interval digraphs and interval bigraphs in polynomial time, *Discrete Appl. Math.*, 78 (1997), 189–205.

[802] H. MÜLLER, On edge perfectness and classes of bipartite graphs, *Discrete Math.*, 149 (1996), 159–187.

[803] A. MUKHOPADHYAY, The square root of a graph, *J. Combin. Theory,* 2 (1967), 290–295.

[804] H.M. MULDER, The interval function of a graph, *Math. Centre Tracts* 132 (Math. Centrum, Amsterdam) (1980).

[805] H.M. MULDER, Interval-regular graphs, *Discrete Math.*, 41 (1982), 253–269.

[806] J.H. MULLER, *Local structure in graph classes*, Ph.D. thesis, Georgia Institute of Technology, Atlanta, GA (1988).

[807] J.H. MULLER, J. SPINRAD, Incremental modular decomposition, *J. Assoc. Comput. Mach.*, 36 (1989), 1–19.

[808] W. NAJI, Reconnaissance des graphes de cordes, *Discrete Math.*, 54 (1985), 329–337.

[809] P.S. NEERALAGI, E. SAMPATHKUMAR, The neighbourhood number of a graph, *Indian J. Pure Appl. Math.*, 16 (1985), 126–132.

[810] F. NICOLAI, *Strukturelle und algorithmische Aspekte distanz-erblicher Graphen und verwandter Klassen*, Dissertation thesis, Gerhard-Mercator-Universität, Duisburg (1994).

[811] F. NICOLAI, A hypertree characterization of distance-hereditary graphs, manuscript, Gerhard-Mercator-Universität, Duisburg (1996).

[812] M.V. NIRKHE, *Efficient algorithms for circular-arc containment graphs*, Masters thesis, Tech. Report SRC 87–211, University of Maryland, Systems Research Center, College Park, MD (1987).

[813] T. NISHIZEKI, *Topological study on network interconnections*, Ph.D. thesis, (1974).

[814] T. NISHIZEKI, N. CHIBA, *Planar Graphs: Theory and Algorithms*, *Ann. Discrete Math.* 32, North–Holland, Amsterdam (1988).

[815] T. NISHIZEKI, N. SAITO, Necessary and sufficient condition for a graph to be three-terminal series-parallel, *IEEE Trans. Circ. Syst.*, CAS–22–8 (1975), 648–653.

[816] T. NISHIZEKI, N. SAITO, Necessary and sufficient condition for a graph to be three-terminal series-parallel-cascade, *J. Combin. Theory B*, 24 (1978), 344–361.

[817] R. NOWAKOWSKI, I. RIVAL, The smallest graph variety containing all paths, *Discrete Math.*, 43 (1983), 223–234.

[818] R. NOWAKOWSKI, P. WINKLER, Vertex-to-vertex pursuit in a graph, *Discrete Math.*, 43 (1983), 235–239.

[819] O. OELLERMANN, J.P. SPINRAD, A polynomial algorithm for testing whether a graph is 3-Steiner distance hereditary, *Inform. Process. Lett.*, 55 (1995), 149–154.

[820] S. OLARIU, *Results on perfect graphs*, Ph.D. thesis, School of Computer Science, McGill University, Montreal (1986).

[821] S. OLARIU, No antitwins in minimal imperfect graphs, *J. Combin. Theory B*, 45 (1988), 255–257.

[822] S. OLARIU, All variations on perfectly orderable graphs, *J. Combin. Theory B*, 45 (1988), 150–159.

[823] S. OLARIU, Paw-free graphs, *Inform. Process. Lett.*, 28 (1988), 53–54.

[824] S. OLARIU, On the strong perfect graph conjecture, *J. Graph Theory*, 12 (1988), 169–176.

[825] S. OLARIU, Coercion classes in unbreakable graphs, *Ars Combin.*, 25 B (1988), 153–180.

[826] S. OLARIU, Weak bipolarizable graphs, *Discrete Math.*, 74 (1989), 159–171.

[827] S. OLARIU, Wings and perfect graphs, *Discrete Math.*, 80 (1990), 281–296.

[828] S. OLARIU, A decomposition for strongly perfect graphs, *J. Graph Theory*, 13 (1989), 301–311.

[829] S. OLARIU, A generalization of Chvátal's star-cutset lemma, *Inform. Process. Lett.*, 33 (1989/90), 301–303.

[830] S. OLARIU, On the structure of unbreakable graphs, *J. Graph Theory*, 15 (1991), 349–373.

[831] S. OLARIU, On the homogeneous representation of interval graphs, *J. Graph Theory*, 15 (1991), 65–80.

[832] S. OLARIU, An optimal greedy heuristic to color interval graphs, *Inform. Process. Lett.*, 37 (1991), 65–80.

[833] S. OLARIU, On sources in comparability graphs, with applications, *Discrete Math.*, 110 (1992), 289–292.

[834] S. OLARIU, Quasi-brittle graphs, a new class of perfectly orderable graphs, *Discrete Math.*, 113 (1992), 143–153.

[835] S. OLARIU, J. RANDALL, Welsh–Powell opposition graphs, *Inform. Process. Lett.*, 31 (1989), 43–46.

[836] E. OLARU, Über die Überdeckung von Graphen mit Cliquen, *Wiss. Zeitschr. Techn. Hochschule Ilmenau*, 15 (1969), 115–121.

[837] E. OLARU, Zur Charakterisierung perfekter Graphen, *Elektron. Inf. verarb. u. Kybern.*, 9 (1973), 543–548.

[838] E. OLARU, On strongly perfect graphs and the structure of critically-imperfect graphs, *Anal. St. Univ. Iasi T. II Informatica* (1993), 45–59.

[839] E. OLARU, The structure of imperfect critically strongly-imperfect graphs, *Discrete Math.*, 156 (1996), 299–302.

[840] E. OLARU, H. SACHS, Contributions to a characterization of the structure of perfect graphs, in *Topics on Perfect Graphs*, C. Berge, V. Chvátal, eds., *Ann. Discrete Math.*, 21 (1984), 121–144.

[841] O. ORE, Theory of Graphs, *Amer. Math. Soc. Colloqu. Publ.*, 38, Providence, RI (1962).

[842] J. O'ROURKE, *Art Gallery Theorems and Algorithms*, Oxford University Press, New York (1987).

[843] J. O'ROURKE, Recovery of convexity from visibility graphs, Tech. Report 90.4.6, Smith College, Northampton, MA (1990).

[844] J. O'ROURKE, Computational geometry column 18, *Internat. J. Comput. Geom. Appl.*, 3 (1993), 107-113; also *SIGACT News*, 24:1 (1993), 20-25.

[845] J. O'ROURKE, Visibility, in *Handbook of Discrete and Computational Geometry*, J.E. Goodman, J. O'Rourke, eds., CRC Press LLC, Boca Raton (1997) 467-480.

[846] M.W. PADBERG, Perfect zero-one matrices, *Math. Programming*, 6 (1974), 180–196.

[847] M.W. PADBERG, Almost integral polyhedra related to certain combinatorial optimization problems, *Linear Algebra Appl.*, 15 (1976), 69–88.

[848] M.W. PADBERG, A characterization of perfect matrices, *Ann. Discrete Math.*, 21 (1984), 169–178.

[849] M.W. PADBERG, Total unimodularity and the Euler subgraph problem, *Oper. Res. Letters*, 7 (1988), 173–179.

[850] R. PAIGE, R.E. TARJAN, Three partition refinement algorithms, *SIAM J. Comput.*, 16 (1987), 973–989.

[851] B.S. PANDA, New linear time algorithms for generating perfect elimination orderings of chordal graphs, *Inform. Process. Lett.*, 58 (1996), 111–115.

[852] B.S. PANDA, S.P. MOHANTY, Recognition algorithm for intersection graphs of edge disjoint paths in a tree, *Inform. Process. Lett.*, 49 (1994), 139–143.

[853] C.H. PAPADIMITRIOU, M. YANNAKAKIS, Scheduling interval ordered tasks, *SIAM J. Comput.*, 8 (1979), 405–409.

[854] A. PARRA ASENSIO, Eine Klasse von Graphen, in der jeder Toleranzgraph ein beschränkter Toleranzgraph ist, *Abh. Math. Sem. Univ. Hamburg*, 64 (1994), 125–129.

[855] A. PARRA ASENSIO, Triangulating multitolerance graphs, *Discrete Appl. Math.*, 84 (1998), 183–197.

[856] A. PARRA ASENSIO, *Structural and algorithmic aspects of chordal graph embeddings*, Dissertation thesis, TU Berlin, FB Mathematik (1996).

[857] A. PARRA ASENSIO, P. SCHEFFLER, How to use the minimal separators of a graph for its chordal triangulation, *Lecture Notes in Comput. Sci.*, 944 (1995), 123–134.

[858] A. PARRA ASENSIO, P. SCHEFFLER, Treewidth equals bandwidth for AT-free claw-free graphs, Tech. Report 436/1995, TU Berlin, FB Mathematik (1995).

[859] A. PARRA, P. SCHEFFLER, Characterizations and algorithmic applications of chordal graph embeddings, *Discrete Appl. Math.*, 79 (1997), 171–188.

[860] K.R. PARTHASARATHY, G. RAVINDRA, The strong perfect graph conjecture is true for $K_{1,3}$-free graphs, *J. Combin. Theory B,* 21 (1976), 212–223.

[861] K.R. PARTHASARATHY, G. RAVINDRA, The validity of the strong perfect graph conjecture for $(K_4 - e)$-free graphs, *J. Combin. Theory B,* 26 (1979), 98–100.

[862] C. PAYAN, A class of threshold and domishold graphs: equistable and equidominating graphs, *Discrete Math.,* 29 (1980), 47–52.

[863] C. PAYAN, Perfectness and Dilworth number, *Discrete Math.,* 44 (1983), 229–230.

[864] I. PE'ER, R. SHAMIR, Interval graphs with side (and size) constraints, in European Sympos. on Algorithms P. Spirakis, ed., *Lecture Notes in Comput. Sci.,* 979 (1995), 142–154.

[865] I. PE'ER, R. SHAMIR, Realizing interval graphs with side and distance constraints, *SIAM J. Discrete Math.,* 10 (1997), 662–687.

[866] U.N. PELED, Matroidal graphs, *Discrete Math.,* 20 (1977), 263–286.

[867] U.N. PELED, B. SIMEONE, Box-threshold graphs, *J. Graph Theory,* 8 (1984), 331–345.

[868] S. PERZ, S. POLEWICZ, Norms and perfect graphs, *Methods Mod. Operat. Research,* 34 (1990), 13–27.

[869] E. PESCH, *Retracts of Graphs,* Athenaeum Verlag, Frankfurt (1988).

[870] D. PETERSON, Gridline graphs and higher-dimensional extension, RUTCOR Research Report, Rutgers University, New Brunswick, NJ, RRR 3–95 (1995).

[871] R. PETRESCHI, A. STERBINI, Recognizing strict 2-threshold graphs in $\mathcal{O}(m)$ time, *Inform. Process. Lett.,* 54 (1995), 193–198.

[872] R.E. PIPPERT, L.W. BEINEKE, Characterizations of 2-dimensional trees, in *The Many Facets of Graph Theory,* G. Chartrand, F. Kapoor, eds., Springer, Berlin (1969), 250–257.

[873] A. PNUELI, A. LEMPEL, S. EVEN, Transitive orientation of graphs and identification of permutation graphs, *Canad. J. Math.,* 23 (1971), 160–175.

[874] T. POSTON, *Fuzzy geometry,* Ph.D. thesis, University of Warwick (1971).

[875] M. PREISSMANN, A class of strongly perfect graphs, *Discrete Math.,* 54 (1985), 117–120.

[876] M. PREISSMANN, Locally perfect graphs, *J. Combin. Theory B,* 50 (1990), 22–40.

[877] M. PREISSMANN, D. DE WERRA, A note on strongly perfectness of graphs, *Math. Programming,* 31 (1985), 321–326.

[878] M. PREISSMANN, D. DE WERRA, N.V.R. MAHADEV, A note on superbrittle graphs, *Discrete Math.,* 61 (1986), 259–267.

[879] E. PRISNER, Tree representation of chordal graphs and the weighted clique graph, manuscript (1986).

[880] E. PRISNER, Convergence of iterated clique graphs, *Discrete Math.,* 103 (1992), 199–207.

[881] E. PRISNER, A common generalization of line graphs and clique graphs, *J. Graph Theory,* 18 (1994), 301–313.

[882] E. PRISNER, Clique covering and clique partition in generalizations of line graphs, *Discrete Appl. Math.,* 56 (1995), 93–98.

[883] E. PRISNER, Line graphs and generalizations — A survey, *Congres. Numer.,* 116 (1996), 193–229.

[884] H.J. PRÖMEL, A. STEGER, Almost all Berge graphs are perfect, *Combin. Probab. Comput.,* 1 (1992), 53–79.

[885] M. QUEST, G. WEGNER, Characterizations of the graphs with boxicity ≤ 2, *Discrete Math.,* 81 (1990), 187–192.

[886] A. QUILLIOT, *Homomorphismes, points fixes, rétractions et jeux de poursuite dans les graphes, les ensembles ordonnés et les espaces métriques,* Ph.D. thesis, Université de Paris VI (1983).

[887] A. QUILLIOT, Circular representation problem on hypergraphs, *Discrete Math.,* 51 (1984), 251–264.

[888] A. QUILLIOT, On the problem of how to represent a graph taking into account an additional structure, *J. Combin. Theory B*, 44 (1988), 1–21.

[889] I. RABINOVITCH, The dimension of semiorders, *J. Combin. Theory A*, 25 (1978), 50–61.

[890] I. RABINOVITCH, An upper bound on the "dimension of interval orders," *J. Combin. Theory A*, 25 (1978), 68–71.

[891] A. RAJARAM, H. BALAKRISHNAN, C. PANDU RANGAN, Modular decomposition techniques for distance-hereditary graphs, manuscript (1994).

[892] S.B. RAO, G. RAVINDRA, A characterization of perfect total graphs, *J. Math. Phys. Sci.*, 11 (1977), 25–26.

[893] T. RASCHLE, K. SIMON, Recognition of graphs with threshold dimension two, Proc. Ann. ACM Sympos. on Theory of Comp. (1995), 68–71.

[894] T. RASCHLE, K. SIMON, On the P_4-components of graphs, Tech. Report ETH, Zürich (1997).

[895] G. RAVINDRA, Strongly perfect line graphs and total graphs. Finite and infinite sets, *Colloquia Math. Soc. János Bolyai*, 37 (1981), 621–633.

[896] G. RAVINDRA, Meyniel graphs are strongly perfect, *J. Combin. Theory B*, 33 (1982), 187–190.

[897] K.T. RAWLINSON, R.C. ENTRINGER, Class of graphs with restricted neighbourhoods, *J. Graph Theory*, 3 (1979), 257–262.

[898] A. RAYCHAUDHURI, On powers of interval and unit interval graphs, *Congres. Numer.*, 59 (1987), 235–242.

[899] A. RAYCHAUDHURI, On powers of strongly chordal and circular arc graphs, *Ars Combin.*, 34 (1992), 147–160.

[900] B. REED, *A semi-strong perfect graph theorem*, Ph.D. thesis, McGill University, Montréal, (1986).

[901] B. REED, A semi-strong perfect graph theorem, *J. Combin. Theory B*, 43 (1987), 223–240.

[902] B. REED, N. SBIHI, Recognizing bull-free perfect graphs, *Graphs Combin.*, 11 (1995), 171–178.

[903] J. REITERMAN, V. RÖDL, E. ŠINAJOVÁ, Geometrical embeddings of graphs, *Discrete Math.*, 74 (1989), 291–319.

[904] P.L. RENZ, Intersection representation of graphs by arcs, *Pacific J. Math.*, 34 (1970), 501–510.

[905] J. RIORDAN, C.E. SHANNON, The number of two-terminal series-parallel networks, *J. Math. and Physics*, 21 (1942), 83.

[906] F.S. ROBERTS, Indifference graphs, in *Proof Techniques in Graph Theory*, F. Harary, ed., Academic Press, New York (1969), 139–146.

[907] F.S. ROBERTS, On the boxicity and cubicity of a graph, in *Recent Progress in Combinatorics*, W.T. Tutte, ed., Academic Press, New York (1969), 301–310.

[908] F.S. ROBERTS, J.H. SPENCER, A characterization of clique graphs, *J. Combin. Theory B*, 10 (1971), 102–108.

[909] N. ROBERTSON, P.D. SEYMOUR, Graph minors. I. Excluding a forest, *J. Combin. Theory B*, 35 (1983), 39–61.

[910] N. ROBERTSON, P.D. SEYMOUR, Graph minors. III. Planar tree-width, *J. Combin. Theory B*, 36 (1984), 49–64.

[911] N. ROBERTSON, P.D. SEYMOUR, Graph width and well–quasi ordering: a survey, in *Progress in Graph Theory*, J. Bondy, U. Murty, eds., Academic Press, New York (1984), 399–406.

[912] N. ROBERTSON, P.D. SEYMOUR, Graph minors — a survey, in *Surveys in Combinatorics*, I. Anderson, ed., Cambridge University Press, Cambridge (1985), 153–171.

[913] N. ROBERTSON, P.D. SEYMOUR, Graph minors. II. Algorithmic aspects of tree width, *J. Algorithms*, 7 (1986), 309–322.

[914] N. ROBERTSON, P.D. SEYMOUR, Graph minors. V. Excluding a planar graph, *J. Combin. Theory B*, 41 (1986), 92–114.

[915] N. ROBERTSON, P.D. SEYMOUR, Graph minors. VI. Disjoint paths across a disc, *J. Combin. Theory B*, 41 (1986), 115–138.

[916] N. ROBERTSON, P.D. SEYMOUR, Graph minors. VII. Disjoint paths on a surface, *J. Combin. Theory B*, 45 (1988), 212–254.

[917] N. ROBERTSON, P.D. SEYMOUR, Graph minors. IV. Tree–width and well–quasi–ordering, *J. Combin. Theory B*, 48 (1990), 227–254.

[918] N. ROBERTSON, P.D. SEYMOUR, Graph minors. IX. Disjoint crossed paths, *J. Combin. Theory B*, 49 (1990), 40–77.

[919] N. ROBERTSON, P.D. SEYMOUR, Graph minors. VIII. A Kuratowski theorem for general surfaces, *J. Combin. Theory B*, 48 (1990), 255–288.

[920] N. ROBERTSON, P.D. SEYMOUR, Graph minors. X. Obstructions to tree-decomposition, *J. Combin. Theory B*, 52 (1991), 153–190.

[921] N. ROBERTSON, P.D. SEYMOUR, Graph minors. XI. Circuits on a surface. *J. Combin. Theory B*, 60 (1994), 72–106.

[922] N. ROBERTSON, P.D. SEYMOUR, Graph minors. XII. Distance on a surface, *J. Combin. Theory B*, 64 (1995), 240–272.

[923] N. ROBERTSON, P.D. SEYMOUR, Graph minors XIII. The disjoint paths problem, *J. Combin. Theory B*, 63 (1995), 65–110.

[924] N. ROBERTSON, P.D. SEYMOUR, Graph minors. XIV. Extending an embedding, *J. Combin. Theory B*, 65 (1995), 23–50.

[925] N. ROBERTSON, P.D. SEYMOUR, Graph minors. XV. Giant steps, *J. Combin. Theory B*, 68 (1996), 112–148.

[926] D.J. ROSE, Triangulated graphs and the elimination process, *J. Math. Analys. Appl.*, 32 (1970), 597–609.

[927] D.J. ROSE, On simple characterizations of k-trees, *Discrete Math.*, 7 (1974), 317–322.

[928] D.J. ROSE, R.E. TARJAN, G.S. LUEKER, Algorithmic aspects of vertex elimination on graph, *SIAM J. Comput.*, 5 (1976), 266–283.

[929] A. ROSENBERG, Interval hypergraphs, *Contemporary Math.*, 89 (1989), 27–44.

[930] I.C. ROSS, F. HARARY, The square of a tree, *Bell System Tech. J.*, 39 (1960), 641–647.

[931] D. ROTEM, J. URRUTIA, Circular permutation graphs, *Networks*, 12 (1982), 429–437.

[932] F. ROUSSEL, I. RUSU, An $\mathcal{O}(m^2+mn)$ algorithm to recognize Meyniel graphs, ODSA'97 workshop, Rostock (1997).

[933] F. ROUSSEL, I. RUSU, Recognizing i-triangulated graphs in $\mathcal{O}(mn)$, Research Report RR 98-10, LIFO, Université d'Orleans, France (1998).

[934] N.D. ROUSSOPOULOS, A $\mathcal{O}\max\{m,n\}$ algorithm for determining the graph H from its line graph G, *Inform. Process. Lett.*, 2 (1973), 108–112.

[935] I. RUSU, A new class of perfect Hoàng graphs, *Discrete Math.*, 145 (1995), 279–285.

[936] I. RUSU, Quasi–parity and perfect graphs, *Inform. Process. Lett.*, 54 (1995), 35–39.

[937] I. RUSU, Building counterexamples, *Discrete Math.*, 171 (1997), 213–227.

[938] H.J. RYSER, Combinatorial configurations, *SIAM J. Appl. Math.*, 17 (1969), 593–602.

[939] G. SABIDUSSI, The composition of graphs, *Duke Math. J.*, 26 (1959), 693–696.

[940] G. SABIDUSSI, Graph multiplication, *Math. Zeitschr.*, 72 (1960), 446–457.

[941] G. SABIDUSSI, Graph derivatives, *Math. Zeitschr.*, 76 (1961), 385–401.

[942] H. SACHS, On the Berge conjecture concerning perfect graphs, in *Combinatorial Structures and their Applications*, Gordon and Breach, New York (1970), 377–384.

[943] H. SACHS, Coin graphs, polyhedra and conformal mapping, *Discrete Math.*, 134 (1994), 133–138.

[944] A. SASSANO, Chair-free Berge graphs are perfect, *Graphs Combin.*, 13 (1997), 369–395.

[945] N. SBIHI, Algorithme de recherche d'un stable de cardinalité maximum dans un graphe sans étoile, *Discrete Math.*, 29 (1980), 53–76.

[946] A.A. SCHÄFFER, Recognizing brittle graphs: remarks on a paper of Hoàng and Khouzam, *Discrete Appl. Math.*, 31 (1991), 29–35.

[947] A.A. SCHÄFFER, A faster algorithm to recognize undirected path graphs, *Discrete Appl. Math.*, 43 (1993), 261–295.

[948] P. SCHEFFLER, The graphs of tree-width k are exactly the partial k–trees, manuscript (1986).

[949] P. SCHEFFLER, Linear-time algorithms for NP-complete problems restricted to partial k-trees, Tech. Report R-MATH-03/87, IMATH, Berlin (1987).

[950] P. SCHEFFLER, *Die Baumweite von Graphen als ein Maß für die Kompliziertheit algorithmischer Probleme*, Dissertation thesis, Akad. d. Wiss. Berlin, Report R-MATH-04/89 (1989).

[951] P. SCHEFFLER, D. SEESE, Graphs of bounded tree–width and linear–time algorithms, manuscript (1986).

[952] E.R. SCHEINERMAN, *Intersection classes and multiple intersection parameters of a graph*, Ph.D. thesis, Princeton University (1984).

[953] E.R. SCHEINERMAN, Characterizing intersection classes of graphs, *Discrete Math.*, 55 (1985), 185–193.

[954] E.R. SCHEINERMAN, On the structure of hereditary classes of graphs, *J. Graph Theory*, 10 (1986), 545–551.

[955] E.R. SCHEINERMAN, A note on planar graphs and circle orders, *SIAM J. Discrete Math.*, 4 (1991), 447–450.

[956] E.R. SCHEINERMAN, A note on graphs and sphere orders, *J. Graph Theory*, 17 (1993) 283–289.

[957] E.R. SCHEINERMAN, A. TRENK, On generalized perfect graphs: bounded degree and bounded edge perfection, *Discrete Appl. Math.*, 44 (1993), 233–245.

[958] E.R. SCHEINERMAN, J.C. WEIRMAN, On circle containment orders, *Order*, 4 (1988), 315–318.

[959] E.R. SCHEINERMAN, D.B. WEST, The interval number of a planar graph: Three intervals suffice, *J. Combin. Theory B*, 35 (1983), 224–239.

[960] W. SCHNYDER, Planar graphs and poset dimension, *Order*, 5 (1989), 323–343.

[961] A. SEBŐ, Forcing colorations, intervals and the perfect graph conjecture, in *Integer Programming and Combinatorial Optimization* II, R. Kannan, E. Balas, G. Cornuéjols, eds., Carnegie Mellon University Press, Pittsburgh (1992).

[962] A. SEBŐ, The connectivity of minimal imperfect graphs, *J. Graph Theory*, 23 (1996), 77–85.

[963] A. SEBŐ, Critical edges in minimal imperfect graphs, *J. Combin. Theory B*, 67 (1996), 62–85.

[964] D. SEESE, Tree–partite graphs and the complexity of algorithms, *Lecture Notes in Comput. Sci.*, 199 (1985), 412–421.

[965] D. SEESE, Tree-partite graphs and the complexity of algorithms, Tech. Report Akad. d. Wiss. R-MATH-8/86, Berlin (1986).

[966] D. SEINSCHE, On a property of the class of n-colorable graphs, *J. Combin. Theory B*, 16 (1974), 191–193.

[967] P.D. SEYMOUR, Decomposition of regular matroids, *J. Combin. Theory B*, 28 (1980), 305–359.

[968] F. B. SHEPHERD, Hamiltonicity in claw-free graphs, *J. Combin. Theory B*, 53 (1991), 173–194.

[969] F.B. SHEPHERD, Near-perfect matrices, *Math. Programming*, 64 (1994), 295–323.

[970] T. SHERMER, Recent results in art galleries, *Proc IEEE*, 80 (1992).

[971] L.N. SHEVRIN, N.D. FILIPPOV, Partially ordered sets and their comparability graphs, *Siber. Math. J.*, 11 (1970), 497–509.

[972] Y. SHIBATA, On the tree representation of chordal graphs, *J. Graph Theory,* 12 (1988), 421–428.

[973] Y. SHIBATA, A. ISHIJIMA, On the minimum tree representation of chordal graphs, *Transact. IEICE,* Vol. E 71, No. 3 (1988), 203–204.

[974] D.R. SHIER, Some aspects of perfect elimination orderings in chordal graphs, *Discrete Appl. Math.,* 7 (1984), 325–331.

[975] S. SHINODA, Y. KAJITANI, K. ONAGA, W. MAYEDA, Various characterizations of series-parallel graphs, in Proc. 1979 ISCHS (1979), 100–103.

[976] R.W. SHIREY, *Implementation and analysis of efficient graph planarity testing algorithms,* Ph.D. thesis, University of Wisconsin (1969).

[977] G. SIERKSMA, H. HOOGEVEEN, Seven criteria for integer sequences being graphic, *J. Graph Theory,* 15 (1991), 223–231.

[978] F.W. SINDEN, Topology of thin film RC-circuits, *Bell System Tech. J.,* (1966), 1639–1662.

[979] D.J. SKRIEN, A relationship between triangulated graphs, comparability graphs, proper interval graphs, proper circular-arc graphs and nested interval graphs, *J. Graph Theory,* 6 (1982), 309–316.

[980] D.J. SKRIEN, J. GIMBEL, Homogeneously representable interval graphs, *Discrete Math.,* 55 (1985), 213–216.

[981] P.J. SLATER, A characterization of SOFT hypergraphs, *Canad. Math. Bull.,* 21 (1978), 335–337.

[982] V.P. SOLTAN, *d*-convexity in graphs, *Soviet Math. Dokl.,* 28 (1983), 419–421.

[983] V.P. SOLTAN, *Introduction to the Axiomatic Theory of Convexity* (in Russian), Stiinţa, Chişinău (1984).

[984] V.P. SOLTAN, V.D. CHEPOI, Conditions for invariance of set diameters under *d*–convexification in a graph, *Cybernetics* (the English translation *of Kibernetika*), 19 (1983), 750–756.

[985] V.P. SOLTAN, V.D. CHEPOI, *d*-convex sets in chordal graphs, *Math. Research* (Chişinău), 78 (1984), 105–124.

[986] L. ŠOLTÉS, Forbidden induced subgraphs for line graphs, *Discrete Math.,* 132 (1994), 391–394.

[987] S. SORG, *Die P_4-Struktur von Kantengraphen bipartiter Graphen,* Diploma thesis, Mathematisches Institut der Universität zu Köln (1997).

[988] J.P. SPINRAD, *Two-dimensional partial orders,* Ph.D. thesis, Dept. of EECS, Princeton University, NJ (1982).

[989] J.P. SPINRAD, On comparability and permutation graphs, *SIAM J. Comput.,* 14 (1985), 658–670.

[990] J.P. SPINRAD, Recognition of circle graphs, *J. Algorithms,* 16 (1994), 264–282.

[991] J.P. SPINRAD, Doubly lexical ordering of dense 0-1 matrices, *Inform. Process. Lett.,* 45 (1993), 229–235.

[992] J.P. SPINRAD, Circular-arc graphs with clique cover number two, *J. Combin. Theory B,* 44 (1988), 300–306.

[993] J.P. SPINRAD, Finding large holes, *Inform. Process. Lett.,* 39 (1991), 227–229.

[994] J.P. SPINRAD, Representations of graphs, book manuscript (1997).

[995] J.P. SPINRAD, A. BRANDSTÄDT, L.K. STEWART, Bipartite permutation graphs, *Discrete Appl. Math.,* 18 (1987), 279–292.

[996] J.P. SPINRAD, J.L. JOHNSON, Brittle, bipolarizable, and P_4-simplicial graph recognition, manuscript (1999).

[997] J.P. SPINRAD, R. SRITHARAN, Algorithms for weakly triangulated graphs, *Discrete Appl. Math.,* 59 (1995), 181–191.

[998] N. SRINIVASAN, J. OPATRNY, V.S. ALAGAR, Bigeodetic graphs, *Graphs Combin.,* 4 (1988), 379–392.

[999] R. SRITHARAN, A linear time algorithm to recognize circular permutation graphs, *Networks,* 27 (1996), 171–174.

[1000] G. STEINER, On the complexity of dynamic programming for sequencing problems with precedence constraints, *Ann. Oper. Res.*, 26 (1990), 103–123.

[1001] A. STERBINI, T. RASCHLE, An $\mathcal{O}(n^3)$ algorithm for recognizing threshold dimension 2 graphs, *Inform. Proc. Lett.*, 67 (1998), 255–259.

[1002] L.K. STEWART, *Permutation graph structure and algorithms*, Ph.D. thesis, TR 185/85, Dept. of Comput. Sci., Univ. of Toronto (1985).

[1003] D.P. SUMNER, *Indecomposable graphs*, Ph.D. thesis, University of Massachusetts, Amherst (1971).

[1004] D.P. SUMNER, Graphs indecomposable with respect to the X-join, *Discrete Math.*, 6 (1973), 281–298.

[1005] D.P. SUMNER, Dacey graphs, *J. Austral. Math. Soc.*, 18 (1974), 492–502.

[1006] D.P. SUMNER, J.I. MOORE, Domination perfect graphs, *Notices Amer. Math. Soc.*, 26 (1979), A-569.

[1007] L. SUN, Two classes of perfect graphs, *J. Combin. Theory B*, 53 (1991), 273–291.

[1008] R. SUNDARAM, K.S. SINGH, C. PANDU RANGAN, Treewidth of circular-arc graphs, *SIAM J. Discrete Math.*, 7 (1994), 647–655.

[1009] L. SURANYI, The covering of graphs by cliques, *Stud. Sci. Math. Hungar. 3* (1968), 345–349.

[1010] M.M. SYSŁO, On characterizations of cycle graphs, Colloq. CNRS, Orsay 1976, *Problèmes Combinatoires et Théorie des Graphes,* (1978), 395–398.

[1011] M.M. SYSŁO, On characterizations of cycle graphs and on other families of intersection graphs, Tech. Report IN-40, Inst. of Comput. Sci., University of Wroclaw, Poland (1978).

[1012] M.M. SYSŁO, NP-complete problems on some tree structured graphs: a review, in Proc. WG'83 Intern. Workshop on Graph–Theoretic Concepts in Comp. Sci. , M. Nagl, I. Perl, eds., Univ.-Verlag Rudolf Trauner Linz (1984), 342–353.

[1013] M.M. SYSŁO, Triangulated edge intersection graphs of paths in a tree, *Discrete Math.*, 55 (1985), 217–220.

[1014] M.M. SYSŁO, A graph-theoretic approach to the jump number problem, in *Graphs and Order*, I. Rival, ed., Reidel, Dordrecht (1985), 185–215.

[1015] G. SZEKERES, H. S. WILF, An inequality for the chromatic number of a graph, *J. Combin. Theory* 4 (1968), 1–3.

[1016] J.L. SZWARCFITER, Recognizing clique-Helly graphs, *Ars Combin.*, 45 (1997), 29–32.

[1017] J.L. SZWARCFITER, C.F. BORNSTEIN, Clique graphs of chordal and path graphs, *SIAM J. Discrete Math.*, 7 (1994), 331–336.

[1018] K. TAKAMIZAWA, T. NISHIZEKI, N. SAITO, Linear-time computability of combinatorial problems on series-parallel graphs, *J. Assoc. Comput. Mach.*, 29 (1982), 623–641.

[1019] R. TAMASSIA, I.G. TOLLIS, A unified approach to visibility representations of planar graphs, *Discrete Comput. Geom.*, 1 (1986), 321–341.

[1020] R.E. TARJAN, Decomposition by clique separators, *Discrete Math.*, 55 (1985), 221–232.

[1021] R.E. TARJAN, M. YANNAKAKIS, Simple linear-time algorithms to test chordality of graphs, test acyclicity of hypergraphs, and selectively reduce acyclic hypergraphs, *SIAM J. Comput.*, 13 (1984), 566–579.

[1022] C. THOMASSEN, Planarity and duality of finite and infinite graphs, *J. Combin. Theory B,* 29 (1980), 244–271.

[1023] C. THOMASSEN, Infinite graphs, in *Selected Topics in Graph Theory* 2, L.W. Beineke, R.J. Wilson, eds.,, Academic Press, New York (1983), 129–160.

[1024] C. THOMASSEN, Interval representations of planar graphs, *J. Combin. Theory B,* 40 (1986), 9–20.

[1025] C. THOMASSEN, The graph genus problem is NP–complete, *J. Algorithms,* 10 (1989) 568–576.

[1026] G. TINHOFER, Strong tree-cographs are Birkhoff graphs, *Discrete Appl. Math.*, 22 (1988/89), 275–288.

[1027] B. TOFT, Coloring, stable sets and perfect graphs, in *Handbook of Combinatorics*, Vol. I, R. Graham, M. Grötschel, L. Lovász, eds., North–Holland, Amsterdam (1995), 233–288.

[1028] A.N. TRENK, On generalized perfect graphs: characterizations and inversion, *Discrete Appl. Math.*, 60 (1995), 359–387.

[1029] A.N. TRENK, k-weak orders, *Discrete Math.*, 181 (1998), 223–237.

[1030] L.E. TROTTER, Line perfect graphs, *Math. Programming*, 12 (1977), 255–259.

[1031] W.T. TROTTER, JR., *Combinatorics and Partially Ordered Sets — Dimension Theory*, Johns Hopkins University Press, Baltimore, London (1992).

[1032] W.T. TROTTER, JR., F. HARARY, On double and multiple interval graphs, *J. Graph Theory*, 3 (1979), 205–211.

[1033] W.T. TROTTER, JR., J. MOORE, Characterization problems for graphs, partially ordered sets, lattices and families of sets, *Discrete Math.*, 16 (1976), 361–381.

[1034] W.T. TROTTER, JR., J. MOORE, The dimension of planar posets, *J. Combin. Theory B*, 22 (1977), 54–67.

[1035] K. TRUEMPER, Testing of matrix total unimodularity, *J. Combin. Theory B*, 49 (1990), 241–281.

[1036] S. TSUKIYAMA, I. SHIRAKAWA, H. OZAKI, H. ARIYOSHI, An algorithm to enumerate all cutsets of a graph in linear time per cutset, *J. Assoc. Comput. Mach.*, 27 (1980), 619–632.

[1037] A.C. TUCKER, Characterizing circular-arc graphs, *Bull. Amer. Math. Soc.*, 76 (1970), 1257–1260.

[1038] A.C. TUCKER, Matrix characterizations of circular-arc graphs, *Pacific J. Math.*, 39 (1971), 535–545.

[1039] A.C. TUCKER, A structure theorem for the consecutive 1's property, *J. Combin. Theory*, 12 (1972), 153–162.

[1040] A.C. TUCKER, The strong perfect graph conjecture for planar graphs, *Canad. J. Math.*, 25 (1973), 103–114.

[1041] A.C. TUCKER, Structure theorems for some circular arc graphs, *Discrete Math.*, 7 (1974), 167–195.

[1042] A.C. TUCKER, Coloring a family of circular arcs, *SIAM J. Appl. Math.*, 29 (1975), 493–502.

[1043] A.C. TUCKER, Critical perfect graphs and perfect 3-chromatic graphs, *J. Combin. Theory B*, 23 (1977), 143–149.

[1044] A.C. TUCKER, An efficient test for circular-arc graphs, *SIAM J. Comput.*, 9 (1980), 1–24.

[1045] A.C. TUCKER, Coloring graphs with stable cutsets, *J. Combin. Theory B*, 34 (1983), 258–267.

[1046] A.C. TUCKER, The validity of the strong perfect graph conjecture for K_4-free graphs, in *Topics on Perfect Graphs*, C. Berge, V. Chvátal, eds., *Ann. Discrete Math.*, 21 (1984), 149–158.

[1047] A.C. TUCKER, Coloring perfect $(K_4 - e)$-free graphs, *J. Combin. Theory B*, 42 (1987), 313–318.

[1048] A.C. TUCKER, A reduction procedure for coloring perfect K_4-free graphs, *J. Combin. Theory B*, 43 (1987), 151–172.

[1049] ZS. TUZA, Graph colorings with local constraints — a survey, *Discussiones Math. Graph Theory*, 17 (1997), 161–228.

[1050] ZS. TUZA, A. WAGLER, On minimally non-preperfect graphs, working paper (1997).

[1051] R.I. TYSHKEVICH, The canonical decomposition of a graph (in Russian), *Doklady Akad. Nauk, BSSR*, 24 (1980), 677–679.

[1052] R.I. TYSHKEVICH, Once more on matrogenic graphs, *Discrete Math.*, 51 (1984), 91–100.

[1053] R.I. TYSHKEVICH, A.A. CHERNYAK, Unigraphs I, *Isv. Akad. Nauk BSSR, Ser. Fiz.-Mat. Nauk*, 5 (1978), 5–11.

[1054] R.I. TYSHKEVICH, A.A. CHERNYAK, Unigraphs II, *Isv. Akad. Nauk BSSR, Ser. Fiz.-Mat. Nauk,* 1 (1979), 5–12.

[1055] R.I. TYSHKEVICH, A.A. CHERNYAK, Unigraphs III, *Isv. Akad. Nauk BSSR, Ser. Fiz.-Mat. Nauk,* 2 (1979), 5–11.

[1056] R.I. TYSHKEVICH, A.A. CHERNYAK, Canonical partition of a graph defined by the degrees of its vertices (in Russian), *Isv. Akad. Nauk BSSR, Ser. Fiz.-Mat. Nauk,* 5 (1979), 14–26.

[1057] R.I. TYSHKEVICH, A.A. CHERNYAK, Box-threshold graphs: The structure and the enumeration, 30. Intern. Wiss. Kolloqu. TH Ilmenau, "Graphen und Netzwerke — Theorie und Anwendungen," Ilmenau, Germany (1985).

[1058] R.I. TYSHKEVICH, A.A. CHERNYAK, ZH. A. CHERNYAK, Graphs and degree sequences I, *Cybernetics* (the English translation *of Kibernetika*), 23 (1988), 734–745.

[1059] R.I. TYSHKEVICH, A.A. CHERNYAK, ZH. A. CHERNYAK, Graphs and degree sequences II, *Cybernetics* (the English translation *of Kibernetika*), 24 (1988), 137–152.

[1060] R.I. TYSHKEVICH, A.A. CHERNYAK, ZH. A. CHERNYAK, Graphs and degree sequences III, *Cybernetics* (the English translation *of Kibernetika*), 24 (1989), 539–548.

[1061] R.I. TYSHKEVICH, O.I. MELNIKOW, V.M. KOTOV, On graphs and degree sequences: the canonical decomposition (in Russian), *Kibernetika,* 6 (1981), 5–8.

[1062] J. URRUTIA, Partial orders and Euclidean geometry, in *Algorithms and Order,* I. Rival, ed., Kluwer Acaddemic, Dordrecht (1989), 327–436.

[1063] J. URRUTIA AND F. GAVRIL, An algorithm for fraternal orientation of graphs, *Inform. Process. Lett.,* 41 (1992), 271–274.

[1064] J. VALDES, R.E. TARJAN, E.L. LAWLER, The recognition of series-parallel digraphs, *SIAM J. Comput.,* 11 (1982), 298–313.

[1065] M. VAN DE VEL, *Theory of Convex Structures,* Elsevier, Amsterdam (1993).

[1066] A. VAN ROOIJ, H. WILF, The interchange graphs of a finite graph, *Acta Math. Acad. Sci. Hung.,* 16 (1965), 263–269.

[1067] V.G. VIZING, The Cartesian product of graphs (in Russian), English translation in *Comp. El. Syst.,* 2 (1966), 352–365.

[1068] V.I. VOLOSHIN, Properties of triangulated graphs (in Russian), *Issled. Operazii i Programmirov.,* (1982), 24–32.

[1069] V.I. VOLOSHIN, *Triangulated graphs and their generalizations* (in Russian), Ph.D. thesis, Kishinev State University (1983).

[1070] H.-J. VOSS, Note on a paper of McMorris and Shier, *Comment. Math. Univ. Carolin.,* 26 (1985), 319–322.

[1071] A. WAGLER, On critically perfect graphs, Preprint SC 96-50, Konrad-Zuse-Zentrum für Informationstechnik Berlin, 1996.

[1072] K. WAGNER, Über eine Eigenschaft der ebenen Komplexe, *Math. Annal.,* 114 (1937), 570–590.

[1073] D. WAGNER, Decomposition of partial orders, *Order,* 6 (1990), 335–350.

[1074] S. WAGON, A bound on the chromatic number of graphs without certain induced subgraphs, *J. Combin. Theory B,* 29 (1980), 345–346.

[1075] J.A. WALD, C.J. COLBOURN, Steiner trees, partial 2-trees, and minimum IFI networks, *Networks,* 13 (1983), 159–167.

[1076] W.D. WALLIS, J. WU, Squares, clique graphs, and chordality, *J. Graph Theory,* 20 (1995), 37–45.

[1077] J.R. WALTER, *Representations of Rigid Cycle Graphs,* Ph.D. thesis, Wayne State University, Detroit (1972).

[1078] G. WEGNER, *Eigenschaften der Nerven homologisch-einfacher Familien im* R^n, Dissertation thesis, , University of Göttingen (1967).

[1079] D.J.A. WELSH, M.B. POWELL, An upper bound on the chromatic number of a graph and its applications to timetabling problems, *Comput. J.*, 10 (1967), 85–87.

[1080] D.B. WEST, D.B. SHMOYS, Recognizing graphs with fixed interval number is NP-complete, *Discrete Appl. Math.*, 8 (1984), 295–305.

[1081] S.H. WHITESIDES, An algorithm for finding clique cut-sets, *Inform. Process. Lett.*, 12 (1981), 31–32.

[1082] S.H. WHITESIDES, A method for solving certain graph recognition and optimization problems, with applications to perfect graphs, *Ann. Discrete Math.*, 21 (1984), 281–297.

[1083] H. WHITNEY, Congruent graphs and the connectivity of graphs, *Amer. J. Math.*, 54 (1932), 150–168.

[1084] R. WILLE, Lexicographic decomposition of ordered sets (graphs), Preprint No. 705, Fachbereich Mathematik, Tech. Univ. Darmstadt (1983).

[1085] H.S. WILF, The eigenvalues of a graph and its chromatic number, *J. London Math. Soc.*, 42 (1967), 330–332.

[1086] T.V. WIMER, *Linear algorithms on k-terminal graphs*, Ph.D. thesis, Dept. of Computer Science, URI-030, Clemson University, Clemson, SC (1987).

[1087] T.V. WIMER, S.T. HEDETNIEMI, k-terminal recursive families of graphs, *Congres. Numer.*, 63 (1988), 161–176.

[1088] P.M. WINKLER, Factoring a graph in polynomial time, *European J. Combin.*, 8 (1987), 209–212.

[1089] S.K. WISMATH, Characterizing bar line-of-sight graphs, in Proc. 1st ACM Symposium on Computational Geometry, Baltimore, MD (1985), 147–152.

[1090] E.S. WOLK, The comparability graph of a tree, *Proc. Amer. Math. Soc.*, 13 (1962), 789–795.

[1091] E.S. WOLK, A note on "The comparability graph of a tree," *Proc. Amer. Math. Soc.*, 16 (1965), 17–20.

[1092] D.R. WOODALL, Forbidden graphs for degree and neighbourhood conditions, *Discrete Math.*, 75 (1989), 387–404.

[1093] Q. XUE, On a class of square–free graphs, *Inform. Process. Lett.*, 57 (1996), 47–48.

[1094] M. YANNAKAKIS, The complexity of the partial order dimension problem, *SIAM J. Alg. Discrete Methods*, 3 (1982), 351–358.

[1095] M. YANNAKAKIS, On a class of totally unimodular matrices, *Math. Oper. Res.*, 10 (1985), 280–304.

[1096] L.S. ZAREMBA, Perfect graphs and norms, *Math. Programming*, 51 (1991), 269–272.

[1097] I.E. ZVEROVICH, V.E. ZVEROVICH, An induced subgraph characterization of domination perfect graphs, *J. Graph Theory*, 20 (1995), 375–395.

[1098] A.A. ZYKOV, *Fundamentals of Graph Theory*, BCS Associates, Moscow, Idaho (1990).

Index